INTELLIGENT TECHNOLOGIES FOR SENSORS

Applications, Design, and Optimization for a Smart World

INTELLIGENT TECHNOLOGIES FOR SENSORS

Applications, Design, and Optimization for a Smart World

Edited by

S. Kannadhasan

R. Nagarajan

Alagar Karthick

AAP | APPLE ACADEMIC PRESS

First edition published 2023

Apple Academic Press Inc.
1265 Goldenrod Circle, NE,
Palm Bay, FL 32905 USA

760 Laurentian Drive, Unit 19,
Burlington, ON L7N 0A4, CANADA

CRC Press
6000 Broken Sound Parkway NW,
Suite 300, Boca Raton, FL 33487-2742 USA

4 Park Square, Milton Park,
Abingdon, Oxon, OX14 4RN UK

Library and Archives Canada Cataloguing in Publication

Title: Intelligent technologies for sensors : applications, design, and optimization for a smart world / edited by S. Kannadhasan, R. Nagarajan, Alagar Karthick.
Names: Kannadhasan, S., editor. | Nagarajan, R. (Control engineer), editor. | Karthick, Alagar editor.
Description: First edition. | Includes bibliographical references and index.
Identifiers: Canadiana (print) 20220481946 | Canadiana (ebook) 20220481997 | ISBN 9781774911853 (hardcover) | ISBN 9781774911860 (softcover) | ISBN 9781003314851 (ebook)
Subjects: LCSH: Intelligent sensors. | LCSH: Detectors.
Classification: LCC TK7871.676 .I58 2023 | DDC 681/.2—dc23

Library of Congress Cataloging-in-Publication Data

CIP data on file with US Library of Congress

ISBN: 978-1-77491-185-3 (hbk)
ISBN: 978-1-77491-186-0 (pbk)
ISBN: 978-1-00331-485-1 (ebk)

About the Editors

S. Kannadhasan is working as an Assistant Professor in the Department of Electronics and Communication Engineering at Study World College of Engineering, Coimbatore, Tamil Nadu, India. He has 12 years of teaching and research experience. He has completed his PhD in the field of Smart Antenna at Anna University in 2022. He earned his BE in ECE at Sethu Institute of Technology, Kariapatti, in 2009 and ME in Communication Systems at Velammal College of Engineering and Technology, Madurai in 2013. He earned his MBA in Human Resources Management at Tamil Nadu Open University, Chennai. India. He has published around 45 papers in reputed indexed international journals indexed by SCI, Scopus, Web of Science, and other major indexing services, and he has presented/published more than 146 papers in national and international journals and conferences. In addition, he has contributed a book chapter. He also serves as a board member, reviewer, speaker, session chair, advisory and technical committee member of various colleges and conferences. He has attended various workshops, seminars, conferences, faculty development programs, short-term training programs, and online courses. His areas of interest are smart antennas, digital signal processing, wireless communication, wireless networks, embedded systems, network security, optical communication, microwave antennas, electromagnetic compatibility and interference, wireless sensor networks, digital image processing, satellite communication, cognitive radio design, and soft computing techniques. He is a member of IEEE, ISTE, IEI, IETE, CSI, IAENG, SEEE, IEAE, INSC, IARDO, ISRPM, IACSIT, ICSES, SPG, SDIWC, IJSPR, and EAI Community.

R. Nagarajan is currently working as Professor of Electrical and Electronics Engineering at Gnanamani College of Technology, Namakkal, Tamil Nadu, India. He has published more than 70 papers in international journals and conferences. He has also worked in industry as an electrical engineer. His research interest includes power electronics, power system, communication engineering, network security, soft computing techniques, cloud computing, big data analysis, and renewable energy sources. Dr. Nagarajan received his BE in Electrical and Electronics Engineering from Madurai Kamarajar University, Madurai, India, in 1997. He received his ME in Power Electronics and Drives from Anna University, Chennai, India, in 2008. He received his PhD in Electrical Engineering from Anna University, Chennai, India, in 2014.

Alagar Karthick is working as Associate Professor in the Electrical and Electronics Engineering Department in KPR Institute of Engineering and Technology, Coimbatore, Tamil Nadu, India. He has published more than 30 international journals and is also reviewer for various journals, such as *Solar Energy, Fuel, Journal of Cleaner Production, Heliyon,* and *Building Services Engineering Research and Technology.* He received his Doctor of Philosophy in the field of Building Integrated Photovoltaic (BIPV) from Anna University, Chennai in 2018. He received a master's degree in Energy Engineering and bachelor 's degree in Electrical and Electronics Engineering. He has received best paper awards for his research articles on biomass conversion. His research area includes solar photovoltaic, bioenergy, zero energy buildings, energy with artificial intelligence, machine learning, and deep learning algorithms.

Contents

Contributors

Deepika Adhikary
Department of Electronics and Communication Engineering, NIST (Autonomous), Berhampur, Odisha, 761008, India

G. Akshaykumar
Department of Electronics and Communication Engineering, NIST (Autonomous), Berhampur 760010, Odisha, India

Anitha G.
Department of Electronics and Communication Engineering, Saveetha School of Engineering, Saveetha Institute of Medical and Technical Sciences (SIMATS), Chennai 602105, Tamil Nadu, India; E-mail: anipsg09@gmail.com

J. Vijitha Ananthi
Department of Biomedical Engineering, Karunya Institute of Technology and Sciences, Coimbatore, India; E-mail: vijithaananthi@karunya.edu.in, vijithaananthij@gmail.com

C. Balaji Ayyanar
Department of Mechanical Engineering, Coimbatore Institute of Technology, Coimbatore 60014, Tamil Nadu, India; E-mail: balajiayyanar2007@gmail.com

R. Banupriya
Department of Electrical and Electronics Engineering, PGP College of Engineering & Technology, Namakkal, Tamil Nadu, India; E-mail: priyarangasamy85@gmail.com

Arpita Bebarta
Department of Electronics and Communication Engineering, National Institute of Science and Technology, Berhampur 761008, Odisha, India

Pradeep Chindhi
Department of Electrical Engineering, SGMCOE, Mahagaon 416503, Maharashtra, India; E-mail: pradeepchindhi.4003@gmail.com

Ashish Kumar Dash
Department of School of Computer Science and Engineering, NIST (Autonomous), Berhampur 760010, Odisha, India

Swastid Dash
Department of Electronics and Communication Engineering, NIST (Autonomous), Berhampur 760010, Odisha, India

S. R. Dash
Department of Electronics Communication Engineering, NIST (Autonomous) Berhampur, Berhampur 760010, Odisha, India

P. Deepak
Department of Electronics and Communication Engineering, NIST (Autonomous), Berhampur 760010, Odisha, India

J. A. Dhanraj
Centre for Automation and Robotics (ANRO), Department of Mechanical Engineering, Hindustan Institute of Technology and Science, Padur, Chennai 603103, Tamil Nadu, India

Chandramohan Dhasarathan
Computer Science and Engineering Department, Thapar Institute of Engineering and Technology, Patiala, Punjab, India

B. Gayathri
Department of Chemistry, Coimbatore Institute of Technology, Coimbatore 60014, Tamil Nadu, India

R. Geetha
Department of Information Technology, Saveetha School of Engineering, Saveetha Institute of Medical and Technical Sciences (SIMATS), Chennai 602105, Tamil Nadu, India

T. K. Giri
Department of Electronics Communication Engineering, NIST (Autonomous) Berhampur, Berhampur 760010, Odisha, India

Bhabani Sankar Gouda
Department of Computer Science and Engineering, NIST (Autonomous), Berhampur 760010, Odisha, India

Sanjeev Gour
Department of Computer Science, Career College Bhopal, MP, India

Subash Chandra Bose Jaganathan
School of Computing Science and Engineering, VIT Bhopal University, Sehore, MP, India

R. Jegan
Department of Biomedical Engineering, Karunya Institute of Technology and Sciences, Coimbatore 641114, India

P. Subha Hency Jose
Department of Biomedical Engineering, Karunya Institute of Technology and Sciences, Coimbatore 641114, India

Apoorva Joshi
Department of Computer Science, Career College Bhopal, MP, India; E-mail: apoorvajoshi16@gmail.com

S. Kannadhasan
Department of Electronics and Communication Engineering, Cheran College of Engineering, Karur, Tamilnadu, India; email: kannadhasan.ece@gmail.com

Alagar Karthick
Electrical and Electronics Engineering Department, KPR Institute of Engineering and Technology, Coimbatore, Tamilnadu, India; email: Karthick.power@gmail.com

Kanishk Kashyap
Department of Electrical and Electronics Engineering, National Institute of Science and Technology, Berhampur 761008, Odisha, India

Geeta Kalkhambkar
Department of Electronics and Telecommunication Engineering, SGMCOE, Mahagaon, Maharashtra, India

R. Kalpana
Dept. of Computer Science and Engineering, Madanapalle Institute of Technology and Science, Angallu, Andhra Pradesh, India

Alagar Karthick
Department of Electrical and Electronics Engineering, KPR Institute of Engineering and Technology, Coimbatore 641024, Tamil Nadu, India

Rajashri Khanai
Department of Electronics and Communication Engineering, KLE'S Dr. MSSCET, Belgaum, Karnataka, India

S. Sovan Kumar
Department of Electronics and Communication Engineering, NIST (Autonomous), Berhampur, Odisha 760010, India

Yashwardhan Kumar
Department of Electronics and Communication Engineering, NIST (Autonomous), Berhampur, Odisha 760010, India

J. Manjushree Kumari
Department of Electrical and Electronics Engineering, Gnanamani College of Technology, Namakkal, Tamil Nadu, India

B. Leelamani
Department of Biotechnology, Aurora's Degree and PG College, Chikkadpally, Hyderabad, Telangana, India

Sanjana Mahapatra
Department of Computer Science and Engineering, NIST (Autonomous), Berhampur, Odisha 760010, India

T. Manimegalai
Department of CSE, Saveetha School of Engineering, Saveetha Institute of Medical and Technical Sciences (SIMATS), Saveetha University, Chennai 600124, Tamil Nadu, India

Sandipan Mallik
Department of Electronics and Communication Engineering, NIST (Autonomous), Berhampur, Odisha, 761008, India; E-mail: sandi.iitkgp@gmail.com; sandipan@nist.edu

K. Marimuthu
Department of Mechanical Engineering, Coimbatore Institute of Technology, Coimbatore 60014, Tamil Nadu, India

Ninoslav Marina
University of Information Science and Technology, "St. Paul the Apostle" Ohrid, North Macedonia

D. Mishra
Department of Computer science and Engineering, NIST (Autonomous) Berhampur, Berhampur 760010, Odisha, India

V. Mohanavel
Centre for Materials Engineering and Regenerative Medicine, Bharath Institute of Higher Education and Research, Chennai 600073, Tamil Nadu, India

Azath Mubarakali
College of Computer Science, Department of CNE, King Khalid University, Abha, Saudi Arabia

R. Nagarajan
Department of Electrical and Electronics Engineering, Gnanamani College of Technology, A. K. Samuthiram, Namakkal, Tamilnadu, India; email: krrajan71@gmail.com

Sonalika Nayak
Department of Electronics Communication Engineering, NIST (Autonomous), Berhampur 760010, Odisha, India

W. S. Nimi
Department of Biomedical Engineering, Karunya Institute of Technology and Sciences, Coimbatore 641114, India; E-mail: nimiwsrec@gmail.com

R. R. Padhi
Department of Electronics Communication Engineering, NIST (Autonomous) Berhampur, Berhampur 760010, Odisha, India

D. Panda
Department of Electronics Communication Engineering, NIST (Autonomous) Berhampur, Berhampur 760010, Odisha, India

Karuna Nidhi Pandagre
Department of Computer Science, Career College Bhopal, MP, India

Lokesh Singh Panwar
Department of Electronics and Communication, Polymer Sensor and Actuator Lab, Graphic Era Deemed to be University, Dehradun, India; E-mail: lokesh31j@gmail.com

Varij Panwar
Department of Electronics and Communication, Polymer Sensor and Actuator Lab, Graphic Era Deemed to be University, Dehradun, India

Sudhir Ranjan Pattanaik
Department of School of Computer Science and Engineering, NIST (Autonomous), Berhampur, Odisha 760010, India; E-mail: sudhir.pattanaik@nist.edu

Sai Swarup Patnaik
Department of Electronics and Communication Engineering, NIST (Autonomous), Berhampur, Odisha 760010, India

Sumanjit Pattanayak
Department of Electrical and Electronics Engineering, National Institute of Science and Technology, Berhampur 761008, Odisha, India

V. R. K. Patro
Department of Computer science and Engineering, NIST (Autonomous) Berhampur, Berhampur 760010, Odisha, India

Bora Pavani
Department of Electronics and Communication Engineering, NIST (Autonomous), Berhampur 760010, Odisha, India

Mădălin-Dorin Pop
Computer and Information Technology Department, Politehnica University of Timişoara, România

R. Thandaiah Prabu
Department of Electronics and Communication Engineering, Saveetha School of Engineering, Saveetha Institute of Medical and Technical Sciences (SIMATS), Chennai 602105, Tamil Nadu, India

Shyam Sundar Pradhan
Department of Information Technology Engineering, NIST (Autonomous), Berhampur 760010, Odisha, India

Priyanka Pratihari
Department of Computer Science and Engineering, NIST (Autonomous), Berhampur 760010, Odisha, India

H. P. Rajani
Department of Electronics and Communication Engineering, KLE'S Dr. MSSCET, Belgaum, Karnataka, India

G. Ramkumar
Department of Electronics and Communication Engineering, Saveetha School of Engineering, Saveetha Institute of Medical and Technical Sciences (SIMATS), Chennai 602105, Tamil Nadu, India; E-mail: pgrvlsi@gmail.com

V. V. R. Raman
Department of Computer Science, Aurora's Degree and PG College, Chikkadpally, Hyderabad, Telangana, India

Velmani Ramasamy
Computer Science and Engineering Department, Adithya Institute of Technology, Kurumbapalayam, Coimbatore, Tamil Nadu, India

S. Ramesh
Department of Electronics and Communication Engineering, Saveetha School of Engineering, Saveetha Institute of Medical and Technical Sciences (SIMATS), Chennai 602105, Tamil Nadu, India

Obbinti Sankar Rao
Department of Electronics Communication Engineering, NIST (Autonomous), Berhampur, Odisha 761008, India

Manoj Kumar Sahoo
Department of School of Computer Science and Engineering, NIST (Autonomous), Berhampur, Odisha 760010, India; E-mail: mksahoo@nist.edu; mksahoo@hotmail.com

Monik Raj Sahu
Department of Electronics and Communication Engineering, NIST (Autonomous), Berhampur, Odisha 760010, India

Anand Kumar Satapathy
Department of Computer Science and Engineering, NIST (Autonomous), Berhampur 760010, Odisha, India

K. Pandi Selvi
Thamirabharani Engineering College, Tirunelveli, Tamil Nadu, India

Swadhin Kumar Senapati
Department of Electronics and Communication Engineering, NIST (Autonomous), Berhampur 760010, Odisha, India

Ambrish Kumar Sharma
Department of Computer Science, NRI College, Bhopal, MP, India

S. A. Shifani
Department of ECE, Jeppiaar Maamallan Engineering College, Chennai 602108, Tamil Nadu, India

Madhvi Singh
Department of Electronics and Communication Engineering, NIST (Autonomous), Berhampur, Odisha 761008, India

L. Suganya
Department of Electrical and Electronics Engineering, PGP College of Engineering & Technology, Namakkal, Tamil Nadu, India

M. Suganthi
AP/CSE Department, Thamirabharani Engineering College, Tirunelveli, Tamil Nadu, India; E-mail: sugi.mp@gmail.com

M. Suresh
Department of Electronics Communication Engineering, NIST (Autonomous), Berhampur, Odisha, 761008, India; E-mail: msuresh73@gmail.com

Kunjabihari Swain
Department of Electrical and Electronics Engineering, National Institute of Science and Technology, Berhampur 761008, Odisha, India

M. Tamilselvi
Department of Mechatronics Engineering, T.S. Srinivasan Centre for Polytechnic College and Advanced Training, Vanagaram, Chennai 600095, Tamil Nadu, India

Murugesh Veerasamy
Department of CSE, Koneru Lakshmaiah Education Foundation, Vaddeswaram, AP, India

Renjith V. S.
Department of Biomedical Engineering, Karunya Institute of Technology and Sciences Coimbatore, India; E-mail: notify_renjithvs@yahoo.com

Abbreviations

ABS	automatic blocking system
ACK	acknowledge
ADC	analog to digital converters
ADS	advanced design system
AES	Advanced Encryption Standard
AI	artificial intelligence
ALPR	automatic license plate recognition system
ANPR	automatic number plate recognition system
APTS	advanced public transportation system
ATCC	automatic traffic counter cum classifier
ATIS	advanced traveler information system
ATMS	advanced transportation management system
ATPS	advanced transportation pricing system
ATR	attenuated total internal reflection
AWS	Amazon Web Services
BCI	brain–computer interface
BF	bamboo flour
BG	blood glucose
BP	blood pressure
BPM	beats per minute
BSHDT	brainwave-based smart home device triggering
BSN	body sensor network
BT	body temperature
CAGR	compound annual growth rate
CAIDI	customer average interruption duration index
CEHS	cooperative-based energy harvesting scheme
C-ITS	cooperative-ITS
CMOS	complementary metal–oxide semiconductor
CNN	convolutional neural network
CoRE	constrained restful environment
CSEE	clustered-based security and energy-efficient
CT	current transformer
CVD	cardio vascular disease
CVS	cooperative vehicle system

DApps	distributed applications
DAS	driver assistant system
DCM	discontinuous current mode
DEs	differential equations
DNN	deep neural networks
DOTs	departments of transportation
DSP	digital signal processors
DSRC	dedicated short-range communications
ECB	emergency calling box
ECG	electrocardiogram
EDA	electro-dermal activity
EEG	electroencephalograph
ELHMS	enhanced learning-based healthcare monitoring system
EMD	empirical mode decomposition
EMG	electromyogram
ERP	event-related potential
ETH	ethereum
EXI	efficient XML interchange
E2E	end-to-end
FIA	flow injection analysis
FIR	finite impulse response
FL	fault location
FN	future network
FPGA	Field Programmable Gate Array
GC	gas chromatography
GDPR	general data protection regulation
GHG	greenhouse gas
G-IoT	green internet of things
GIS	Geographical Information System
GPS	global positioning System
GPSR	general-packet-radio-service
GSM	global system for mobile
HB	harmonic balance
HCI	human–computer interaction
HD	Hodgkin's disease
HPF	high-pass filter
HPLC	high-performance liquid chromatography
HR	heart rate
IaaS	infrastructure as a service
ICO	initial coin offering

IETF	internet engineering task force
IIR	infinite impulse response
IoE	internet of energy
IoE	internet of everything
IoS	internet of services
IoT	Internet of Things
IoTaaS	Internet of Things as a service
IPTV	internet protocol television
IR	infrared
ISM	industrial scientific and medical band
ITS	intelligent transportation systems
JS	JavaScript
LeNN	single layer legendre neural network
MaaS	Mobility-as-a-Service
MDS	meteorological data station
MEMS	micro-electro-mechanical-systems
MIMO	multiple-input and multiple-output
MRCS	mobile radio communication system
MS	mass spectrometry
NDVI	Normalized Difference Vegetable Index
NHL	non-Hodgkin lymphoma
NHSR	National Health Statistics Reports
ODE	ordinary differential equations
OCR	optical character recognition
PaaS	platform as a service
PD	photodetector
PDE	partial differential equations
PP	polypropylene
PLI	power line interference
PV	photovoltaic
QoS	quality of service
RF	radio frequency
RFID	radio frequency identification
RHD	resistive heat detectors
RHF	rice husk fiber
ROI	return on investment
ROLL	routing over low-power and lossy networks
ROM	read-only memory
RSA	Rivest–Shamir–Adleman
RSSI	receiver signal strength indicator

SAIDI	system average interruption duration index
SAIFI	system average interruption frequency index
SCS	*Syzygium Cuminiseed*
SDGs	sustainable development goals
SFMU	smart field monitoring unit
SG	smart grid
SHCMK	smart healthcare monitoring kit
SNR ratio	signal to noise ratio
SOC	system on a chip
SPDEs	singularly-perturbed differential equations
SpO_2	oxygen saturation
S-RAM	static-random access memory
SRV	ServiceCoin
SSL	secure sockets layer
STWS	single term Walsh series
TCP/IP	Transmission Control Protocol/Internet Protocol
TCR	temperature coefficient of resistance
TDMS	time-division multiple access
TMCs	traffic management centers
TMCs	transportation management centers
TxDOT	Texas Department of Transportation
ULP	ultra-low power
VANET	vehicle ad-hoc networks
VIDS	video incident detection system
VMS	variable message sign
VMT	vehicle miles travelled
VR	virtual reality
VRP	vehicle-routing problem
VSH	virtual smart home
V2I	vehicle-to-infrastructure
V2IoT	vehicle-to-IoT
V2V	vehicle-to-vehicle
V2X	vehicle-to-everything
WBAN	wireless body area network
WF	waste fish
WHO	World Health Organization
WSNs	wireless sensor network
6LoWPAN	IPv6 over low-power wireless personal area networks
W3C	World Wide Web Consortium

Preface

This book provides an overview of systems and machines that are intelligent. It is intended for anybody who is interested in future advancements in these domains or wants to be informed on the current state of these multidisciplinary technologies.

Sensor devices that are flexible and printable have received a lot of interest in recent years. New techniques like as printing and additive manufacturing are being developed to realize a wide range of readily deployable systems such as displays, sensors, and RFID tags. The needs of the growing area of modular and writable sensors are being met by repurposing silicon-based planar electronics and solid-state sensing technology. This book brings together leading academics, architects, and scientists who are experts in the subject from across the globe. Engineers discuss about their research projects, experiments, discoveries, innovative ideas and principles, contributions, and advancements in the fields of inventions and software, measurement theories and applications, and instrumentation theories and applications.

Pervasive, reliable, robust, and streamlined positioning technologies will benefit a wide range of resources, including personal navigation, search and rescue, robot and fleet control, and health care. Despite the fact that there are mature GNSS solutions for outside areas, more than 10 years of sensor technology research and development have failed to provide a widely available offer of generic and cheap standard solutions for inside. A new technology is the requirement for sophisticated computing technologies to evolve in order to satisfy the increasing needs of knowledge and communication technologies in smart real-time world applications. There are a number of clever technologies that may help the educational system significantly.

We would like to take this opportunity to thank our family members and friends, who encouraged us a lot during the preparation of this book. First and most obviously, we give all the glory and honor to our almighty Lord for his abundant grace that sustained us for successful completion of this book. We would like to thank the authors for their contribution in this edited book. We would also like to thanks Apple Academic Press, CRC Press, a Taylor & Francis Group and its whole team for facilitating the work and providing us the opportunity to be a part of this work.

PART I

Applications of Intelligent Technologies for Sensors

CHAPTER 1

Product Details Identification for Visually Impaired Persons

M. SUGANTHI[1*] and K. PANDI SELVI[2]

[1]*AP/CSE Department, Thamirabharani Engineering College, Tirunelveli, Tamil Nadu, India*

[2]*Thamirabharani Engineering College, Tirunelveli, Tamil Nadu, India*

[*]*Corresponding author. E-mail: sugi.mp@gmail.com*

ABSTRACT

Android is a smartphone and tablet operating system that is free and open-source. Smartphones are used for the majority of activities, such as e-commerce and commercial processing. People with visual impairment face difficulties to read the details of the product. So to help the visually impaired people in purchasing the projects, we developed an application in Android and PHP. As most of the supermarkets sell products of all brands in a single place, visually impaired people can shop easily with the assistance of QR code generated by shop owners through the web application. The product information for the QR code produced will be registered by the business owner. Then the customer with the "VIP Helper" Android application can easily scan the QR code and the application speaks aloud the product details like product name, brand, price, expiry date of the product which is embedded. So this greatly helps the visually impaired people by eliminating the third person's assistance in shopping at ease.

Intelligent Technologies for Sensors: Applications, Design, and Optimization for a Smart World.
S. Kannadhasan, R. Nagarajan, & Alagar Karthick (Eds.)

1.1 INTRODUCTION

A customer looks through the available items or services from one or more merchants with the intention of purchasing a suitable option. Scholars have developed a shopper typology that categorizes one sort of shopper as recreational shoppers, or those who like shopping as a leisure activity. Consumers can now look up product information and place purchases across several platforms, making internet shopping a major disruptor in the retail industry. Customers' purchases are delivered to their homes, offices, or other locations by online retailers. Thanks to the business to consumer approach, customers may now choose any product from a retailer's website and have it delivered quite quickly. By not having to travel to actual stores, consumers who utilize online shopping strategies save time and energy. They may be able to save both time and money as a result of this. A retailer, often known as a shop, is a business that exhibits a wide range of goods and offers to trade or sell them to customers in return for money or other goods. Customers' purchasing experiences may vary. They are influenced by a variety of factors, including how the customer is treated, the ease with which the transaction is completed, the items purchased, and the consumer's mood.

1.2 ONLINE SHOPPING

One of the earliest sorts of internet commerce was IBM's online transaction processing, which was developed in the 1960s and allowed for the real-time processing of financial transactions. The Semi-Automatic Business Research Environment, a computerized ticket reservation system for American Airlines, was one of its uses. A massive IBM mainframe computer linked computer terminals at multiple travel agencies, processing, and coordinating transactions so that all travel agents had access to the same information at the same time. The growth of online shopping[1] as we know it now began with the introduction of the Internet. Initially, this platform was just a tool for companies to promote and share information about their products. It quickly advanced from this fundamental utility to true online purchase transactions thanks to the creation of dynamic Web sites and secure connections. With the initial sales of Sting's album "Ten Summoner's Tales" in 1994, the internet started to grow as a secure purchasing channel. Wine, chocolates, and flowers were among the first retail categories to promote the expansion of online shopping, and they were among the pioneering retail categories. The presence of e-commerce-ready items is a strong predictor of Internet success, according to researchers. Many of these things worked well because they were generic items that customers did not have to touch or feel before buying. But, more importantly, there were few internet users in the early days, and those that did were from a certain demographic: rich guys in their 30s.

1.3 RETAIL SHOPPING

Retail is the business of earning money by selling consumer goods or services to customers via different distribution channels. A supply chain has identified a requirement that retailers must satisfy. The word "retailer" is used when a service provider fills modest orders for a large number of end-users rather than large purchases for a small number of wholesale, corporate, or government customers. Shopping refers to the act of buying items. This is done to get final things, such as food and clothing, and it is also done as a recreational activity. Window shopping and browsing are popular leisure shopping activities that may not always result in a purchase. Markets and retail businesses have a long and famous history that dates back to antiquity. The earliest retailers were itinerant peddlers.[7] Throughout the years, retail businesses have grown from basic "rude booths" to elaborate shopping malls of the modern day.

1.4 PRODUCT DETAILS IDENTIFICATION

As we progress into the contemporary era of technology, we may discover that many engineering-related applications are very helpful to society's development. This is the technological world, where individuals utilize cellphones to do everyday activities such as shopping, job management, and so on. The specifics of a product are difficult to read for those with vision impairment. By scanning the QR code of the goods, this initiative assists visually challenged individuals in learning about the product description. This programme may also be used to create new QR codes. This application is for visually impaired individuals who want to provide product information via voice. It focuses on shopping facilities.

1.4.1 EXISTING SYSTEM

Visually challenged individuals used to shop with the assistance of others under the current arrangement. They need regular people's assistance. The checkout operator software in an existing system produces a table containing the product name and expiry date, which is subsequently uploaded to the cloud. The customer scans a single Quick Response number written[5] on the purchase receipt using his or her smartphone. The table is then instantly downloaded from the cloud to the smart phone.

1.4.2 PROBLEM DEFINITION

The existing system gives the notification to display product details. User must depend on third person to know the details of product. Locating the barcode using mobile is difficult.

1.5 PROPOSED SYSTEM

In the proposed system, to help the visually impaired people in purchasing the project, we develop the application in Android. This application reads the product information from the QR code. The PHP language is used to create the QR code. The product information is incorporated in the QR code when it is being generated. An event like QR scanning and reading product information does not involve the use of the internet is shown in Figure 1.1.

1.5.1 SYSTEM ARCHITECTURE

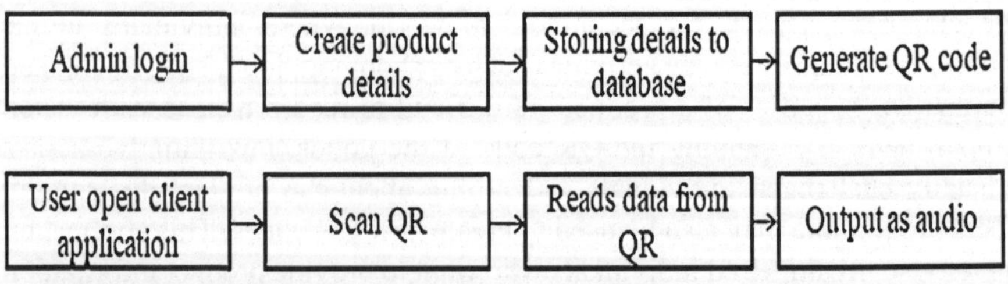

FIGURE 1.1 Block diagram of product details identification.

1.5.2 IMPLEMENTATION

When the theoretical design is transformed into a working system, it is called implementation. This is the most important stage in creating a new, effective system. It can only be deployed once all testing has been completed and the system has been confirmed to fulfill the criteria. The implementation phase includes a number of tasks. The purchase of hardware and software is completed. It's possible that the system will need the creation of certain software. The compilation and execution of the planned system are both part of software implementation. During this stage, modular and subsystem programming code will be completed. The developers do unit and module testing at this stage.

The proposed system has three modules. The modules are:

- QR generation
- QR reading
- Output as speech

QR Generation

The administrator module is where this is done. This QR code for available items is generated by the business or shopping center owner. The site's controller is the Category Administrator. He is in charge of product development. The product's category is created by the first administrator. It is then presented in the module when it has been constructed. The unwelcome category may then be deleted from the website. The primary division of items such as Cookies, Cosmetics, Hair oil, and Dairy Products are examples of categories.

New Product

In this module, the administrator creates a new product. First administrator selects the categories that were created earlier, and then the product name, brand, units, and price are entered. Then the product is created. The created product is displayed in this module and unwanted products can be removed.

Edit Product

If the administrator needs to modify the product information such as price, brand, and name, then in this module it can be changed. After selecting the product the product information is shown in the screen. Then the administrator can modify any information of the product. Then on submitting the form the product information is modified and stored in the server.

Print QR Code

The administrator can print QR codes by categories. On selecting the categories the corresponding QR code and the name of the product are printed in the screen. Then the administrator can take print using the printer. This QR code can be placed in the products in the shop, so that the customers can read the QR code with the android application.

QR Reading

This is done using android-based mobile phones. Using the camera in the mobile phone the user can scan the available QR code. Then it fetches the information that is embedded inside the QR code given by the administrator who generated the QR code.

Output as Speech

The last job is to read out the information derived from the QR code. The QR code carries product information such as the product name, brand, price, and so on. The QR code is then scanned by the Android application and spoken to the user. As a result, blind individuals may shop for things without the need for an aid.

FIGURE 1.2 Home page.

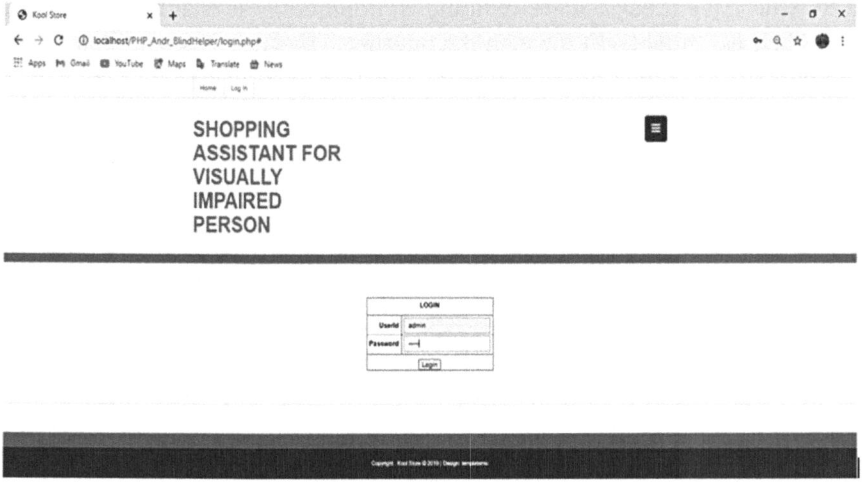

FIGURE 1.3 Login page.

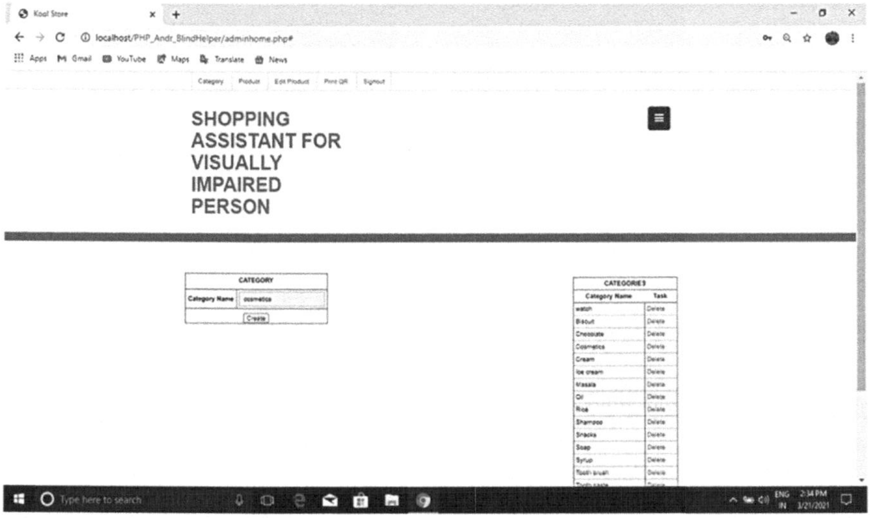

FIGURE 1.4 Create category.

1.6 CONCLUSION

The application's creation provides us with a nice experience while delivering great outcomes. The goal of the project "VIP Helper" is to identify users who have a basic understanding of computers. It enables individuals to do their essential tasks through the internet. The project is completed on a computer. It is a quick procedure that saves both time and money.

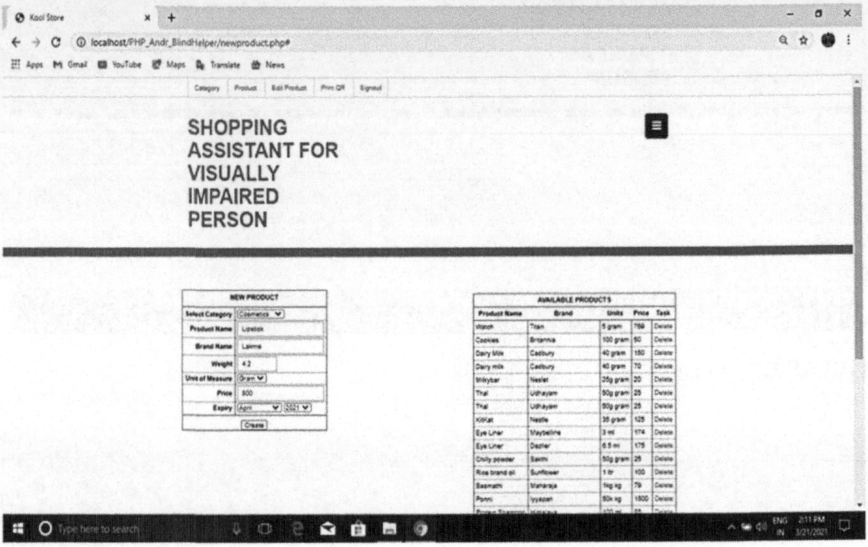

FIGURE 1.5 Create new product.

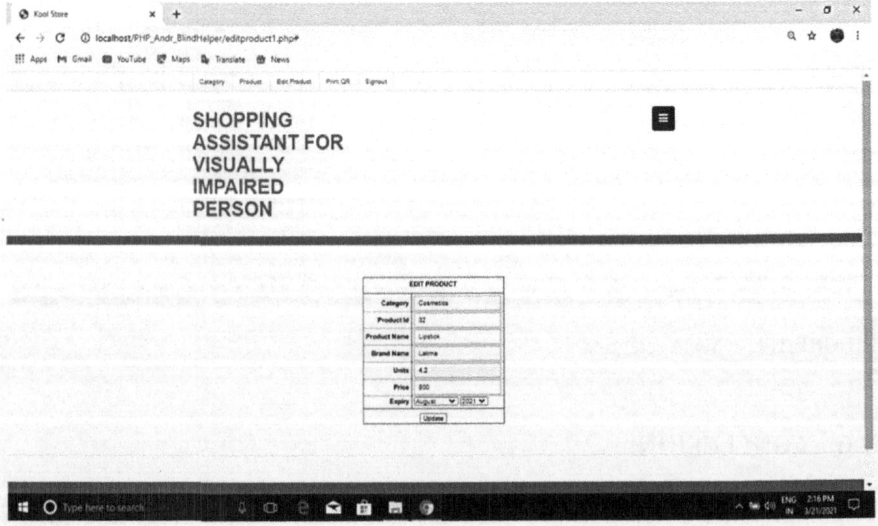

FIGURE 1.6 Edit product details.

FIGURE 1.7 Generate QR code.

1.7 FUTURE ENHANCEMENT

We know that much information cannot be stored in a QR code under the present system. So, in the future, more information should be stored in QR codes so that more information may be retrieved even when the QR code is used by visually impaired individuals. We'll create a sound in several more languages in the future.

KEYWORDS

- **android**
- **QR code**
- **shopper**
- **authetication**
- **customer**

REFERENCES

1. Mohammed, R. H.; Khan, T. Automatic Expiry Date Notification System Interfaced with Smart Speaker. *Int. J. Eng. Sci. Invent.* **2020,** ISSN: 2319-6734, 515–525.
2. Padmapriya, V.; Suresh, R.; Nithyasri, B.; Pavithra, L. Expiry Date and Cost Tracking in Medicine for Visually Impaired. *Int. Res. J. Eng. Technol.* **2020,** *07*(03).
3. Shahnoor, A.; Syeda, A. K.; Amulya, K. S.; Monisha, K. S. Product Details and Its Expiry Date Recognition through Speech. *Int. J. Eng. Res. Technol.* **2019,** ISSN: 2278-0181.
4. Devipriya, D.; Sushma Sri, V.; Mamatha, I. In *Smart Store Assistor for Visually Impaired.* International Conference on Advances in Computing, Communications and Informatics (ICACCI), 2018.
5. Kumar, K.D.; Thangavel, S. K. Assisting Visually Challenged Person in the Library Environment. *Lect. Notes Comput. Vision Biomech.* **2018,** *28*, 722–733.
6. Elgendy, M.; Sik-Lanyi, C.; Kelemen, A. Making Shopping Easy for People With Visual Impairment Using Mobile Assistive Technologies. *Appl. Sco.* **2019,** *9*(6), 1061.
7. Tiwari, S. In *An Introduction to QR Code Technology*, 2016 IEEE International Conference on Information Technology (ICIT), 2016.
8. Singh, V.; Verlekar, P.; Mishra, N. R.; Shaikh, S. Expiry Remainder. *Int. J. Innov. Res. Sci. Technol.* **2016.**
9. Zientara, P.; Advani, S.; Shukla, N.; Okafor, I.; Irick, K.; Sampson, J.; Datta, S.; Narayanan, V. K. A Multitask Grocery Assistance System for the Visually Impaired. *IEEE Consumer Electron. Mag.* **2017**, *6*(1), 73–81.

CHAPTER 2

IoT-Based Teaching Assistant System for Smart Classrooms

MANOJ KUMAR SAHOO[1*], SUDHIR RANJAN PATTANAIK[1*], YASHWARDHAN KUMAR[2], S. SOVAN KUMAR[2], MONIK RAJ SAHU[2], SAI SWARUP PATNAIK[2], and SANJANA MAHAPATRA[1]

[1]School of Computer Science and Engineering, NIST (Autonomous), Berhampur, Odisha 760010, India

[2]Department of Electronics and Communication Engineering, NIST (Autonomous), Berhampur, Odisha 760010, India

[*]Corresponding author. E-mail: mksahoo@nist.edu; mksahoo@hotmail.com; sudhir.pattanaik@nist.edu

ABSTRACT

In the present generation of smart technology, students are expecting colleges and university campus life to be innovative and inclined toward advanced learning methods. Cloud/Mesh computing technologies, IOT, and a robust system of communication transfer data at long distances with a cost-effective method. This leading edge technology is called LoRa (Long Range) communication, which is a secure and low-power technology and can give solutions for a smart and suitable classroom and campus to upgrade the teaching technique of the students as well as the efficiency of classroom. To initiate a smart classroom for teaching, we try to give teachers as well as students the indistinguishable experiences as received in a normal classroom during the lectures. The Smart Classroom could sincerely perceive, listen, and obey the lecturer and the lecturer can also be written on a virtual board by their hands or may take the help of speeches and indication to manage the

Intelligent Technologies for Sensors: Applications, Design, and Optimization for a Smart World.
S. Kannadhasan, R. Nagarajan, & Alagar Karthick (Eds.)

class, for the outlying students. Students can also use Personal Digital Assistance to approach the given assignments and exam appearance by online platforms. Lectures can be shared in the cloud, online lecturing permits students to remotely attend classrooms. Attendance during classroom hours is very important, because it affects the academic achievement of students. Therefore, several institutions impose a minimum percentage of attendance criteria for students to be allowed to seat in examinations. Conventional methods for taking student attendance in the classroom, such as roll-call and sign-in, are time taking and also increase teacher's workload. This system can also track student's attendance and keep track of who leave out classes, send alert notifications that assist students to focus on educational work regularly. This system consists of various hardware components such as sensors, microcontroller, and LoRa technology. We are also designing the software structure to process the data to and from sensors which will be transferred to cloud storage. In this paper, it is described how precisely Cloud, IOT, and low power communication technology restructure the traditional classroom and teaching techniques.

2.1 INTRODUCTION

The continuous up-gradation of modern technology encourages the update of the modern education system. The continuous updating of the education system is integrated with the up-gradation of the classrooms performance. It has established a suitable environment for the operation of the new-age education system in academia.[1] The uses of modern accessories to develop and assist smart learning as the teaching assistant have become a familiar circumstance and then advancement into a smart tutoring mode.[2] Depending upon the thought of energy-saving management, smart accessories management in the smart classroom develops an android application-based IoT device for online/offline education and obtains direct management of classroom circumstance[3–5] detail information, coupled with the advancement of smart educational campus. The circumstances are organized by the smart classroom reviving students' passion in schooling, consequently developing the feature of educating[6] to get the full benefit of the IoT network to increase the information management of the classrooms. This chapter displays the traditional classroom and teaching techniques using the IoT platform and LORA technology. We are also designing the software structure to process the data to and from sensors which will be transferred to cloud storage. This system can also track student's attendance and keep track of who leaves out classes, and send an alert notification that assists students to focus on educational work regularly.

2.2 RELATED WORK

In recent years,[7] many institutes have relied on roll-call and paper-based methods to determine the students' attendance. It is challenging and time-consuming for this roll-call and paper-based system to take attendance. In this manual, papers and a variety of stationery materials are used to work out. There had previously been little work completed on the academic attendance monitoring issue. There has previously been software designed to keep track of attendance.[8,9] This process requires that a teacher or staff member enter the data manually, which can be time-consuming and difficult. As a result, no solution has been found. A follow-up procedure is used to integrate the RFID system with the attendance-monitoring system. The database stores the records that are executed in an MS SQL Server database. SQL Server is fast and easy to use, and it can handle very large records, can be accessed from anywhere, and requires little configuration.[10] Similarly, the database in this system must also be manually updated by the staff. As such, the matter remains unresolved. There are also a lot of new ideas being proposed, such as face recognition technology to keep attendance records. However, this system is expensive, and it does not produce accurate results.

2.3 WORKING PROCESS OF PROPOSED SYSTEM

In this prototype we have proposed a finger print-based attendance system and an android application-based personal teaching assistance system that will give a new experience to an online/offline education system. For the attendance system that we developed, we used fingerprint scanners on Android smartphones because we used biometric methods to identify users during the biometric identification research. So we need to take the fingerprints of 100 volunteers (participants) as well as personal information for the identification test and attendance system. Next, we develop a fingerprint matching algorithm. The hardware equipment is also located outside the classroom door for offline purposes. Before entering a classroom, students must have to give their fingerprints for their identification. The attendance of a student is recorded upon identification.

First take a student's/teacher's fingerprint sample and create a digital copy. It consists of a set of features. The features of an individual fingerprint that make it unique are known as a feature set as shown in Figure 2.1.

FIGURE 2.1 Student making their attendance before entering the classroom.

After that the user will retrieve the templates from the repository (database). Thereafter, it compares the fingerprint with the fingerprint templates stored in the database and makes a match-or-no-match decision. Figure 2.2 illustrates the process of identification.

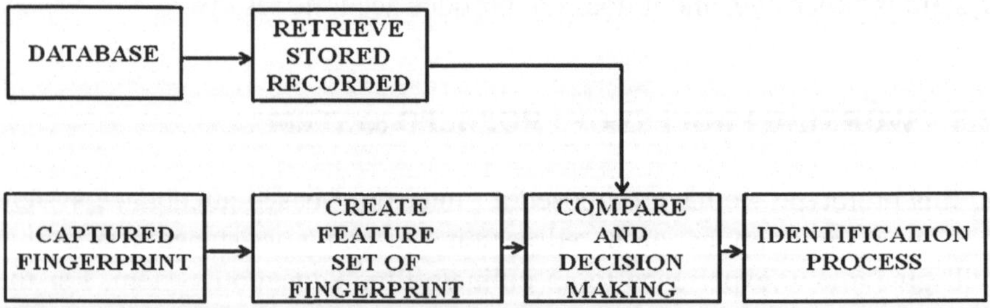

FIGURE 2.2 Represents process of identification.

If the match is found, fetch student information (roll number, department, subjects) from the database. At last we locate the scheduled lecture in the database and mark attendance if the student is present within 30 min of the lecture starting time. (For more information, see the rules). In the event of a verification error or an incorrect fingerprint enrolment, the system returns to its initial state without marking the attendance as shown in Figure 2.3.

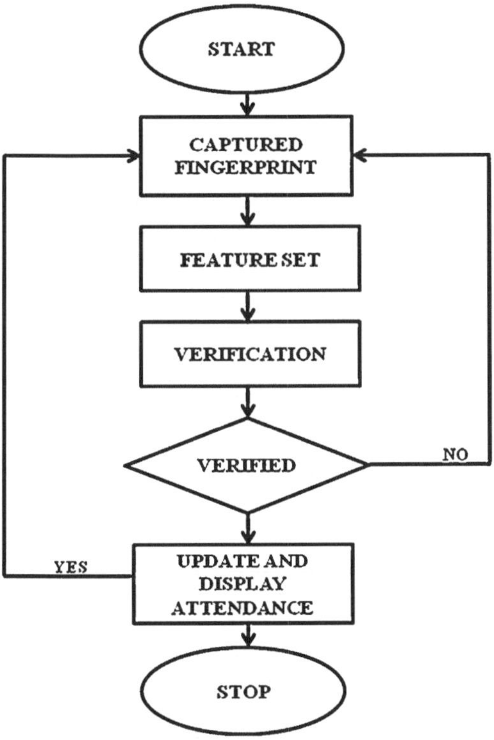

FIGURE 2.3 Flowchart of the attendance system.

2.4 SYSTEM DESIGN

The circuit diagram for the smart classroom is simple. Starting with the power supply, the voltage for Arduino UNO should be within 9 to 12 V, and it is regulated internally by the board to 5 V. All the components are connected to the Arduino UNO with the help of jumper wires. For the attendance system, we have used an optical fingerprint sensor, the Vcc of this sensor connected to a 5 V pin, ground to GND pin, the RX and TX connected to D3 and D2 pins respectively of the Arduino board. The fingerprint sensor takes the input data as a fingerprint, transmitting it to the database with the help of the LoRa SX1278 Transmitter. LoRa SX1278 Transmitter cannot be operated at 5 V, so the Vcc of Lora should be connected to the 3.3 V pin of Arduino. The ground should be connected to the GND pin of the Arduino. Now, connect the RST pin to D9 and the DIO0 to D2. The SPI Pins NSS, MOSI, MISO, SCK are connected to pins D10, D11, D12, D13 of Arduino, respectively, and the same connection goes for the LoRa SX1278 Receiver. The LoRa SX1278 Receiver will receive the data and store it on the Cloud server.

2.5 IMPLEMENTATION

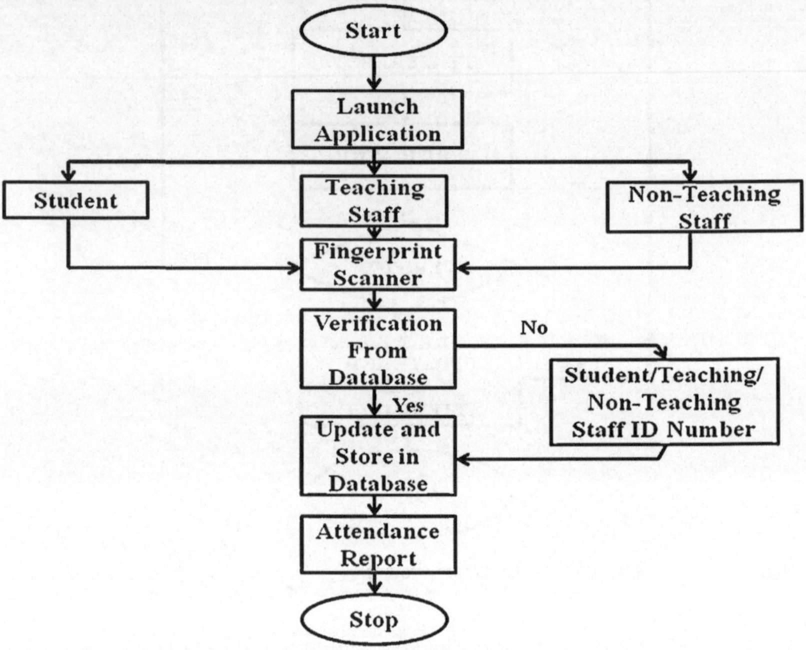

FIGURE 2.4 Flowchart of smart classroom attendance system.

The flowchart shown in Figure 2.4 displays the overall structure of IoT-based Attendance system for smart classrooms. This framework operates basically in two stages. At the first stage, the application turns on, and then the app will show to choose the category (Student/Teaching Staff/ Non-Teaching Staff); then fingerprint will be taken for attendance using the fingerprint scanner. After giving the fingerprint the fingerprint data will be verified. In this process, the verification will work as an interface, by updating and storing the data in the database. At last, the attendance report will be shown where we can see the status of our attendance. The next stage means after the verification of the data will mismatch with the user and some problem in data, in this situation you have to give the manual attendance, where we have to write our ID number; then the data will update and store in the database and then the attendance will be shown.

2.6 APPLICATIONS

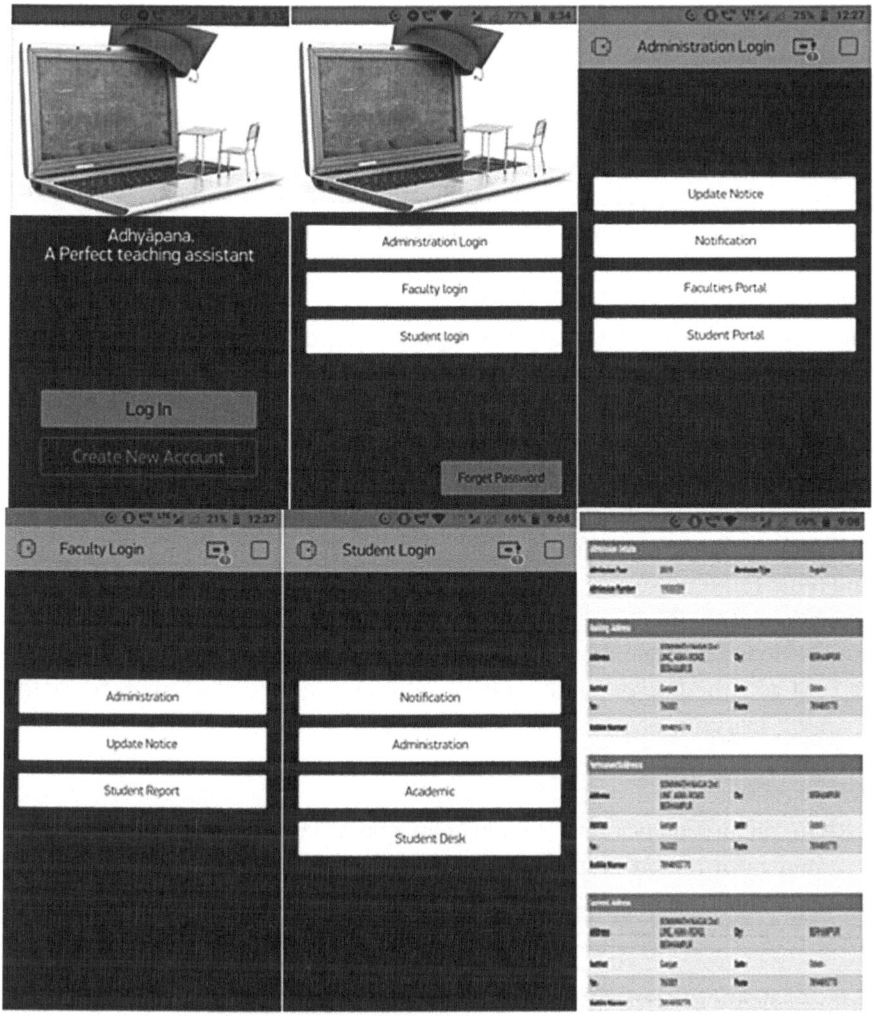

FIGURE 2.5 The overall architecture of smart classroom application.

2.7 CLOUD STORAGE

The main idea behind this chapter is to build a wireless teaching system between the teacher and the student where we need not to use pen and paper for storage of information of students. HTML can be used as an interface between smartphone/pc and cloud server.[11] Smartphones are becoming most essential part of human life as convenient tools for communication irrespective of time and place. That is why we can use mobiles, PDA as teaching instruments. However, these devices are facing many challenges in communications. But these challenges are tackled to some extent by using cloud computing. The integration of cloud computing in PDA offers a betterment to use networks, servers, and other infrastructures, platforms, software etc. It ensures that the applications and data should be available offline also. Smart campus, smart class rooms can be provided by IoT-based cloud computing technology for teaching and management of teacher's and student's activities.[12] Students and staff can use PDA (mobiles, laptops, and touch pads) for easy access of teaching and learning process. It helps teachers for easy monitoring of students and their activities. It allows both online and offline lectures for students which makes a big advantage for students over the manual teaching system.

In our proposed system, HTML5 is used for applications and database. Database covers the information of students and staff which also includes the subjects, assigned teachers, syllabus, identifications, materials, attendance, and other required stuffs. All these information is stored in cloud so that both students and teachers can access the data from anywhere, anytime, and from any device. The classrooms will be provided with laptops or the students can connect themselves with their PDAs and smartphones. It will make the teaching learning process more convenient[13] as shown in Figure 2.5. This system will give an innovative platform for teachers and students to use their pc/smartphones for a smart classroom. Even teachers can take online classes with no inconvenience and students can also access the materials and other required information from any place. Thus data will be available in both online and offline; it will make an advantage for students who skip a class.

2.8 LORA TECHNOLOGY

The LoRa protocol is a secure, low-power, long-range, and low-cost wireless communication technology that can be used as a foundation for IoT. It uses chirp spread spectrum modulation, which has many of the same characteristics as FSK modulation but is able to handle long-distance communications.[14] Sensors, gateways, devices, animals, people, machines, and other objects can be connected wirelessly to the cloud using LoRa. LoRa uses different bands of frequency in different areas.[15] LoRa transmits over megahertz radio frequency bands without a license. Its communications can reach distances of ten miles or more in rural areas using low power. To accomplish this, LoRa's bandwidth should be narrowed.

By developing an internet-of-things platform that provides real-time tracking and management of many educational buildings and their different characteristics,[14,16] the objective of this technology is to establish IoT networks with the help of LoRa so that various kinds of sensors are embedded in IoT networks that can operate in a lot of buildings and spread in several rooms inside each building. LoRa gateways should be installed on the roof of the building or at a height of at least 15 m above the ground.[15] The data collected from various sensors are now uploaded to the cloud and later on, the data can be accessed from the web server by anyone as shown in Figure 2.6.

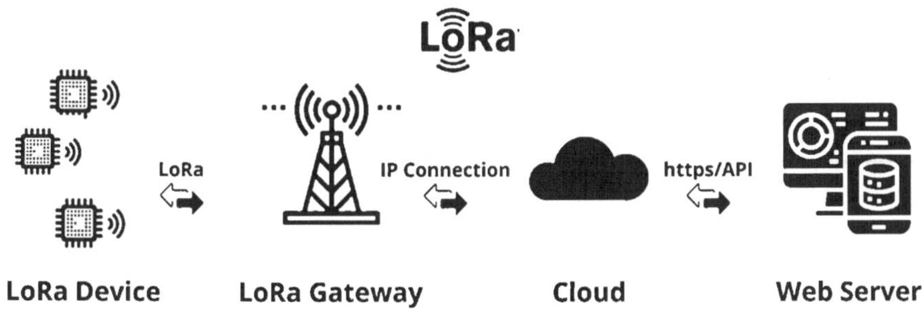

FIGURE 2.6 Shows the working process of LoRa technology.

LoRa technology offers many benefits for an educational institute's operation such as interactive learning: In[17] this era of digital textbooks,[18] Students can engage in learning by providing them with materials, assignments, and recorded class videos for proper understanding of the subject with the Internet of Things. Learners with disabilities: LoRa technology is useful as it can be used to teach and enable the disabled to work and support them. Attendance monitoring system: Educational institutes can benefit from an attendance monitoring system in many ways. For example, it will send an alert message to a student if their attendance is below 75%.

As a result of LoRa technology, e-learning services benefit from an expanded learning ecosystem that integrates a physical and virtual component.

2.9 RESULT

The proposed IoT-based teaching assistant system for smart classroom is being set up and tested over prototype in our college in which we have used finger print sensor for attendance and android application-based personal digital assistance for student as well as for faculty for online teaching.

2.10 CONCLUSION AND FUTURE WORKS

IoT technology will turn out and exist in the real world in the near future. Especially, IoT in the education sector is regarding the ability to learn new things. The IoT-based teaching assistant for the smart classroom for a smart campus creates an evolution in education technique results in high efficiency in classroom teaching methodology. Experimentally, IoT has been proven that a smart classroom with a teaching assistant system is functioning correctly by connecting devices that are being effectively controlled. This system will assist the teacher and students in time-saving and focus on studies. The architecture of our device can be extended to smart home implementation and also in a smart office. Two applications can be converted into a single application for the desirable use of the user. We are planning to implement text-to-speech system work using multiple languages in near future.

KEYWORDS

- **cloud computing**
- **intelligent education system**
- **IoT**
- **LoRa technology**
- **smart classroom**
- **teaching assistant**
- **innovation**

REFERENCE

1. Memos, V.A.; Minopoulos, G.; Stergiou, C.; Psannis, K. E.; Ishibashi, Y. In *A Revolutionary Interactive Smart Classroom (RISC) With the use of Emerging Technologies*, 2020 2nd International Conference on Computer Communication and the Internet (ICCCI), 2020, pp 174–178.

2. Lee, J.; Park, Y.; Cha, M. S. Smart Classroom: Converging Smart Technologies, Novel Content and Advanced Pedagogies for Future of Education. *J. Educ. Vocat. Res.* **2013,** *4*(1).
3. Li, S.; Da Xu, L.; Zhao, S. The Internet of Things: A Survey. *Inf. Syst. Front.* **2015,** *17*(2), 243–259.
4. Marry, W. Disruptive Civil Technologies Six Technologies With Potential Impacts on us Interests Out to 2025, 2013.
5. Hepp, M.; Siorpaes, K.; Bachlechner, D. Harvesting Wiki Consensus: Using Wikipedia Entries as Vocabulary for Knowledge Management. *IEEE Internet Comput.* **2007,** *11*(5), 54–65.
6. Subbarao, V.; Srinivas, K.; Pavithr, R. S. In *A Survey on Internet of Things Based Smart, Digital Green and Intelligent Campus*, 2019 4th International Conference on Internet of Things: Smart Innovation and Usages (IoT-SIU), 2019, pp 1–6.
7. Augusto, J. C. Ambient Intelligence: Opportunities and Consequences of its use in Smart Classrooms. *Innov. Teach. Learn. Inf. Comput. Sci.* **2009,** *8*(2), 53–63.
8. Ghosh, S.; Mohammed, S. K. P.; Mogal, N.; Nayak, P. K.; Champaty, B. In *Smart Attendance System*, 2018 International Conference on Smart City and Emerging Technology (ICSCET), 2018, pp 1–5.
9. Soewito, B.; Gaol, F. L.; Simanjuntak, E.; Gunawan, F. E. In *Smart Mobile Attendance System Using Voice Recognition and Fingerprint on Smartphone*, 2016 International Seminar on Intelligent Technology and Its Applications (ISITIA), 2016, pp 175–180.
10. Talaviya, G.; Ramteke, R.; Shete, A. K. Wireless Fingerprint Based College Attendance System Using Zigbee Technology. *Int. J. Eng. Adv. Technol.* **2013,** *2*(3), 201–203.
11. Jiang, C.; Shi, Y.; Xu, G.; Xie, W. Classroom in the Era of Ubiquitous Computing Smart Classroom. In *Wireless Lans And Home Networks: Connecting Offices and Homes*; World Scientific: Singapore, 2001; pp 14–26.
12. Jansen, W. A.; Grance, T. Guidelines on Security and Privacy in Public Cloud Computing. 2011.
13. Shi, Y.; Qin, W.; Suo, Y.; Xiao, X. Smart Classroom: Bringing Pervasive Computing into Distance Learning. In *Handbook of Ambient Intelligence and Smart Environments*; Springer: New York, 2010; pp 881–910.
14. Gkamas, A. LoRa Technology Benefits in Educational Institutes. In *Interactive Mobile Communication, Technologies and Learning*; Springer: New York, 2019; pp 413–424.
15. Pocero, L.; Tsampas, S.; Mylonas, G.; Amaxilatis, D. In *Experiences From Using LoRa and IEEE 802.15. 4 for IoT-Enabled Classrooms*, European Conference on Ambient Intelligence, 2019, pp 186–202.
16. Workgroup, T. M. A Technical Overview of LoRa and LoRaWAN. *no. November*, 2015.
17. Petäjäjärvi, J.; Mikhaylov, K.; Hämäläinen, M.; Iinatti, J. In *Evaluation of LoRa LPWAN Technology for Remote Health and Wellbeing Monitoring*. 2016 10th International Symposium on Medical Information and Communication Technology (ISMICT), 2016, pp 1–5.
18. Stočes, M.; Vaněk, J.; Masner, J.; Pavlík, J. Internet of Things (IoT) in Agriculture-Selected Aspects. *Agris. On-line Pap. Econ. Inform.*, **2016,** *8* (1), 83–88.

CHAPTER 3

Legendre Neural Network Method for Solving Nonlinear Singular Systems

MURUGESH VEERASAMY[1], SUBASH CHANDRA BOSE JAGANATHAN[2*], CHANDRAMOHAN DHASARATHAN[3], AZATH MUBARAKALI[4], VELMANI RAMASAMY[5], R. KALPANA[6], and NINOSLAV MARINA[7]

[1]*Department of CSE, Koneru Lakshmaiah Education Foundation, Vaddeswaram, AP, India*

[2]*School of Computing Science and Engineering, VIT Bhopal University, Sehore, MP, India*

[3]*Computer Science and Engineering Department, Thapar Institute of Engineering and Technology, Patiala, Punjab, India*

[4]*College of Computer Science, King Khalid University, Abha, Saudi Arabia*

[5]*Computer Science and Engineering Department, Adithya Institute of Technology, Kurumbapalayam, Coimbatore, Tamil Nadu, India*

[6]*Madanapalle Institute of Technology and Science, Angallu, Andhra Pradesh, India*

[7]*University of Information Science and Technology, "St. Paul the Apostle", Ohrid, North Macedonia*

[*]*Corresponding author. E-mail: jsubashme@gmail.com*

ABSTRACT

Legendre Neural Network Method was utilized in this research work to investigate time invariant and time varying nonlinear singular systems. The researchers attained the results from different methods such as RK-Butcher

Intelligent Technologies for Sensors: Applications, Design, and Optimization for a Smart World.
S. Kannadhasan, R. Nagarajan, & Alagar Karthick (Eds.)

algorithms, Legendre Neural Network Method and Singe-Term Walsh Series (STWS) and compared it against the analytical solution of time invariant and time varying nonlinear singular systems. From the comparison, it was inferred that the solution produced by Legendre Neural Network Method is nearby the analytical solution of nonlinear singular systems. One can easily simulate Legendre Neural Network Method in MATLAB. Further, one can obtain the solution for any length of time in time-invariant and time-varying nonlinear singular systems

3.1 INTRODUCTION

Differential equations (DEs) are algebraic relations that exist between functions and their derivatives. These DEs are crucial in the functioning of all kinds of physical systems. The solution for the majority of the problems raised in selected streams such as physics, chemistry, math, and engineering is modeled by either Partial Differential Equations (PDE) or Ordinary Differential Equations (ODE). However, it is not easy to obtain analytic solution for differential equations. Therefore, different numerical methods are considered to find the approximate solution for DEs. However, numerical methods have few limitations, for instance high computational cost.

Numerical methods have been widely used in resolving differential equations and evolved since the first differential equations were introduced. Spectral methods, finite elements, fine volumes, and finite difference are some of the classical approaches followed in spatial discretization of Partial Differential Equations (PDEs).[11] On the other hand, classical methods for discretizing Ordinary Differential Equations (ODEs) include Euler's Method,[12] Runge-Kutta Method, RK-Gill Method,[12] and RK-Butcher Algorithm.[1, 6, 7, 13, 14, 15, 16]

The emergence of artificial intelligence in recent years kindled interest among the researchers toward neural network methods. Various applications make use of neural networks, for instance, control systems,[16] image processing,[13,14] and pattern recognition[14] that possess huge scope for the improvement in the future. Having been loaded with advantages, these neural network method functions with approximation capabilities resulted in massive development of neural network models aimed at resolving differential equations.

In the study conducted by Susmita Mall and Chakravarthy,[11] a novel method was introduced by the researchers to resolve ODE with the help of Legendre Neural Network. In line with this, Legendre neural network was proposed by Hongliang Liu et al.[18, 10] as a solution for Linear Coefficients Delay Differential-Algebric Equations and Singularly Perturbed Differential Equations. Yunlei Yang et al.[20] solved Legendre Neural Network-based algorithm for elliptic PDEs. Yinghao Chen et al.[2] found a solution for Ruin Probability in continuous Time Model on the basis of Block Trigonometric Exponential. A new algorithm was proposed by Toni Schneidereit et al.[8] on the basis of Artificial Neural Network. The intention of the study was to resolve ODEs.

The current study considered Nonlinear Singular system used by Evans et al.[3] However, a different approach was proposed by the author in which Legendre Neural Network Method is incorporated. Here, high accuracy is maintained while the author considered both time-invariant and time-varying cases.

3.2 LEGENDRE NEURAL NETWORKS

In the presence of one input node and one output node, the Single-layer Legendre Neural Network (LeNN) undergoes functional expansion depending on Legendre polynomials. Legendre polynomials constitute a set of orthogonal polynomials obtained as a solution of Legendre differential equations.[11] Legendre polynomials are simply denoted as $L_n(u)$ where n is the order of polynomial and u lies between −1 and 1. These are nothing but a set of orthogonal polynomials that are attained as a solution to resolve Legendre differential equations. Figure 3.1 shows the structure of Legendre Neural Network.

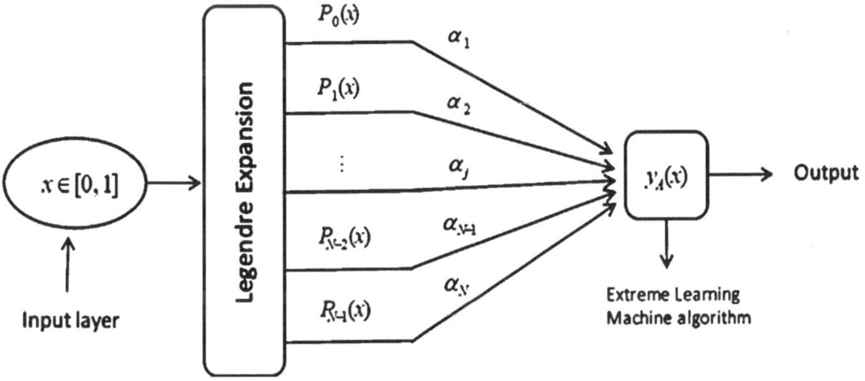

FIGURE 3.1 General structure of Legendre neural network.

Single-layer Legendre Neural Network possesses one input and one output, while its functional expansion occurs on the basis of Legendre polynomial $P_n(x)$.[11] The mathematical model for Legendre Neural Network for "N" nodes polynomial $P_n(x)$ is given herewith

$$y_A(x) = \sum_{j=1}^{N} \alpha_j p_{j-1}\left(w_j x + b_j\right) \tag{3.1}$$

Here, the input value of the network is denoted by x, while the output is denoted by y_A. Further, w_j denotes the input node's weight for the jth hidden node and b_j corresponds to the threshold of the jth hidden node. Here, α_j corresponds to the weight vector of the jth hidden node. To simplify equation (3.1), let us take $w_j = 1$ and $b_j = 0$, then the model in equation (3.1) is calculated as herewith

$$y_A(x) = \sum_{j=1}^{N} \alpha_j p_{j-1}(x) \tag{3.2}$$

On the basis of the universal approximation theorem, the Singularly-Perturbed Differential Equations (SPDEs) represent the analytical solution while $y_A(x)$ represents its approximate solution

$$\left\| y(x) - y_A(x) \right\| = \left\| y(x) - \sum_{j-1}^{N} \alpha_j p_{j-1}(x) \right\| \leq \in \tag{3.3}$$

that is,

$$L_{\in} y_A(x) = C \ on \ \partial I \tag{3.4}$$

utilizes the discretization of intervals $I = \{x_1, x_2, x_3 ......x_n\}$, $f_i = f(x_i)$, and x_B denotes the boundary points and the weight α_j can be solved

$$\begin{bmatrix} L_{\in}\left(\sum_{j=1}^{N} p_{j-1}(x_1')\right) \\ L_{\in}\left(\sum_{j=1}^{N} p_{j-1}(x_2)\right) \\ L_{\in}\left(\sum_{j=1}^{N} p_{j-1}(x_3)\right) \\ \vdots \\ L_{\in}\left(\sum_{j=1}^{N} p_{j-1}(x_B)\right) \end{bmatrix} (\alpha) = \begin{bmatrix} f_1 \\ f_2 \\ f_3 \\ \vdots \\ c \end{bmatrix} \tag{3.5}$$

It is simplified as follows:

$$H\alpha = F \tag{3.6}$$

H matrix is the first left term of equation (3.4) that corresponds to the output matrix of the neural network only next to linear $L\in$ operator and *B*. Here, *f* denotes the right term of equation (3.4). To mitigate the error that occurs between true solution $y(x)$ and approximate solution $y_A(x)$, extreme learning machine algorithm[6] should be incorporated as given herewith for optimization.

$$\min\left\|H(\alpha)-f\right\| \tag{3.7}$$

3.3 NONLINEAR SINGULAR SYSTEMS

Here, the researcher considered the time invariant nonlinear singular system

$$K\dot{x}(t) = Ax(t) + f(x(t)) \tag{3.8}$$

While the value $x(0) = x_0$, K denotes $n \times n$ singular matrix and A corresponds to $n \times n$ matrix. In this equation, $x(t)$ is the n-state vector while "f" is the "n" vector function. To convert the system mentioned above (3.8) into time-varying one, few system components are converted as time-varying ones. There is no need to convert all the elements. Then, the system attains the form as given below.

$$K(t)\dot{x}(t) = A(t)x(t) + f(x(t)) \tag{3.9}$$

In the above equation, $x(0) = x_0$ and $K(t)$ corresponds to $n \times n$ singular matrix. Further, $A(t)$ is a $n \times n$ matrix and $x(t)$ is an n-state vector. The researcher considered f as the "n" vector function. The time-varying singular nonlinear systems are generally deemed to be challenging to resolve in comparison with time-invariant systems. To overcome this characteristic, a number of studies have been conducted earlier with different transform methods. The current study introduces Legendre Neural Network Method with high accuracy so that time-invariant and time-varying singular nonlinear systems are resolved.

3.4 SIMULATION EXAMPLES

The authors considered two examples in this section such as time-invariant and time-varying cases. Here, three methods mentioned earlier, that is,

Legendre neural network method, STWS, and RK-Butcher algorithm were utilized to obtain numerical solutions.

3.4.1 FIRST EXAMPLE

The time-invariant nonlinear singular system is used in this study in line with Campbell[19] and Lin and Yang.[20]

$$K = \begin{bmatrix} 0 & 1 \\ 0 & 0 \end{bmatrix}, A = \begin{bmatrix} -1 & 0 \\ 0 & 2 \end{bmatrix} \text{ and } f(x(t)) = \begin{bmatrix} 0 \\ -x^2 \end{bmatrix} \tag{3.10}$$

with initial condition $x(0) = [0 \quad 0]^T$.

The analytical solutions are

$$x_1(t) = -t$$
$$x_2(t) = \frac{t^2}{2} \tag{3.11}$$

Tables 3.1 and 3.2 provide detailed information on the results (discrete solutions) attained from three methods such as Legendre neural network method, STWS, and RK-Butcher Algorithms (with step size time $t = 0.25$). The table provides information on analytical solutions as well as absolute errors calculated among these methods.

TABLE 3.1 Time-Invariant System (3.10) Solution for Different Values of "x_1."

		$x_1(t)$						
S.No.	**Time**	**Analytical solutions**	**STWS-Solutions**	**STWS-Error**	**RK-Butcher solutions**	**RK-Butcher error**	**LeNN Method**	**LeNN Error**
1	0	0	0	0	0	0	0	0.00000
2	0.25	−0.25	−0.254	0.004	−0.25002	0.00002	−0.25000	0.00000
3	0.5	−0.50	−0.504	0.004	−0.50007	0.00007	−0.50000	0.00000
4	0.75	−0.75	−0.754	0.004	−0.75009	0.00009	−0.75000	0.00000
5	1	−1.00	−1.004	0.004	−1.00014	0.00014	−1.00000	0.00000
6	1.25	−1.25	−1.254	0.004	−1.25017	0.00017	−1.25000	0.00000
7	1.5	−1.50	−1.504	0.004	−1.50019	0.00019	−1.50000	0.00000
8	1.75	−1.75	−1.754	0.004	−1.75022	0.00022	−1.75000	0.00000
9	2	−2.00	−2.004	0.004	−2.00026	0.00026	−2.00000	0.00000

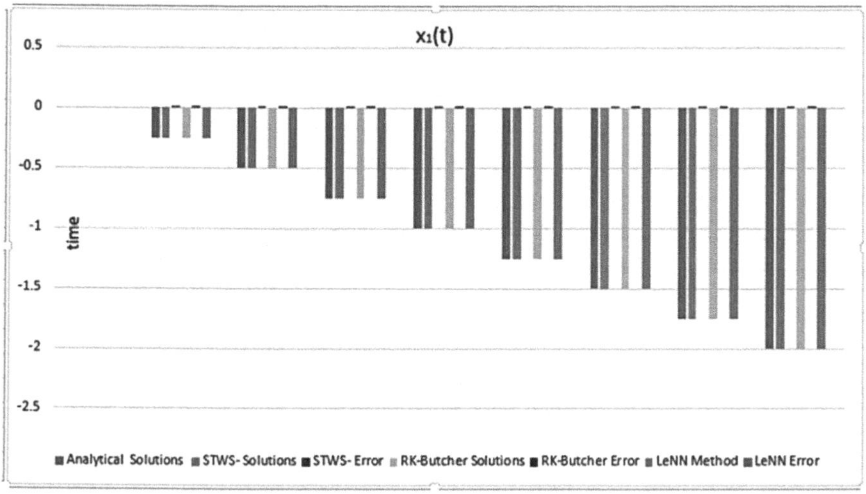

FIGURE 3.2 Time-invariant system (3.10) solution for different values of "x_1."

TABLE 3.2 Time-Invariant System (3.10) Solutions for Different Values of "x_2."

S.No.	Time	$x_2(t)$						
		Analytical solutions	**STWS-Solutions**	**STWS-Error**	**RK-Butcher solutions**	**RK-Butcher error**	**LeNN Method**	**LeNN Error**
1	0	0	0	0	0	0	0	0.00000
2	0.25	0.031	0.032	0.001	0.031002	0.000002	0.03100	0.00000
3	0.5	0.125	0.127	0.002	0.125007	0.000007	0.12500	0.00000
4	0.75	0.281	0.285	0.004	0.281009	0.000009	0.28100	0.00000
5	1	0.500	0.505	0.005	0.500014	0.000014	0.50000	0.00000
6	1.25	0.781	0.787	0.006	0.781017	0.000017	0.78100	0.00000
7	1.5	1.125	1.132	0.007	1.125019	0.000019	1.12500	0.00000
8	1.75	1.531	1.540	0.009	1.531022	0.000022	1.53100	0.00000
9	2	2.000	2.010	0.010	2.000026	0.000026	2.00000	0.00000

3.4.2 SECOND EXAMPLE

The time-varying nonlinear singular system is used in the study as per the literature Hsiao and Wang[18] and Sepehrian and Razzaghi.[18]

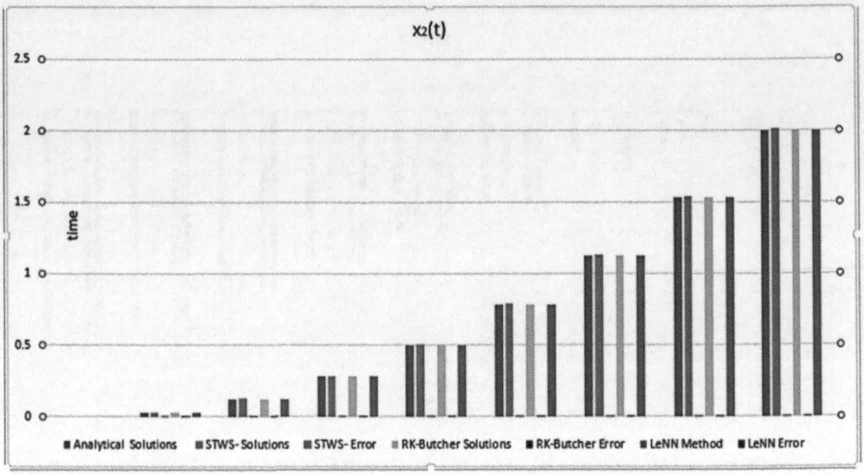

FIGURE 3.3 Time-invariant system (3.10) solutions for different values of "x_2."

$$\begin{bmatrix} 0 & 1 & 0 \\ 0 & 0 & t^2 \\ 0 & 0 & 0 \end{bmatrix} \dot{x}(t) = \begin{bmatrix} tx_1(t) + x_2(t) \\ \exp(t)x_1(t)x_2(t) \\ x_2(t)\left(x_1(t) + x_3(t)\right) \end{bmatrix} x(t) + \begin{bmatrix} 0 \\ 2t^2\exp(-t) \\ 0 \end{bmatrix}, \tag{3.12}$$

with initial condition $x(0) = \begin{bmatrix} 2 \\ 0 \\ -2 \end{bmatrix}$.

The analytical solutions are $x(t) = \begin{bmatrix} 2\exp(-t)(1-t) \\ t^2\exp(-t) \\ -2\exp(-t)(1-t) \end{bmatrix}$ (3.13)

Tables 3.3-3.5 tabulate the results attained from different methods such as Legendre Neural Network Method, RK-Butcher algorithm, and STWS method (step size time $t = 0.25$). The tables also provide the calculated information on analytical solutions and its absolute errors.

Majority of the time intervals tabulated in Tables 3.1–3.5 and Figures 3.2–3.6 postulate that the absolute error in the Legendre Neural network method was negligible (almost absence of error) in comparison with Single Term Walsh Series (STWS) Technique and RK-Butcher algorithm.

TABLE 3.3 Time-Varying System (3.12) Solution Obtained for Different Values of "x_1."

		$x_1(t)$						
S.No.	**Time**	**Analytical solutions**	**STWS-Solutions**	**STWS-Error**	**RK-Butcher solutions**	**RK-Butcher error**	**LeNN Method**	**LeNN Error**
1	0	2	2	0	2	0	2	0.00000
2	0.25	0.778801	0.783458	0.00466	0.778805	4E−06	0.778801	0.00000
3	0.5	0	−0.09452	0.09452	0.000006	6E−06	0	0.00000
4	0.75	−0.47236	−0.49452	0.02216	−0.47239	3E−05	−0.47236	0.00000
5	1	−0.73575	−0.76504	0.02929	−0.73579	4E−05	−0.73575	0.00000
6	1.25	−0.85951	−0.89451	0.035	−0.85956	5E−05	−0.85951	0.00000
7	1.5	−0.89252	−0.94892	0.0564	−0.89258	6E−05	−0.89252	0.00000
8	1.75	−0.86886	−0.92765	0.05879	−0.86893	7E−05	−0.86886	0.00000
9	2	−0.81201	−0.88310	0.07109	−0.81209	8E−05	−0.81201	0.00000

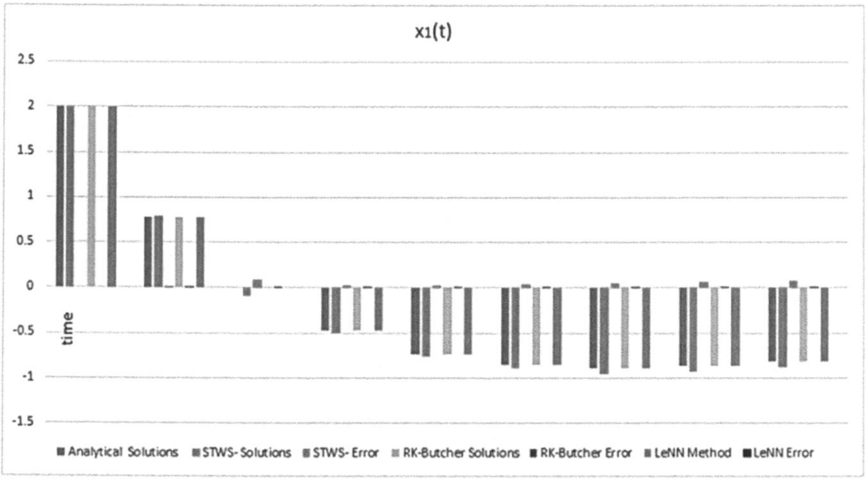

FIGURE 3.4 Time varying system (3.12) solution obtained for different values of "x_1."

3.5 CONCLUSIONS

The results obtained from the study (discrete solutions) in case of time-invariant and time-varying nonlinear singular systems established the fact that the Legendre neural network method is excellent in identifying the state vector. Majority of the time intervals tabulated in Tables 3.1–3.5 postulate that the absolute error in the Legendre Neural network method was negligible (almost absence of error) in comparison with Single Term Walsh

TABLE 3.4 Time-Varying System (3.12) Solutions Obtained for Different Values of "x_2."

S.No.	Time	$x_2(t)$ Analytical solutions	STWS-Solutions	STWS-Error	RK-Butcher solutions	RK-Butcher error	LeNN Method	LeNN Error
1	0	0	0	0	0	0	0	0.00000
2	0.25	0.048675	0.054867	0.00619	0.048676	1E−06	0.048675	0.00000
3	0.5	0.151632	0.175132	0.0235	0.151634	2E−06	0.151632	0.00000
4	0.75	0.265706	0.296506	0.0308	0.265709	3E−06	0.265706	0.00000
5	1	0.367879	0.403879	0.036	0.367883	4E−06	0.367879	0.00000
6	1.25	0.447663	0.494763	0.0471	0.447667	4E−06	0.447663	0.00000
7	1.5	0.502042	0.552042	0.05	0.502048	6E−06	0.502042	0.00000
8	1.75	0.532182	0.605382	0.0732	0.532189	7E−06	0.532182	0.00000
9	2	0.541341	0.635141	0.0938	0.541349	8E−06	0.541341	0.00000

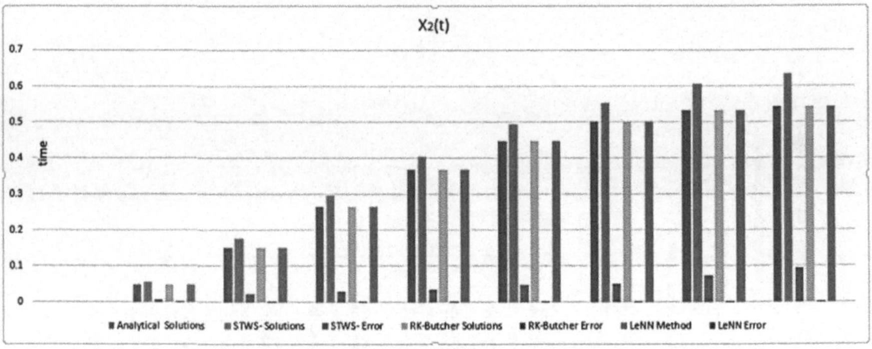

FIGURE 3.5 Time-varying system (3.12) solutions obtained for different values of "x_2."

TABLE 3.5 Time-Varying System (3.12) Solution Obtained for Different Values of "x_3."

S.No.	Time	$x_3(t)$ Analytical solutions	STWS-Solutions	STWS-Error	RK-Butcher solutions	RK-Butcher error	LeNN Method	LeNN Error
1	0	−2	−2	0	2	0	−2	0.00000
2	0.25	−0.77880	−0.78880	0.01	−0.77881	1E−05	−0.77880	0.00000
3	0.5	0	0.000009	9E−06	0	0	0	0.00000
4	0.75	0.472366	0.472367	1E−06	0.472367	1E−06	0.472366	0.00000
5	1	0.735758	0.735760	2E−06	0.735760	2E−06	0.735758	0.00000
6	1.25	0.859514	0.859517	3E−06	0.859517	3E−06	0.859514	0.00000
7	1.5	0.892521	0.892525	4E−06	0.892525	4E−06	0.892521	0.00000
8	1.75	0.868869	0.868875	6E−06	0.868876	7E−06	0.868869	0.00000
9	2	0.812011	0.812019	8E−06	0.812019	8E−06	0.812011	0.00000

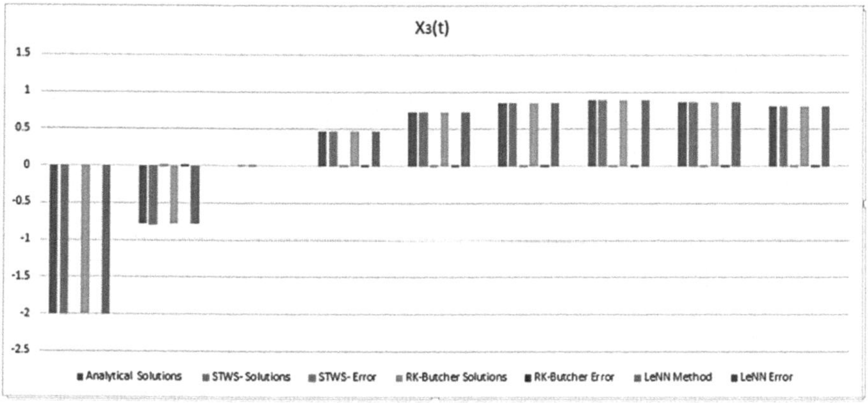

FIGURE 3.6 Time-varying system (3.12) solution obtained for different values of "x_3."

Series (STWS) Technique and RK-Butcher algorithm. The latter methods accomplished low errors in addition to analytical solutions. So, it can be concluded that the Legendre Neural Network Method is the most optimum and suitable method to study time-invariant and time-varying nonlinear singular systems.

KEYWORDS

- **nonlinear singular systems**
- **time invariant and time varying**
- **STWS**
- **RK-Butcher algorithms**
- **Legendre neural network method**

REFERENCES

1. Murugesh, V.; Murugesan, K. Computational Science - ICCS 2006: 6th International Conference, Reading, UK, May 28–31, 2006, Proceedings, Part I (Lecture Notes in Computer Science, 3991) (2006th ed.). Springer, 2006.
2. Chen, Y.; Yi, C.; Xie, X.; Hou, M.; Cheng, Y. Solution of Ruin Probability for Continuous Time Model Based on Block Trigonometric Exponential Neural Network. *Symmetry*. **2020**, *12*(6), 876. https://doi.org/10.3390/sym12060876.

3. Evans, D. J.; Murugesan, K.; Sekar, S.; Kim, H. M. Non-Linear Singular Systems Using RK–Butcher Algorithms. *Int. J. Comput. Math.* **2006**, *83*(1), 131–142. https://doi.org/10.1080/00207160500069888.
4. Hsiao, C. H.; Wang, W. J. State Analysis of Time-Varying Singular Nonlinear Systems Via Haar Wavelets. *Math. Comput. Simul.* **1999**, *51*(1–2), 91–100. https://doi.org/10.1016/s0378-4754(99)00107-x.
5. Huang, G. B.; Zhu, Q. Y.; Siew, C. K. Extreme Learning Machine: Theory and Applications. *Neurocomputing.* **2006**, *70*(1–3), 489–501. https://doi.org/10.1016/j.neucom.2005.12.126.
6. Lee, G.; Howard, D.; ŚLęzak, D. Convergence and Hybrid Information Technology: 5th International Conference, ICHIT 2011, Daejeon, Korea, September 22–24, 2011. Proceedings (Communications in Computer and Information Science, 206) (2011th ed.). Springer, 2011.
7. Lee, G.; Howard, D.; ŚLęzak, D. Convergence and Hybrid Information Technology: 5th International Conference, ICHIT 2011, Daejeon, Korea, September 22–24, 2011, Proceedings (Lecture Notes in Computer Science, 6935) (2011th ed.). Springer, 2011.
8. Schneidereit, T.; Breu, M. Solving Ordinary Differential Equations Using an Optimization Technique Based on Training Improved Artificial Neural Networks. **2020**, *25*(5). Springer Science and Business Media LLC. https://doi.org/10.1007/s00500-020-05401-w
9. Lin, J. Y.; Yang, Z. H. Existence and Uniqueness of Solutions for Non-Linear Singular (Descriptor) Systems. *Int. J. Syst. Sci.* **1988**, *19*(11), 2179–2184. https://doi.org/10.1080/00207728808964111.
10. Liu, H.; Xing, B.; Wang, Z.; Li, L. Legendre Neural Network Method for Several Classes of Singularly Perturbed Differential Equations Based on Mapping and Piecewise Optimization Technology. *Neural Process. Lett.* **2020**, *51*(3), 2891–2913. https://doi.org/10.1007/s11063-020-10232-9.
11. Mall, S.; Chakraverty, S. Application of Legendre Neural Network for solving ordinary differential equations. *Appl. Soft Comput.* **2016**, *43*, 347–356. https://doi.org/10.1016/j.asoc.2015.10.069.
12. Murgesh, V.; Murugesan, K. Comparison of Numerical Integration Algorithms in Raster CNN Simulation. Springer Berlin Heidelberg, 2004. https://doi.org/10.1007/978-3-540-30176-9_15.
13. Murugesh, V. Raster Cellular Neural Network Simulator for Image Processing Applications with Numerical Integration Algorithms. *Int. J. Comput. Math.* **2009**, *86*(7), 1215–1221. https://doi.org/10.1080/00207160701798772.
14. Murugesh, V. Image Processing Applications via Time-Multiplexing Cellular Neural Network Simulator with Numerical Integration Algorithms. *Int. J. Comput. Math.* **2010**, *87*(4), 840–848. https://doi.org/10.1080/00207160802217219.
15. Murugesh, V.; Batri, K. An Efficient Numerical Integration Algorithm for Cellular Neural Network Based Hole-Filler Template Design. *Int. J. Comput. Commun. Control.* **2007**, *2*(4), 367. https://doi.org/10.15837/ijccc.2007.4.2367.
16. Murugesh, V.; Murugesan, K. RK–Butcher Algorithms for Singular System-Based Electronic Circuit. *Int. J. Comput. Math.* **2009**, *86*(3), 523–536. https://doi.org/10.1080/00207160701652912.
17. Liu, H.; Song, J.; Liu, H.; Xu, J.; Li, L. Legendre Neural Network for Solving Linear Variable Coefficients Delay Differential-Algebraic Equations with Weak Discontinuities. *Adv. Appl. Math. Mech.* **2021**, *13*(1), 101–118. https://doi.org/10.4208/aamm.oa-2019-0281.

18. Sepehrian, B.; Razzaghi, M. Solution of Time-Varying Singular Nonlinear Systems by Single-Term Walsh Series. *Math. Probl. Eng.* **2003**, *2003*(3), 129–136. https://doi.org/10.1155/s1024123x03202027.
19. Wood, W. L. Singular Systems of Differential Equations, S. L. Campbell (176 pp.). Research Notes in Mathematics 40, Pitman Advanced Publishing Program 1980. *Int. J. Numer. Methods Eng.* **1980**, *15*(9), 1421. https://doi.org/10.1002/nme.1620150916.
20. Yang, Y.; Hou, M.; Sun, H.; Zhang, T.; Weng, F.; Luo, J. Neural Network Algorithm Based on Legendre Improved Extreme Learning Machine for Solving Elliptic Partial Differential Equations. *Soft Comput.* **2019**, *24*(2), 1083–1096. https://doi.org/10.1007/s00500-019-03944-1.

CHAPTER 4

Characterization of *Syzygium cumini* Particulates-Filled Epoxy Composites

C. BALAJI AYYANAR[1*], K. MARIMUTHU[1], and B. GAYATHRI[2]

[1]*Department of Mechanical Engineering, Coimbatore Institute of Technology, Coimbatore 60014, Tamil Nadu, India*

[2]*Department of Chemistry, Coimbatore Institute of Technology, Coimbatore 60014, Tamil Nadu, India*

[*]*Corresponding author. E-mail: balajiayyanar2007@gmail.com*

ABSTRACT

Different weight percentages of the *Syzygium cumini* waste particles, which degrade naturally, were used to fill E-glass fiber laminate-reinforced composites. Currently being worked on is creating a new composite out of massive amounts of *S. cumini* seed debris. Through the use of open layup molding processes, the two distinct composite combinations with varying reinforcement concentrations (10, 20, 30, 40, and 50 wt%) were created. Functional groups, X-ray diffraction, surface morphology, carbon, hydrogen, nitrogen, and sulfur (CHNS) elemental analyses, and mechanical properties (tensile, flexural strength, and hardness) of constructed composites were assessed. The composite's maximum tensile strength, flexural strength, and Shore D hardness were determined to be 14.04 MPa, 1603 MPa, and 92 SHN, respectively. In comparison to *S. cumini* particles loaded epoxy composites, mechanical characteristics were improved by adding E-glass fiber laminate on both sides of the composite. According to the findings, these composites may be used as panels for applications requiring minimal strength.

Intelligent Technologies for Sensors: Applications, Design, and Optimization for a Smart World.
S. Kannadhasan, R. Nagarajan, & Alagar Karthick (Eds.)

4.1 INTRODUCTION

The degradable and low-density *Syzygium Cumini* Seed (SCS) particulates filled epoxy composites were casted with different weight percentages (10, 20, 30, 40, and 50 wt.%). Current work is focused on making a new composite from vast amount of SCS waste.

Different concentrations of SCS were used to make different combinations of SCS filled epoxy composites; the composites were fabricated by open layup molding methods.

Presence of functional groups and orientation of crystal plane were evaluated through X-ray diffraction. Also, mechanical properties such as hardness, tensile, and flexural strength of fabricated composites were analyzed. The newly fabricated composites can be utilized as low-strength applications. Particulates filled composites are used in many fast-growing industries such as automobiles, aircraft, marine, and biomedical applications due to their low density, higher specific strength, low wear rate, good corrosion resistance, biocompatibility, and biodegradability.[1,2]

The combination of peat ash and HDPE blended to composites was characterized. It was found that tensile strengths and flexural modulus were higher than virgin HDPE,[3] Bagasse Fiber (BF), Waste Fish (WF), and Rice Husk Fiber (RHF) were used as reinforcing biodegradable agents for thermoplastic composites. In general, fillers such as RHF and BF enhance the existing mechanical property of Polypropylene composites when compared to the neat Polypropylene (PP). Some research reported that composites significantly improved the biodegradation and led to a higher degradation rate.[4] The recycled paper blended in the HDPE composite was fabricated through injection molding or by turbo mixing found a tensile strength of 17 MPa. The fibers have been added with maleated polyethylene in different weight percentages such as 1, 3, and 5 wt% by the turbo mixing method.[5]

Wood flour fillers have been added to make composites using either recycled or virgin polypropylene. Adding maleated polypropylene by 3–5 wt.% in the composite considerably enhanced both mechanical properties and stability of the composites. It has been reported that tensile strength of the HDPE-wood flour composite was found to be 9.5 MPa and 23.2 MPa.[6] Meleated ethylene/propylene elastomers were used to make a modified HDPE/Bamboo Flour (BF) composites were also investigated. Without the addition of modifiers, the tensile strength of the BF-filled HDPE composite was found to be 17.5 MPa, which is less than virgin HDPE (18. 9 MPa).[7]

Ayyanar et al. synthesized hydroxyapatite (HAp) from fish scale and filled with HDPE. The cell viability, cell morphology, and levels of cytotoxicity of the HAp-HDPE specimen were studied as per ISO 10993:5 and ISO 10993:12 standards. It was reported after 24 h contact that 30 wt.% HAp-HDPE with MG -63 cells showed none to slight cytotoxicity reactivity. The results of 30 wt.% HAp-HDPE have been compared with the standard reactivity level and confirmed that it showed low cytotoxic level.[8] The cell viability and morphology of the composite specimens were examined and results were compared with (control) reference fresh cell culture medium. Further, cell adhesion and cell growth were investigated on HAp-HDPE composites through FESEM. Since it has no toxicity in the specimens it opens a wide opportunity for the development of byproducts from the fish scale wastes.[9]

There is domestic and industrial use of huge quantities of jamun fruits, especially to produce juice and wine.[10] To date, some parts of jamun fruits have been used as cattle feed due to their high fiber content.[11] The influence of several parameters (extraction temperature, extraction duration, and liquid-to-solid ratio), as well as the application of the RSM technique to jamun seeds, allowed for the quantification of specific phenolic components in the jamun seed extract that was prepared under ideal conditions.[12] Jamun seed accounts about 20% of the fruit weight, which can be a good and cheap source of polyphenols having antioxidant capacity. Jamun seeds are traditionally used in Ayurveda medicine due to their medicinal properties such as antibacterial, antidiabetic, and anti-inflammatory effects.[13]

Large quantities of waste SCS that were available on premises in nature motivated us to carry out this research work. The novelty of these studies was that fillers were extracted from SCS and reinforced with epoxy composite. The aim of this study was to develop new composites material by using SCS particulates filled epoxy composite with different wt.%. The composite was fabricated and mechanical characterizations were carried out, that is, FTIR, XRD, tensile, hardness, and flexural and results were compared.

4.2 EXPERIMENTAL DETAILS

4.2.1 MATERIALS AND FABRICATION OF COMPOSITE

The SCS was collected from local area Coimbatore. The epoxy resin (LY 556), hardener HY-951 was used for this experiment. The SCS particulates filled epoxy composites were fabricated through open layup techniques with low-temperature cured matrix epoxy resin and the corresponding hardener was mixed in a proportion of 10:1 by weight percentage. The casted specimens of the required size as per ASSTM standards of dimension were prepared using a suitable cutter for different characterization studies.

4.3 RESULTS AND DISCUSSION

4.3.1 STRUCTURAL CHARACTERIZATION

The FTIR spectrum revealed the presence of different functional groups in the SCS seed given in Table 4.1. Also the XRD spectrum SCS particulates revealed that the 2θ value for the SCS particulates was found as 17, 33, 47, and 63 corresponding to the crystal plane.

TABLE 4.1 Results of FTIR.

Sl. No.	Wavelength (cm^{-1})	Functional groups assignment
1	2848 to 2915	C—H
2	1603 to 1504	C—C
3	1293	N—O
4	1177	C—H
5	1021	C—N
6	821	C—Cl
7	722	C—H
8	561	C—Br

DSC analysis showed that SCS particulates filled epoxy composite10:3 ratio with a function of increasing temperature. The results showed the melt peak temperature was 75.2 °C. At this transition zone, the composite was changed from a relatively hard and glassy solid to more stretchy or rubbery material when the temperature was increased above 75.2 °C. The TGA revealed the decreasing weight of the SCS particulates filled epoxy composite 10:3 with a function of increasing temperature and the percentage of mass reduction (%). The temperature increases steadily from 0 to 375 °C, the weight is approximately stable and started decreasing when the temperature was increasing from 375 to 450 °C mg. When temperature increases further from 450 to 525 °C and a huge quantity of weight was decreased.

Finally, increasing the temperature further from the entire remaining weight is exhausted. The peak temperature of weight degradation was at 547.8 °C. The major loss of weight is attained between the temperatures of 545 to 575 °C. After this range, to increase the temperature, the materials become oxidized and finally get decomposed.

4.3.2 MECHANICAL CHARACTERIZATION

Tensile strength tests were performed on the naturally occurring SCS particulates reinforced epoxy composites (10, 20, 30, 40, and 50 weight percent specimens) and on SCS particulates filled and reinforced epoxy composites with various weight percentages (10, 20, 30, 40, and 50 weight percent specimens) in accordance with ASTM 638. The tensile strength was high at 10 wt.% and it was gradually decreasing by varying the particulate contents of 20 wt.%, 30 wt.%, 40 wt.%, and 50 wt.% respectively. The strength was gradually decreased by increasing the particulates content which was observed in the test. The flexural strength of the SCS particulates reinforced composites has been decreasing. There may be two reasons for this decline in the flexural strength of these SCS reinforced composites. The first reason may be due to chemical reaction at the interface between the particulates and the matrix may be too weak to transfer the flexural strength. And the second reason may be because of the fact that the irregular corner points of the particulates result in stress concentration in the epoxy matrix. The Shore D hardness values were also measured at five different positions at each sample. It can be inferred that the measured hardness of the composites was improved with the addition of different weight percentages of reinforcement.

4.4 CONCLUSION

The structural properties and functional group of newly developed composites were carried out by FT-IR and XRD analysis. The thermal stability of the composites was carried out through DSC and TGA and found its meting point, energy observation, and mass degradation of the composite. Mechanical properties such as tensile strength, shore D hardness, and flexural strength were found better compared with pure epoxy. The newly composite would be used in low load-carrying structural applications such as panels, cabins, cupboards, and containers.

KEYWORDS

- **jamun seeds**
- **epoxy**
- **composites**
- **DSC**
- **TGA**
- **FTIR**
- **XRD**

REFERENCES

1. Sanjay, M. R.; Madhu, P.; Jawaid, M.; Senthamaraikannan, P.; Senthil, S.; Pradeep, S. *J. Clean. Prod.* **2018**, *172*, 566–581.
2. Jagadeesh, P.; ThyavihalliGirijappa, Y. G.; Puttegowda, M.; Rangappa, S. M.; Siengchin, S. *J. Natur. Fibers.* **2020**, 1–16.
3. Cao, Z.; Daly, M.; Geever, L. M.; Major, I.; Higginbotham, C. L.; Devine, D. M. *Compos. Part B.* **2016**, *94*, 312–321.
4. Nourbakhsh, A.; Ashori, A.; Tabrizi, A. K. *Compos. Part B.* **2014**, *56*, 279–283.
5. Valente, M.; Tirillò, J.; Quitadamo, A.; Santulli, C. In *Electronic proceeding of: 5th International Conference on Innovative Natural Fibre Composites for Industrial Applications,* 2015. *ISBN 9788890924002.*
6. Adhikary, K. B.; Pang, S.; Staiger, M. P. *Compos. Part B.* **2008**, *39*(5), 807–815.
7. Liu, H.; Wu, Q.; Han, G.; Yao, F.; Kojima, Y.; Suzuki, S. *Compos. Part A.* **2008**, *39*(12), 1891–1900.
8. BalajiAyyanar, C.; Marimuthu, K. *Polym. Polym. Compos.* **2020**, *28*(4), 285–296.
9. Balaji Ayyanar, C.; Marimuthu, K.; Gayathri, B.; Sankarrajan. *Polym. Polym. Compos.* **2020**, 0967391120981551.
10. Patil, S. S.; Thorat, R. M.; Rajasekaran, P. *J. Adv. Lab. Res. Biol.* **2012**, *3*(3), 200–203.
11. Sagar, N. A.; Pareek, S.; Sharma, S.; Yahia, E. M.; Lobo, M. G. *Compreh. Rev. Food Sci. Food Saf.* **2018**, *17*(3), 512–531.
12. Balyan, U.; Sarkar, B. *Int. J. Food Proper.* **2017**, *20*(2), 372–389.
13. Shrikan Baslingappa, S.; Nayan Singh J, T.; Meghatai M, P.; Parag M, H.. *Food Nutr. Sci.* **2012**, *2012*.

CHAPTER 5

HCI: Designing a Smart Tool for Analyzing Human Brain Signals and Operating Smart Home Devices

M. TAMILSELVI[1], R. GEETHA[2], ANITHA G.[3*], J. A. DHANRAJ[4], and V. MOHANAVEL[5]

[1]*Department of Mechatronics Engineering, T.S. Srinivasan Centre for Polytechnic College and Advanced Training, Vanagaram, Chennai 600095, Tamil Nadu, India*

[2]*Department of Information Technology, Saveetha School of Engineering, Saveetha Institute of Medical and Technical Sciences (SIMATS), Chennai 602105, Tamil Nadu, India*

[3]*Department of ECE, Saveetha School of Engineering, Saveetha Institute of Medical and Technical Sciences (SIMATS), Chennai 602105, Tamil Nadu, India*

[4]*Centre for Automation and Robotics (ANRO), Department of Mechanical Engineering, Hindustan Institute of Technology and Science, Padur, Chennai 603103, Tamil Nadu, India*

[5]*Centre for Materials Engineering and Regenerative Medicine, Bharath Institute of Higher Education and Research, Chennai 600073, Tamil Nadu, India*

[*]*Corresponding author. E-mail: anipsg09@gmail.com*

ABSTRACT

The main motto of this article is to provide a smart home automation system using Human Brainwave Signals, in which it is helpful to elder people,

Intelligent Technologies for Sensors: Applications, Design, and Optimization for a Smart World.
S. Kannadhasan, R. Nagarajan, & Alagar Karthick (Eds.)

patients and other related people. To provide an intelligent working nature to residence to automate the home based on brain based Electroencephalogram (EEG) signals acquired by using the Neurosky Brainwave analyzing tool. This system provides ability to patients or elder people and others can operate electronic devices based on their human brain signals. The logic of Internet of Things (IoT) is associated into the proposed Smart Home to operate the electronic devices from remote regions as well without any range limits. To provide a high range of efficiency to operate the smart home operations using Brainwave Signals.

5.1 INTRODUCTION

Human Computer Interaction (HCI) is considered to be the most significant and effective area in the information technology and communication industry. Machine-to-machine communication provides an effective strategy till now to operate the devices based on human or systematic triggers. However, some patients and disabled persons cannot operate such electronic devices based on triggers due to their illness. Another constraint is to take care of remote actions, in which the signals from such a kind of local communication scheme provide an ability to operate the devices within a few meters of range such as 10 to 20 feet. To avoid the issues such as range restrictions, instant triggering for disabled persons, patients, as well as others to operate the electronic devices easily without any delay, a new Smart Home Automation scheme is required with respect to HCI logic called Brainwave based Smart Home Device Triggering (BSHDT). This kind of Human Computer Interfacing logic is helpful in several situations as well as it establishes a link between the human brainwave signals and the hardware device associated with the smart device. The primary objective is to enable human-free connection with the surrounding environment, which could be the sole means for individuals in a locked state to communicate with the outside environment. Numerous methodologies are described in this article where peoples' better and healthier brain signals are registered and evaluated to regulate some smart home devices as well as a mechanical arm for self-feeding to assist people with disabilities who cannot move or speak in living autonomously in a brain-controlled platform. The Neurosky Brainwave Sensor acquires brainwave signals from the human brain and processes it based on the ThinkGear module connected into it.[1] The ThinkGear module accumulates the EEG signals and operates the electronic devices over the smart home. The brainwave signals are collected into the computer in a wireless manner and operate the electronic devices based on the triggers. The Brainwave Sensor is connected into the computer via Bluetooth, in which the computer end brain signals are processed with respect to different levels of degree variations such as sleep state, alert state, sedation state, eye blink state, and eye movements including left, right, up, and down. Based on these signal variations triggers are generated in a numerical manner and passed to the hardware unit to control the electronic devices in an intelligent manner.

5.1.1 MAJOR OBJECTIVES OF THIS STUDY

The major objective of this smart health monitoring system is to provide a sufficient health care details management scheme without any flaws over server handling to the users. This provision allows several patients to avoid keeping the physical records everywhere as well as this scheme allows the respective guardian to view the details of the patients from anywhere at any time. The adaptation of Internet of Things (IoT) allows the Smart HealthCare Monitoring Kit (SHCMK) on the patient end to connect with the server end and store the health details instantly without any delay. The security concerns are the major concentration over this study as well as the respective security perceptions are carried out in two ways such as Access Control and the Health Data Security.

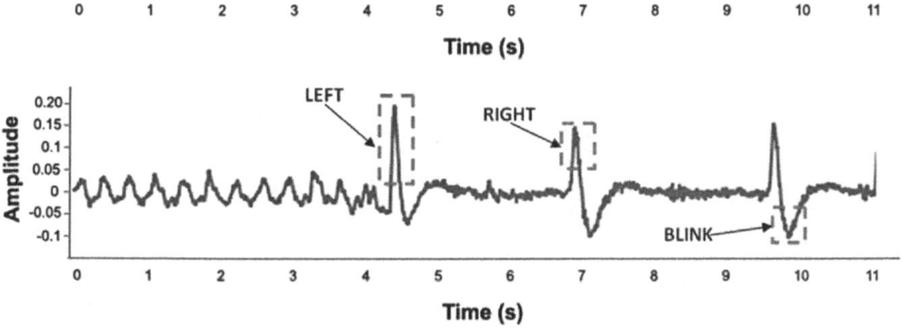

FIGURE 5.1 Brainwave signals with respective triggers.

The Access Control logics are handled in association with the adaptation of RFID and the data security is achieved by means of the cipher mechanism called Advanced Encryption Standard (AES). A supervised learning scheme is utilized to provide a good healthcare-assisted prediction to the patient and doctors with respect to the analysis of the existing health records. The existing patient health records are maintained as a dataset and preserved separately over the server unit for processing. Those records are processed according to the machine learning principles and created a model based on the training samples. Based on the created model the health prediction for the present patient is analyzed as well as this testing record is also appended to the dataset for further reference. With respect to all these specialized sensors, latest technologies, supervised learning abilities, and the security principles, a novel smart healthcare monitoring system is introduced with intellectual skill set. The proposed approach is termed Enhanced Learning based Healthcare Monitoring System (ELHMS), in which it associates all the specified techniques in a proper way to provide an efficient healthcare surveillance scheme to the people. The following summaries will illustrate all these details in a clear manner with graphical representations. Figure 5.1 illustrates the clear perception of EEG brainwave signals in a graphical manner, in which the x-axis indicates the time variations and the y-axis indicates the amplitude.

The states acquired from the brainwave signals are categorized into several types such as: (1) Delta: for identifying the unconscious state, (2) Theta: for identifying an imaginary or dream state, (3) Low-Beta: for identifying the relaxed, focused, or both integrated state, (4) Mid-Beta: for identifying the Thinking or Analyzing State, (5) High-Beta: for identifying the alert state, (6) Gamma: for identifying the information processing state, and (7) Mid-Gamma: for identifying the eye movement states. Table 5.1 illustrates the view of all the mentioned brainwave types, respective state conditions, and the associated frequency ranges in detail with hertz unit specification.

5.2 BRAINWAVE-BASED SMART HOME DEVICE TRIGGERING

Nowadays, smart technologies and the associated devices are more popular as well as everyone likes that kind of adaptation due to its simplicity and easiness. Similarly, the concept of Smart Home Automation is introduced to ease the operations of triggering the electronic devices in the residence automatically instead of manual operations. The purpose of home automation is to make the device automatically switched on when the user requires;

similarly, it will get off when it is not required. In this case, the user needs to operate the devices according to their convenience. Many people's quality of life has improved as a result of the ability to control home equipment such as light/lamp, fan, window doors, and heaters through the usage of smart home automation systems. Regrettably, a segment of the population is unable to benefit from technological improvements. All these cases of automating the residence by means of manual triggering to normal individuals are not an issue, but consider the case of people who cannot speak and move. For example, consider the patients with diseases such as Arthritis, Palsy, and Neural Diseases. These kinds of patients cannot operate the smart home automation system triggers through the application or any other operation modes. So, a novel and intelligent Smart Home Automation system is required to provide an intelligent solution to such a category of people.

TABLE 5.1 Brainwave Signal Types and Their Associated Specifications.

Brainwave types	Conditions/state	Frequency range
Delta	Unconscious state	0.1–3 Hz
Theta	Imaginary or dream state	4–7 Hz
Low-beta	Relaxed, focused or both integrated state	13–15 Hz
Mid-beta	Thinking or analyzing state	16–20 Hz
High-beta	Alert state	21–30 Hz
Gamma	Information processing state	31–40 Hz
Mid-gamma	Eye movements	41 Hz and above

This article is intended to design a new Smart Home Automation architecture with Human Computer Interfacing (HCI) features such as Brain Computer Interface (BCI), in which it provides an ability to acquire the signals from the human brain and control the electronic devices in the residence in an intelligent manner. This article proposed a new methodology called Brainwave based Smart Home Device Triggering (BSHDT) to identify the brain wave signals of the people and operate the electronic devices residing in residence in an intelligent manner. Based on this technology, an individual can operate up to eight numbers of electronic devices in an intelligent manner based on eight different triggers acquired by using brainwaves. The following are the signal states used to operate the devices in an intelligent manner, such as sleep state, alert state, sedation state, eye blink state, and eye movement states including left, right, up, and down.

TABLE 5.2 Device Operations Based on Brainwave Signals.

Brainwave state signals	Electronic devices	Operations
Sleep state	Door lock or curtain motor	Closing the door lock or closing the window curtain based on DC motor rotation controls.
Alert state	Door lock or curtain motor	Opening the door lock or opening the window curtain based on DC motor rotation controls.
Sedation state	Fan	ON/OFF
Eye blink state	Light	ON/OFF
Left movement of the eye	Motor pump	ON/OFF
Right movement of the eye	Air conditioner	ON/OFF
Upside movement of the eye	Air conditioner	Increasing the A.C. level
Downside movement of the eye	Air conditioner	Decreasing the A.C. level

The operating frequency and the associated specifications are clearly mentioned in Table 5.1. Table 5.2 illustrates the perception of electronic device operations according to the brainwave signals, in which the following electronic device operations are just a sample for understanding, but the person can set any devices according to their convenience. Given that Electroencephalography signals are not relying on the typical neural system and muscular connections,[2] they might be utilized to construct a smart home regulation system. The smart home automation design incorporates a Brain signal to collect Electroencephalography data, transform them to queryable information, as well as subsequently transform them to usable signals that may be utilized to manage a simple smart home automation system that is not contingent on customer gesture controls.

Figure 5.2 illustrates the process of Brainwave-based Smart Home Device Triggering with respect to the logic of Application Programming Interface. The logic of brain waves is controlled by using Electroencephalography signals and these signals are accumulated by using computer interfacing entity. Once the signals are acquired, the brainwave processing unit estimates the signal ranges and provides a trigger according to the associated devices over the residence. The EEG signals are handled by means of either one of the two application programming interface tools such as C# or MATLAB. Both these tools support the EEG signal processing by means of the Think-Gear library associated with it.

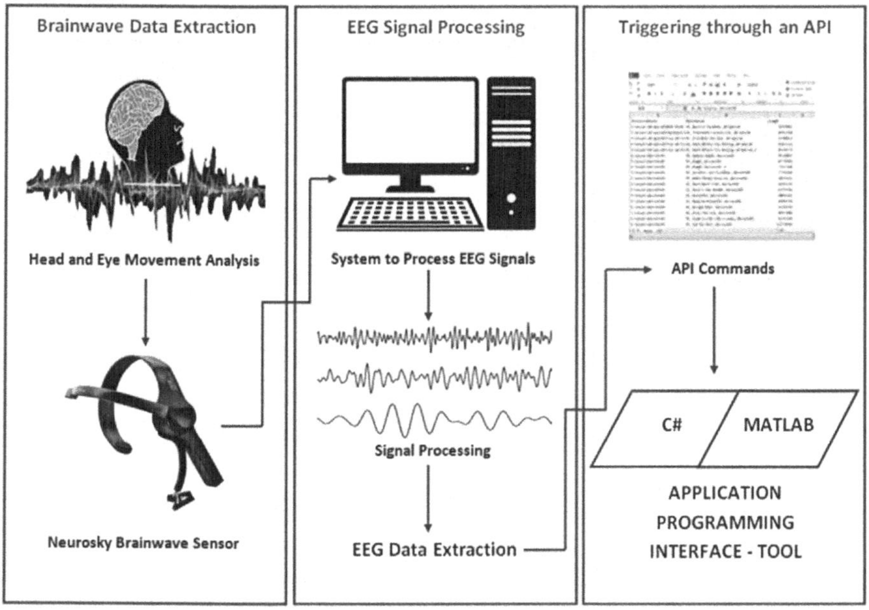

FIGURE 5.2 BSHDT signal processing diagram.

5.3 BRAINWAVE PROCESSING INTERFACE

The Human Computer Interface (HCI)-based brainwave signal analysis logic is a technology for controlling a computer only by brain activity without the use of muscular activity or gesture-related activity.[3] While there are various types of Human Computer Interface, they all rely on either motor imagery or P300 detection.[4] The P300 event-related potential was generated in the same subjects using audio stimulation, while their eyes were open or closed.[5] Visually elicited potential-based Human Computer Interface devices are now the quickest way to develop noninvasive HCI management. A Human Computer Interface-based brainwave monitoring is one of the quickest emerging areas of intensive studies, encompassing neurology, software engineering, system architecture, and clinical reintegration.[6] Two distinct forms of Human Computer Interface are used and there are two types of electrode placements presented into it: (1) Invasive, which needs medical attention to implant sensors within the brain and (2) Non-invasive, in which it just involves sensors to be implanted on the head.[7] Richard Canton developed functional magnetic resonance imaging based on Electroencephalography signals in 1875 using exposed monkey and rabbit brain[8] as well as it is a technique that is utilized to monitor the electromagnetic activity in the brains of animals. The procedure uses small metal plates with a device connected to the head to monitor the brainwave signal pattern, in which it is subsequently transmitted as waveforms. There are various sophisticated equipment accessible that can quantify the Electroencephalography from the brain using electrodes as sensing devices. Several gadgets employ a variety of numerous techniques and frameworks when developing headsets capable of recording electrical impulses.

Mechref et al.[9] proposed a paper related to providing a smart living environment for people by using a brain–computer interface. In this paper,[9] the authors illustrated that the Human Computer Interface establishes a strong link between the human brain and the computational program. Its primary objective is to enable nonmuscular connection with the outside environment which may have been the sole means for individuals in a closed state to communicate with the outside universe. Different methodologies are discussed in this paper[9] where customers' better and healthier brainwave patterns are documented as well as analyzed to regulate several compatible smart home devices and a mechanical arm for self-feeding in order to aid people with disabilities who really cannot move or speak in living individually in a brainwave operating environment.

Furthermore, the application utilizes the Steady-State Visual-Evoked-Potentials neural activity caused by visual information to determine the wavelength at which individual is gazing but also then transmits the appropriate permission to change the mechanical system to the proper plates to serve themselves. Additionally, a WiFi-assisted smart home management system is proposed,[9] in which intensity of light and speed of the fan are changed autonomously or voluntarily via the suggested approach,[9] dependent on the person's individual physical response. Eventually, this is a convenient method for allowing the customer to select among two alternatives, in which it is provided using electroencephalogram-assisted RGB-based color identification.

Hasan et al.[10] proposed a paper related to acquisition, interpretation, and deployment of brain signals using Brain Computer Interfacing options. In this paper,[10] the authors illustrated that Machine-to-Machine and Device-to-Device communication paradigm provide quick, efficient, and independent interaction, which benefits gadgets, individuals, computers, and materials. Furthermore, other implementations, especially those in a Smart Home

Network, necessitate a propulsion signal from a person. Individuals with disabilities may have additional challenges when it comes to generating an actuating signal for Internet of Things enabled household equipment. This article[10] studies the collection of human brainwave signals utilizing an emotionally charged Epoch headset as well as the interpretation and manipulation of a computational track pad. Pressing and pointer movements are examples of computer mouse working movements. This work[10] has the potential to be expanded to manage other gadgets, household utilities, and also to regulate power consumption through giving accurate appliances with processing abilities for the pleasure of a resource-constrained person with a disability in a home automation based on Home Area Network.

The procedure to construct a Human Computer Interface system consists of three major processes that are essential as seen in Figure 5.3. The three major processes needed to develop a HCI are listed further such as: (1) Signal Acquisition, (2) Signal Processing, and (3) Execution.

i) **Signal Absorption:** This process of signal collection is used to obtain an electromagnetic representation of the brainwave signals.
ii) **Signal Processing:** This mechanism is used to process the obtained signal, in which it is next deciphered by using filtration algorithms as well as interpretation-based computations.
iii) **Execution:** This execution phase is used to process the categorized output signals, in which it is the outcome of the process that can be viewed using an instrument called oscilloscope as well as it is used as an interface to the appropriate electrical portable gadget.

According to the specifications of Figure 5.3, the Electroencephalography signals are accumulated and processed by using the proposed Brainwave-based Smart Home Device Triggering logic. Initially, the brainwave signals are absorbed by using the Neurosky brainwave sensor and that signals are transmitted into the processing unit by using the inbuilt communication medium called Bluetooth. The accumulated features are evaluated under the following norms such as preprocessing, feature extraction, signal evaluation, and classification. The preprocessing stage is an important stage to accumulate the brainwave signals and optimize the collected signals into a common standard for evaluating further. The feature extraction process removes the noise level presented into the brainwave signals and provides noiseless signals for processing. The signal evaluation block analyzes the processed signals with respect to the corresponding values as presented in Table 5.1. The classification phase analyzes the exact category of the signal and sends an appropriate trigger to operate the respective electronic device in an intelligent manner.

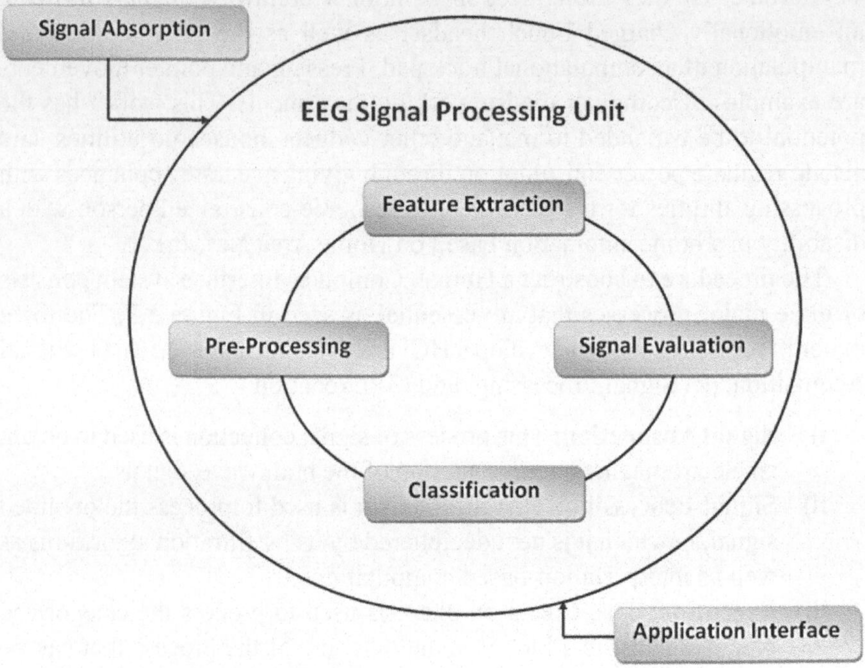

FIGURE 5.3 Processing nature of HCI-based brainwave processing.

5.4 NERVOUS SYSTEM ASSOCIATION WITH BRAINWAVE SENSOR

The nervous system has five senses that allow it to interact with its environment, and each of these sensations is intimately related to the brain. The primary role of certain central nervous system activities, such as (certain instances), is universal in all humans. The entire human body is composed of trillions of neuron cells and this power grid regulates all internal functions. To transport data between the peripheral and fully independent nervous systems, to operate the brain as well as spinal cord to operate muscular movements, and to operate body parts, trillions of electrical impulses must always be produced in the brain and nervous framework. These electromagnetic pulses are transmitted by electrical and chemical potentials in neurons. The purpose of this article is to explore how a computer mouse operates when a human head movement and facial gestures are used, such as wining, smiling, and so on. And according to preliminary presentation of human neural networks, the brain may generate electrochemical stimulation that can be collected and analyzed into an electronic format to be sent to a computer as a command. To explore the process of brain signal extraction, it is necessary to have a better understanding of the functions that generate electric pulses within the human body. Neurons transfer information by passing messages to other neurons or different types of cells, such as muscles. Neurons communicate by electrical signals, beginning with one component of the neuron and progressing to the next.

The neural network is composed of three distinct types of neurons, of which perception neurotransmitters control facial muscles. Qu et al.[11] proposed a paper related to a simulated smart home automation system powered by electroencephalogram signals. In this paper,[11] the authors illustrated that VR (Virtual-Reality) is a comparatively recent innovation that uses a computer to generate a virtual world, providing consumers with a 3D comprehensive learning process. Additionally, using Brain Computer Interface (BCI) technology, users can communicate and engage with gadgets without using a handle, keyboard, or mouse. By combining virtual reality and brain–computer interface (BCI), this article suggests and constructs a revolutionary platform dubbed the virtual smart home, which is controlled by electroencephalograph (EEG) (VSH-EEG). P300 is a type of event-related potential (ERP) that can be produced in response to a random input. The paradigm display of the P300 is included in the virtual smart home via the VSH-EEG system. The conscience Bayes nonlinear discriminate analysis algorithm is used to extract, evaluate, and classify user's P300. Additionally, a sequential BCI methodology predicated on P300 can provide a sufficient quantity of controlling commands and require consumers to freely operate appliances in a virtualized home. Eight subjects are requested to manipulate the respective items to determine the system's accuracy and effectiveness. Finally, the experiment results demonstrate that the VSH-EEG system is acceptable, allowing consumers to experience a novel form of entertainment.

Rajmohan et al.[12] proposed a paper related to Smart Home Automation using Brian Computer Interface logic. In this paper,[12] the authors presented that Smart Home Automation is reshaping lifestyles through the use of numerous sensors and interfaces. The Brain Computer Interface (BCI) technology is critical for combining the embedded processor and the household appliance in an aesthetic manner. Our study intends to establish a prominent mode of interaction between BCI and processors to make common tasks such as controlling home appliances considerably easier. As brain signals are collected using a sensor worn on the head, the sensor module does the fundamental preprocessing and data refining before transmitting it to the intermediate interface, which is an Android Smartphone. The smartphone application receives the signal and attempts to decode it according to the predefined criteria. Additionally, if the brain reading matches the expected reaction, it generates the controller signal. This system[12] incorporates numerous features, including a novel coding technique that enables control of a larger number of devices and protects against illegal control. Other notable features include auto sleep on inadequate signal feeds, device programming to minimize false toggles, three-way control of the system, and scalable architecture. It is highly beneficial for individuals with a high level of impairment.

5.5 CONCLUSION AND FUTURE DIRECTIONS

In this article, a brain-controlled smart home automation system is presented with the help of proposed approach called Brainwave-based Smart Home Device Triggering (BSHDT), in which it is capable of activating various home devices such as lights, switches, and ceiling fans and this was presented in a clear manner. Devices are controlled exclusively via aural impulses acquired from the client. Different types of brainwave signals enabled smart home electronic gadgets were utilized to construct the system: a smart lamp, window curtains, window/room doors, fans, and so on connected to a smart BSHDT socket. Different people were used in the testing, which comprises five trials for each of these two states. Two parameters were evaluated: initial and final setting efficiency and the platform's fast response to finish each task. The task for the smart plug was to switch things upside down. The intelligent light bulb required activities to switch on reduce, brighten, and switch off. Subjects were able to turn on and off the intelligent fan with a reliability of 100 and 91.75%, respectively. Furthermore, task correctness was 92% for turning on the lamp, 67% for dimming, 58.33% for brightening, and 50% for turning off. The experimental results demonstrate that smart home automation equipment can be effectively and successfully operated by utilizing an individual's steady-state response.

Technological advancements have resulted in the development of highly effective smart gadgets and services that can be operated via a home automation system. The suggested system can be expanded to control an increasing number of home devices and applications, with an emphasis on disabled users. Two distinct auditory tones were employed to operate two different home gadgets in the proposed system. It can be developed to incorporate a broader spectrum of aural tones, enabling control of a broader variety of household devices, including garage door openers, automatic door locks, thermostats, and security systems, as well as many emerging smart gadgets. Accuracy is the second factor that can be improved. Enhancing the signal-to-noise ratio and peak thresholds base for a particular individual will result in an increase in the system's accuracy. The suggested system in this study is intended for use with persons who have impairments, but it can easily be expanded to a broader user basc, including individuals who do not have disabilities. Additionally, an actual system may be implemented with a single Arduino, and a computer equipped with a processing interface might be utilized to improve accuracy. Accuracy would be improved by transmitting more information, such as the sort of tone being played, to minimize false alarms.

KEYWORDS

- **IoT**
- **HCI**
- **EEG**
- **BSHDT**
- **GPS**
- **sensor**

REFERENCES

1. http://neurosky.com/biosensors/eeg-sensor/ultimate-guide-to-eeg/
2. Ali, M. M.; Griffiths, A.; Hasan, M. S. In *Study of Brain Signal Extraction, Processing and Implementation*, 2018 24th International Conference on Automation and Computing (ICAC), September 2018, pp 1–5; IEEE.
3. Schalk, G.; Allison, B. Z. Noninvasive Brain–Computer Interfaces. In *Neuromodulation*; Academic Press: New York, 2018; pp 357–377.
4. Mudgal, S. K.; Sharma, S. K.; Chaturvedi, J.; Sharma, A. Brain computer Interface Advancement in Neurosciences: Applications and Issues. *Interdisciplinary Neurosurg.* **2020,** *20*, 100694.
5. Kirasirova, L.; Bulanov, V.; Ossadtchi, A.; Kolsanov, A.; Pyatin, V.; Lebedev, M. A P300 Brain-Computer Interface With a Reduced Visual Field. *Frontiers Neurosci.* **2020,** *14*, 1246.

6. Šumak, B.; Špindler, M.; Pušnik, M. In *Design and Development of Contactless Interaction With Computers Based on the Emotiv EPOC+ Device*, 2017 40th International Convention on Information and Communication Technology, Electronics and Microelectronics (MIPRO), May 2017, pp 576–581; IEEE.
7. Kanaga, E. G. M.; Kumaran, R. M.; Hema, M.; Manohari, R. G.; Thomas, T. A. In *An Experimental Investigations on Classifiers for Brain Computer Interface (BCI) Based Authentication*, 2017 International Conference on Trends in Electronics and Informatics (ICEI), May 2017, pp 1–6; IEEE.
8. Zeigler-Hill, V.; Shackelford, T. *Encyclopedia of Personality and Individual Differences*; Springer: New York, 2020.
9. Tabbal, J.; Mechref, K.; El-Falou, W. In *Brain Computer Interface for Smart Living Environment,* 2018 9th Cairo International Biomedical Engineering Conference (CIBEC), December, 2018, pp 61–64; IEEE.
10. Ali, M. M.; Griffiths, A.; Hasan, M. S. In *Study of Brain Signal Extraction, Processing and Implementation*, 2018 24th International Conference on Automation and Computing (ICAC), September, 2018, pp 1–5; IEEE.
11. Zhao, W.; Zhang, X.; Qu, J.; Xiao, J.; Huang, Y. In *A Virtual Smart Home Based on EEG Control*, 2019 IEEE 9th International Conference on Electronics Information and Emergency Communication (ICEIEC), July 2019, pp 85–89; IEEE.
12. Rajmohan, M.; Vali, S. C. H.; Raj, A.; Gogoi, A. In *Home Automation Using Brain Computer Interface (BCI)*, 2020 International Conference on Power, Energy, Control and Transmission Systems (ICPECTS), December 2020, pp 1–7; IEEE.

CHAPTER 6

Internet of Things Enabled Energy-Efficient Flying Robots for Agricultural Field Monitoring Using Smart Sensors

M. TAMILSELVI[1], T. MANIMEGALAI[2], G. RAMKUMAR[3*], S. A. SHIFANI[4], and V. MOHANAVEL[5]

[1]Department of Mechatronics Engineering, T.S. Srinivasan Centre for Polytechnic College and Advanced Training, Vanagaram, Chennai 600095, Tamil Nadu, India

[2]Department of CSE, Saveetha School of Engineering, Saveetha Institute of Medical and Technical Sciences (SIMATS), Saveetha University, Chennai 600124, Tamil Nadu, India

[3]Department of ECE, Saveetha School of Engineering, Saveetha Institute of Medical and Technical Sciences (SIMATS), Chennai 602105, Tamil Nadu, India

[4]Department of ECE, Jeppiaar Maamallan Engineering College, Chennai 602108, Tamil Nadu, India

[5]Centre for Materials Engineering and Regenerative Medicine, Bharath Institute of Higher Education and Research, Chennai 600073, Tamil Nadu, India

[*]Corresponding author. E-mail: pgrvlsi@gmail.com

ABSTRACT

The major objective of this study is to enhance the performance of agricultural productivity by associating some intelligent and latest technologies into the respective field. To provide a smart agricultural monitoring system

Intelligent Technologies for Sensors: Applications, Design, and Optimization for a Smart World.
S. Kannadhasan, R. Nagarajan, & Alagar Karthick (Eds.)

to farmers, in which it leads their lifestyle to the next level as well as safe-guard their lives in a great manner. To eliminate the difficulties to monitor the crops in a manual way and protect them from diseases by using the latest technologies. To provide sufficient support to farmers to enhance their way of cultivation and improve their way of crop monitoring using intellectual applications. To provide a facility to monitor the agricultural field from anywhere in the globe using Internet of Things support. To produce the new agricultural monitoring device, in which it associates several smart sensors over the process and predicts the crop growth in fine manner. A new flying robot assistance is grasped over this approach to enhance the cultivation and identify the diseases over the crops instantly using powerful drone assistance.

6.1 OBJECTIVES

The major objective of this study is to enhance the performance of agricultural productivity by associating some intelligent and latest technologies with the respective field. To provide a smart agricultural monitoring system to farmers, in which it leads their lifestyles to the next level as well as safeguards their lives in a great manner. To eliminate the difficulties to monitor the crops in a manual way and protect them from diseases by using the latest technologies. To provide sufficient support to farmers to enhance their way of cultivation and improve their way of crop monitoring using intellectual applications. To provide a facility to monitor the agricultural field from anywhere in the globe using Internet of Things support. To produce the new agricultural monitoring device, in which it associates several smart sensors over the process and predicts the crop growth in a fine manner. A new flying robot assistance is grasped over this approach to enhance the cultivation and identify the diseases over the crops instantly using powerful drone assistance.

6.2 INTRODUCTION

In this modern world each and every individual is growing according to the development of the latest communication technology and its associated devices. The most important communication technology nowadays is internet-enabled services, in which it interfaces a local entity to the global medium through the resistance and quality of the service. This is done by the power of smart technology called Internet of Things (IoT), in which it establishes a bridging connectivity between the local entity hardware unit to the remote IoT server end. Today, all the applications used in both commercial and noncommercial sectors are acquiring the benefits of smart technologies such as IoT. The most important fields of adapting such IoT-based technologies are hospitality, digital record maintenance, agricultural field monitoring, and so on. However, compared to all these fields agriculture is the most important and required area to include such new things to enhance productivity. This article aims to provide a new innovative robotic technology to maintain and monitor the agricultural crops in a fine manner with 24/7 surveillance. Indeed, developing and validating an optimization technique for an unmanned agriculture robotic design is complex, in which it is subjected to a variety of interference circumstances in an outdoor setting which is difficult.

Numerous assessment techniques have been employed recently to undertake mechanical evaluation as well as confirmation of intelligent robot control methodologies. Though these analytical techniques offer the benefits of shortening development time and reducing risk, it must be supported by precise dynamic modeling approaches. Due to these issues, a new robotic technology-based approach is required to monitor the agricultural field in an intelligent manner. In this article, a new drone-based agricultural field monitoring system is designed, in which it consists of three sectors such as: (1) Drone Unit, (2) Smart Field Monitoring Unit, and (3) IoT Server Unit. This proposed drone section is manipulated and handled by using the most advanced flying drone called EVO,[1] which includes a remote controller for wireless operations and display unit for monitoring the activity in a clear manner. This drone has an ability to record the videos and capture the plant and crop images with respect to 720 pixel clarity in a high-definition format. A Smart Field Monitoring Unit (SFMU) consists of sensors such as temperature estimation sensor, humidity monitoring sensor, soil water level indicator, Global Positioning System (GPS) module, and ESP8266 WiFi interfacing controller. With respect to all these sensors the smart device SFMU accumulates the agricultural field details in a clear manner and sends those details to the IoT remote server end for further manipulation. In the server end, a novel deep learning-based classification strategy is utilized called Convolutional Neural Network (CNN). This algorithm acquires the captured images from the SFMU and processes it for identifying the diseases over the crops in the agricultural field. For all the proposed approaches it is more significant to monitor the agricultural field in an intelligent manner and report the details properly to the respective farmer instantly without any delay.

6.2.1 AGRICULTURAL FIELD AND ITS INNOVATIONS

The agricultural field is the most important concern nowadays because most of the farmers and agri-related laborers left the field and work in some other industries for their convenience. An association of latest technologies such as Internet of Things (IoT) and other related smart applications makes the process of agriculture as well as provides good yield to farmers. Many of the farmers and interested youngsters get more and more benefits in such fields of agri-related works. These kinds of technologies are associated with the agri-field to attract the people to come forward to do such business and save the nation in an intelligent manner. The objective of this work is to construct an intelligent robotic kit and an algorithm to classify the affected crops in the field for use in the field of farming. Indeed, developing and validating an optimization method for an independent agricultural field monitoring robot that is subjected to a variety of disturbance circumstances in an outdoor setting is challenging. Numerous analysis techniques have been employed recently to undertake factors that reflect as well as validation of robotics engineering systems. While latest methodological approaches offer the advantages of shortening production time and minimizing risk, they must be supported by precise flexible classification methods. The purpose of this study is to construct a precise estimation of a remote agricultural field monitoring robot's confined vibrant characteristics as well as an automated navigation mechanism depending on the energetic paradigm. These kinds of intelligent robotic designs can navigate into the crop area and analyze the crop details as well as store it into the temporary memory for manipulations. However, the issues raised in this case to analyze the records with respect to the physical operations of data processing are a complex task as well as the farmer or the respective individual needing to grasp the records from the robot.

Kulkarni et al.[2] proposed a paper related to an importance of multiple purpose robot design for monitoring the agricultural field to help farmers to boost up their growth and production level of the crops. In this paper[2] the authors illustrated that the goal of such an article is to plan, establish, and fabricate the agricultural field-monitoring robot called a multiple purpose robot capable of performing all agricultural activities such as burrowing the land of the agricultural field, seeding crops in the plough area, clearing the field with a leveler, irrigating and fertilizing the plants, and monitoring the agricultural field-monitoring robot with a video camera. Conventional agricultural practices require a great deal of physical labor and some procedures are performed physically, whereas others are carried out by hand-controlled equipment. As a result, there are no equivalent robots capable of performing all of these tasks automatically. Additionally, once the principal agriculture is completed, the farmer must maintain an eye on the agricultural field for a variety of purposes. The surveillance mechanism accomplishes this and monitoring technology will take care of several aspects such as irrigation systems, fertilizing, preserving climate levels (for greenhouses), removing excess water, and tracking crop production via connected camera.

This article is intended to design a flying robot with several interesting features to monitor the overall crop details such as temperature, humidity, soil wet ratio, and exact location. Apart from these scenarios, the flying robot accumulates the crop leaf pictures and sends those to the control unit for manipulation. These kinds of transmissions and communication norms are handled by using Internet of Things (IoT) assistance. Once the crop details are accumulated, the prediction process begins to estimate the accumulated values with respect to the threshold ratio of the respective sensors and based on that the alert will be triggered to the respective farmers. These actions are accurately done by using the latest technologies as well as the smart sensors association into the field of agriculture. By using such innovations farmers can operate the agricultural business efficiently without any struggles and monitor their agricultural land details from anywhere in the globe at any time.[3]

Figure 6.1 illustrates the perception of the proposed agricultural field monitoring system with Smart Field Monitoring Unit (SFMU) and the defined drone for capturing the crop leaf pictures and passes it to the SFMU for processing and pushes it to remote server end using Internet of Things. This image shows the performance of EVO drone unit to capture the crop details over the agricultural field with the pixel resolution of 750 and these captured images are sending to the SFMU. The details collected from the agricultural field such as temperature, humidity, and soil irrigation level will be accumulated over the same SFMU and pass all these details to the remote server unit for processing. In the server end the powerful classification algorithm is utilized to manipulate the images and the accumulated agricultural data to provide a proper prediction to the respective farmer in an innovative way.

Table 6.1 shows the clear specification of the proposed EVO drone unit with different parameters.

6.2.2 SMART FIELD MONITORING UNIT

The Smart Field Monitoring Unit (SFMU) is a device that consists of several smart sensors interlinked together with a microcontroller called ESP8266 with integrated WiFi options. The proposed SFMU device is placed into the agricultural field for accumulating the digital image oriented details from the EVO drone for processing it in accordance with the prediction of leaf diseases as well as the sensors associated with the smart device provide the required details regarding the agricultural field. The ESP8266 WiFi-enabled microcontroller is a self-contained System On a Chip (SOC) with an inbuilt Connection-oriented application layer capable of providing connectivity to a Wireless connection network area to any control unit. The ESP8266 may access the network or delegate all WiFi network communications to some other programming unit. All these will be sent to the Internet of Things enabled remote server in an organized way and in that server all the manipulations take place to provide proper predictions to the farmer.

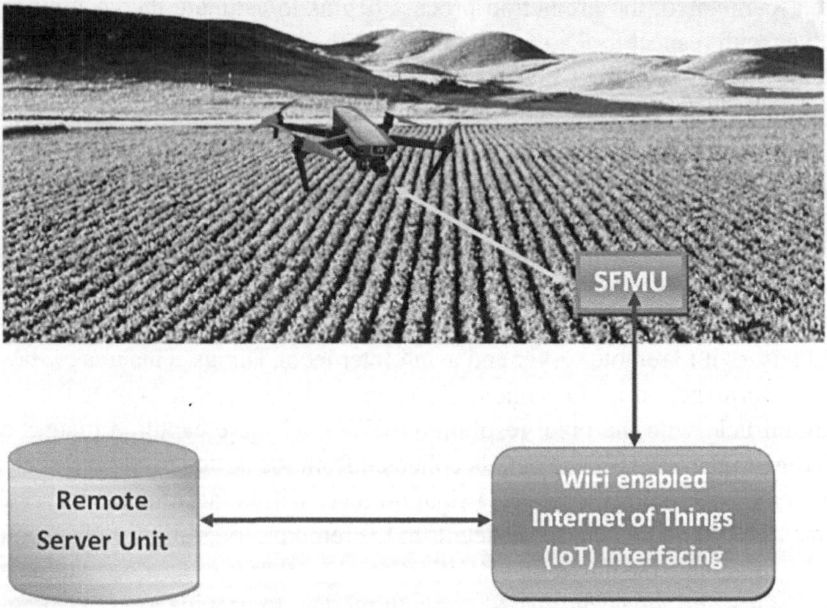

FIGURE 6.1 IoT-based agricultural field monitoring system.

Figure 6.2 illustrates the perception of the SFMU block diagram, in which it portrays the sensor unit and the related modules in detail. The SFMU consists of three major sensor units such as Temperature Sensor, Humidity Sensor, and the Soil Moisture Level Identification sensor. Along with these sensors it contains the Global Positioning System (GPS) module to identify the exact location of the field with respect to latitude and longitude coordinates. The Global System for Mobile (GSM) Communications module is also utilized in this approach to send relevant alert notifications to the farmer based on the emergency needs over the farm land. These all portions are united together to make robust device for monitoring the agricultural field in a clear manner with all specifications.

TABLE 6.1 EVO Drone Specifications.[4]

Drone specifications/ requirements	Description with ranges
Authorization requirements	Customers are not required to enroll the drone with someone, but the concern strongly recommends that really do so that the customer could support their drone information to the cloud data center via the Explorer Application. Additionally, accessing the drone might assist customers in the horrible tragedy of damaged or destroyed drones.
Maximum range of travel	4.3 miles (7 km) for EVO Model-I
	5.5 miles (9 km) for EVO Model-II
	Both these drones have an exceptional transmission distance, allowing them to be used for a broad spectrum of aviation applications. These are just the highest unimpeded communication distances that could be achieved with optimum situations. The inclination of the transmitter, the meteorological and climatic circumstances, as well as the Radio Waves surroundings, all play a significant impact in determining the distance of the drones. Be always aware of and adhere to local aviation rules, standards and operate the drone with precaution and security.
Battery specification	This drone is operated with Smart Lithium Polymer Batteries, in which it provides higher density and energy efficiency. The batteries will systematically deplete to a reasonable temperature and this is the standard position. The operation of discharging requires approximately 2 to 3 days. Though there is no sign that perhaps the batteries are completing a self-discharge period, minor warmth of the battery seems usual and the Autel Explorer Application allows for customizing the discharge limit.
Battery type	LiPo-3S
Battery capacity	4300 mAH
Voltage	11.4 Volts
Maximum charging voltage	13.05 Volts
Watts (energy)	49 Watt h
Overall weight	345 g
Charging temperature ratio	5 to 40 °C
Maximum charging power	49 Watt h

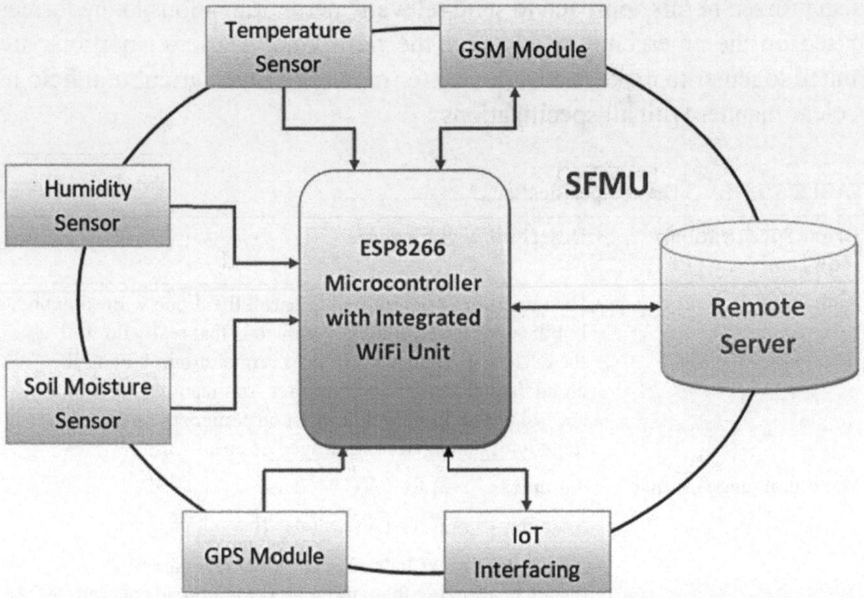

FIGURE 6.2 Smart field monitoring unit block diagram.

Table 6.2 portrays the temperature and humidity ratios of the proposed approach, in which it is attained based on the real-time agricultural field monitoring dataset. The proposed SFMU is designed and placed into the crop unit for 10 days and report the details in a clear manner.

Figure 6.3 illustrates the graphical perception of the temperature and humidity values as mentioned in Table 6.2 in a clear manner with proper high, moderate, and low range specifications. In this, the x-axis shows the number of days and the y-axis shows the temperature and humidity ratios in proper unit specifications.

6.2.3 CROP DISEASE IDENTIFICATION

This article is intended to design a novel flying robot-based crop monitoring system with the help of a drone unit over an agricultural field. In this approach, a power EVO drone[1] is utilized to capture the crop images by means of its digital camera as well as picture quality is 720 pixels. This kind of excellent picture quality grasps the image of crops and sends them to the Smart Field Monitoring Unit by using Internet of Things support. The crop image details are accumulated into the SFMU and evaluate the picture details by means of a server end scripting tool called MATLAB. This tool is significant enough to process the digital image processing and provides the proper leaf disease predictions to the user in an efficient manner. In past research several authors proposed numerous techniques to classify the leaf diseases and report to the user accordingly. However, the logic of such classification schemes is probabilistic as well as provides prediction with respect to maximized supportive threshold ratio instead of confidence threshold level.[8–10]

TABLE 6.2 Temperature and Humidity Measurements

Days	Temperature (°C)	Humidity (%)
1	**Moderate:** 23.6	**Moderate:** 62
	Low: 21	**Low:** 58
	High: 26	**High:** 71
2	**Moderate:** 22.7	**Moderate:** 59
	Low: 19	**Low:** 52
	High: 26	**High:** 67
3	**Moderate:** 20.8	**Moderate:** 75
	Low: 19	**Low:** 62
	High: 26	**High:** 90
4	**Moderate:** 30	**Moderate:** 44
	Low: 22	**Low:** 23
	High: 35	**High:** 80
5	**Moderate:** 29	**Moderate:** 60
	Low: 33	**Low:** 52
	High: 31	**High:** 82
6	**Moderate:** 28	**Moderate:** 50
	Low: 23	**Low:** 31
	High: 30	**High:** 70
7	**Moderate:** 25	**Moderate:** 65
	Low: 20	**Low:** 54
	High: 27	**High:** 80
8	**Moderate:** 24	**Moderate:** 58
	Low: 17	**Low:** 42
	High: 29	**High:** 60
9	**Moderate:** 23.4	**Moderate:** 59
	Low: 17	**Low:** 51
	High: 27	**High:** 70
10	**Moderate:** 28	**Moderate:** 60
	Low: 25	**Low:** 57
	High: 32	**High:** 70

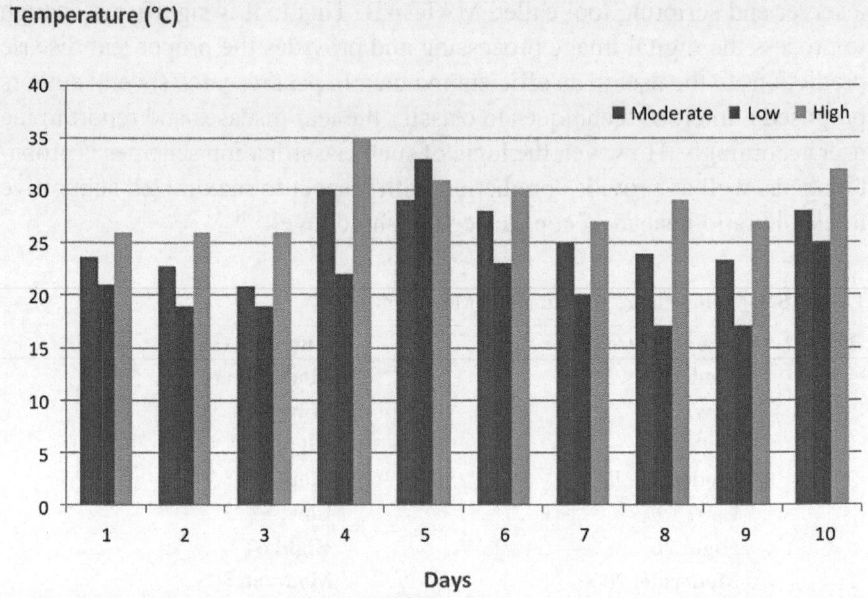

(a)

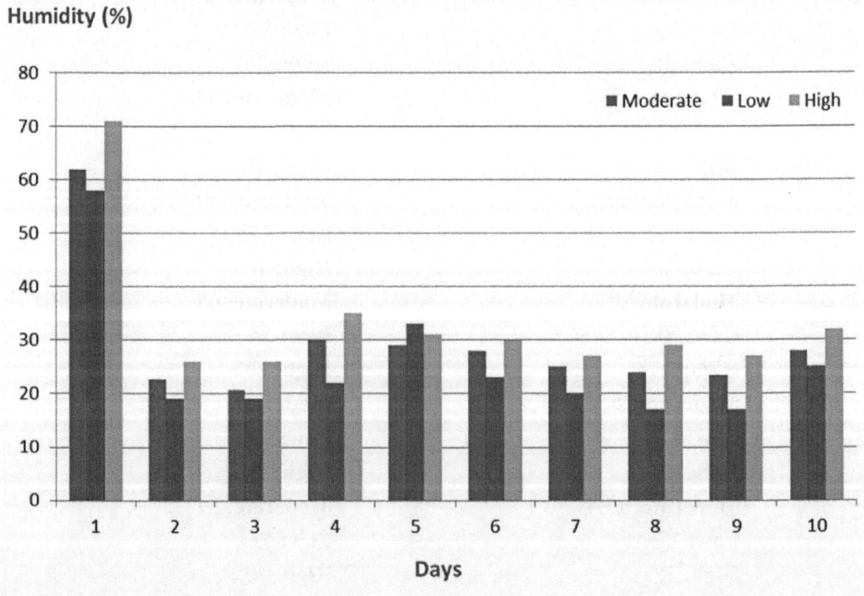

(b)

FIGURE 6.3 (a) Temperature ratio and (b) humidity ratio.

Zhang et al.[5] proposed a paper related to an enhancement of object identification for tomato diseases using classification techniques. In this paper,[5] the authors illustrated such as an enhanced Accelerated R-CNN for detecting normal tomato plants including four infections: downy mildew, blight and plant mold fungal. To begin with, VGG16 is substituted with a layer-based fully connected network infrastructure for dimensionality reduction to acquire more detailed illness characteristics. Next, the convolution layers are clustered using the k-means methodology. An anchoring based on the clustering results is optimized and the revised anchoring structure approaches the dataset's true feature vector. Furthermore, a k-means analysis is conducted using three distinct types of dimensionality reduction frameworks. The empirical outcomes indicate that the upgraded technique for crop leaf diseases had a detection performance of 2.71% and data was conducted that was faster than the conventional accelerated R-CNN.

In this paper,[6] the authors illustrated that quick detection of citrus illnesses is critical for avoiding yield reduction and implementing infection mitigation strategies on fields in a reasonable time. Given the scarcity of labeled afflicted cases, implementing artificial intelligence-based techniques for reliable diagnosis of numerous citrus infections is hard.

Additionally, a resource-constrained technology, including a smartphone, requires a scalable framework with a low computing overhead for citrus illness categorization. This demonstrates[6] the design and architecture useful properties in terms of disease surveillance by farmers who use their individual smartphones on the field. As a result, a lightweight, quic, and accurate architecture is offered based on deep feature engineering for citrus diseases classification from small datasets. A patch-based classification network is presented that consists of an encapsulation component, a clustering prototype subsystem, and a basic neural network classification for the purpose of effectively detecting citrus disorders. An approach is suggested to demonstrate a logic, in which it is effective in reliably diagnosing various illnesses from plant photos using a publicly released citrus fruits as well as leaves dataset. Additionally, our approach's generalization logic is proved by just using a simple dataset, namely the tea leaves dataset. This technique[5] outperforms existing state-of-the-art machine learning in contexts of prediction performance (95.04%), the quantity of tuning functions provided (somewhere around 2.3 million), and the time necessary to analyze citrus infections (somewhere around 10 milliseconds using the classification). Additionally, the capacity to acquire with limited efforts but without compromising quality increases the suggested system's economic applicability on resource-constrained systems, including such cell devices. Yu et al.[7] proposed a paper related to a deep learning-assisted classification model for tomato plant leaves disease prediction.

In this paper,[7] the authors illustrated that the identification of illnesses in the plant sections of crops during the cultivation of crops is an essential component in the control and prevention of plant disease. This article[7] uses tomato plants as research models and employs a deep learning approach to retrieve diseased traits from the plant leaves, focusing on three of the main prevalent varieties. The system can estimate the classification of every illness image with continual experimentation. Thousand images were chosen for every one of the three disorders, separated into 900 images for the training dataset as well as 100 images for the testing dataset. The research makes use of the Resnet-50 communication channel as a starting point. For reference, the network's operational amplifier was modified to Leaky-ReLU and the first segmentation element's linear function was increased to 1111. Following the enhancement, the accuracy rate in the training dataset is 98.3%, while the classification accuracy in the testing dataset is 98.0%.

The proposed crop disease identification concept utilizes the logic of Convolutional Neural Network (CNN) as a classification model to identify the diseases over the leaf of the plant. The received plant leafs are undergone into several classification stages such as preprocessing, feature extraction, and segmentation. Based on these functions the input image is classified and reports the disease details to the farmer in an intelligent manner. The preprocessing stage is helpful to scale the input image according to the size factors of 256 × 256 pixels. The same process assists the system to convert the image into a grayscale format. Figure 6.4 illustrates the perception of preprocessed input images with scaling and feature extraction in a clear manner with graphical representation.

(a) (b) (c)

FIGURE 6.4 Acquired image processing details (a) input image, (b) scaled image with contrast enhancements, and (c) feature extraction.

Figure 6.5 illustrates the perception of image filtration logic, in which it portrays the background and foreground filtration view of the processed image in a clear manner. This view highlights the affected portion of the image in detail with hiding all backgrounds with black color. This provision allows the system to easily identify the affected portion and report the respective affected range ratio to the farmer via respective alert message by using the GSM module connected within the SFMU. The above-mentioned input image is processed properly and attains the prediction accuracy of 97.6% in outcome as well as this proposed approach shows the prediction ratio of the affection level of image is 17.9%. Once these details are clearly noticed by the system with exact location specifications using the GPS module, it can be immediately notified to the farmers to take an appropriate action to preserve the crops using respective functions.

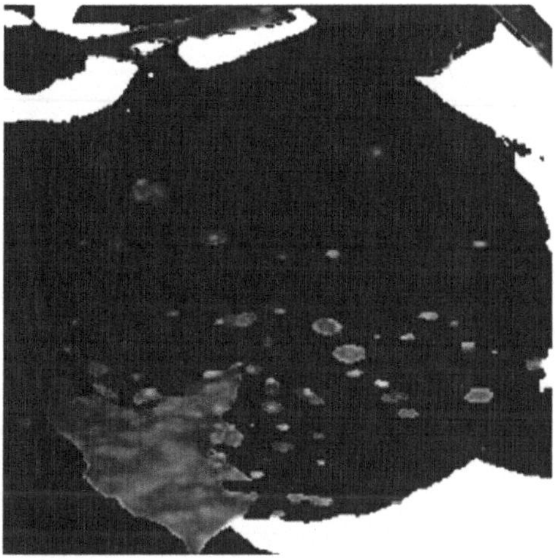

FIGURE 6.5 Segmentation of affected portion of a leaf.

6.3 CONCLUSION AND FUTURE SCOPE

In this article, a new drone-based flying robot design is introduced with three entity factors such as drone unit, Smart Field Monitoring Unit, and the Internet of Things-enabled server unit. Based on these factors the agricultural field is monitored properly with dynamic principles and classification logic. The classification logic is used to analyze the disease ratio of the leaf in the crop field in a clear manner as well as Figure 6.4 and Figure 6.5 show the resulting nature of the classification process in detail with graphical representations. The drone unit accumulates the features based on the EVO drones and the features are sufficient to monitor the agricultural fields using the digital camera connected into it. The smart device unit is placed into the agricultural field for continuous ten days and monitors the details properly to check the working nature of the entire hardware unit. The resulting ratios are given properly over Figure 6.3 in a graphical manner. By using the logic of CNN, this proposed approach can easily identify the disease ratio and report properly to the respective farmer to avoid further affections. For the entire system it is more feasible to manage the agricultural business in a successful manner without any troubles. In the future, the work can further be enhanced by means of adding some crypto features[11, 12] to make security to the remote agricultural data to avoid attacks as well as the crypto logics are reducing the size of storage and providing cost efficiency in terms of storage space reduction.

KEYWORDS

- **IoT**
- **flying robot**
- **crop monitoring**
- **smart agriculture**
- **SMFU**
- **EVO drone**

REFERENCES

1. https://auteldrones.com/products/evo
2. Gupta, S.; Devsani, R.; Katkar, S.; Ingale, R.; Kulkarni, P. A.; Wyawhare, M. In *Iot Based Multipurpose Agribot with Field Monitoring System*, 2020 International Conference on Industry 4.0 Technology (I4Tech), February, 2020, pp 65–69; IEEE.
3. Krishnan, A.; Swarna, S. In *Robotics, IoT, and AI in the Automation of Agricultural Industry: A Review*, 2020 IEEE Bangalore Humanitarian Technology Conference (B-HTC), October, 2020, pp 1–6; IEEE.
4. https://auteldrones.com/pages/evo-support
5. Zhang, Y.; Song, C.; Zhang, D. Deep Learning-Based Object Detection Improvement for Tomato Disease. *IEEE Access* **2020,** *8*, 56607–56614.

6. Janarthan, S.; Thuseethan, S.; Rajasegarar, S.; Lyu, Q.; Zheng, Y.; Yearwood, J. Deep Metric Learning Based Citrus Disease Classification With Sparse Data. *IEEE Access*, **2020,** *8*, 162588–162600.
7. Jiang, D.; Li, F.; Yang, Y.; Yu, S. In *A Tomato Leaf Diseases Classification Method Based on Deep Learning*, 2020 Chinese Control And Decision Conference (CCDC), August, 2020, pp 1446–1450; IEEE.
8. Albuquerque, C. K.; Polimante, S.; Torre-Neto, A.; Prati, R. C. In *Water Spray Detection for Smart Irrigation Systems With Mask R-CNN and UAV footage*, 2020 IEEE International Workshop on Metrology for Agriculture and Forestry (MetroAgriFor), November, 2020, pp 236–240; IEEE.
9. Ayaz, M.; Ammad-Uddin, M.; Sharif, Z.; Mansour, A.; Aggoune, E. H. M. Internet-of-Things (IoT)-Based Smart Agriculture: Toward Making the Fields Talk. *IEEE Access* **2019,** *7*, 129551–129583.
10. Ju, C.; Son, H. I. Modeling and Control of Heterogeneous Agricultural Field Robots Based on Ramadge–Wonham Theory. *IEEE Robot. Autom. Lett.* **2019,** *5*(1), 48–55.
11. Mankar, R. Y.; Gade, A.; Babu, R. In *Association Rules Generation of Outsourced Transaction Data with Privacy-Preserving using Paillier Encryption*, 2020 International Conference on Smart Electronics and Communication (ICOSEC), September, 2020, pp 898–903; IEEE.
12. Holzbaur, L.; Kruglik, S.; Frolov, A.; Wachter-Zeh, A. In *Secrecy and Accessibility in Distributed Storage*, GLOBECOM 2020-2020 IEEE Global Communications Conference, December, 2020, pp 1–6; IEEE.

CHAPTER 7

Medical Devices and Sensor Application

APOORVA JOSHI[1*], AMBRISH KUMAR SHARMA[2],
KARUNA NIDHI PANDAGRE[1], and SANJEEV GOUR[1]

[1]*Department of Computer Science, Career College Bhopal, MP, India*

[2]*Department of Computer Science, NRI College, Bhopal, MP, India*

[*]*Corresponding author. E-mail: apoorvajoshi16@gmail.com*

ABSTRACT

Sensors in medical devices have a lot of potential in terms of lowering patient expenditure and enhancing patient care. Wearable and nonwearable technologies are helping to revolutionize healthcare and health outcomes in the mobile health era, as well as offering real-time advice on better health management and tracking. Researchers and medical practitioners require safe and cost-effective methods for conducting research, maintaining public safety, and providing patients with personalized health alternatives. Medical sensors may be used to quickly implement one such approach. Biomedical diagnostic investigations are becoming increasingly important in the modern medical area. Infectious disease screening, chronic therapies, health management, and tracking are all applications for medical sensors. In this chapter we will look at different types of sensor used in healthcare sector.

7.1 INTRODUCTION

The application of medical sensors in patient treatment, pharmaceutical, biomedical, and healthcare industries has sparked a lot of interest in the field. Disease detection, prevention, patient health surveillance, and healthcare

Intelligent Technologies for Sensors: Applications, Design, and Optimization for a Smart World.
S. Kannadhasan, R. Nagarajan, & Alagar Karthick (Eds.)

management have all been effectively applied with medical sensors. Bacteria, pathogens, and viral germs may all be detected with medical sensors. Physical activity and step trackers enable people to develop and enhance healthier habits. These devices open opportunities for people working in clinical science as well. These sensors detect substances without requiring the human body to take blood. They are always searching for several factors at the same time. People are generally required to stay in a hospital to track their body chemistry on a continuous basis.

People may use this gadget to get information about the substances in our bodies from anyplace. They will use chemicals to track various physiological processes. This information may be collected on a mobile app and shared with a healthcare worker, caregiver, or anybody else.[10,21] As a result, comprehensive quality assurance and testing procedures are required. Training must be conducted at many stages, including climate, customers, and algorithms, to obtain the needed performance and a long-term marketing strategy. The environment's real accuracy is then assessed and confirmed, and better product design and performance are recognized. Biosensors are used to increase the quality of human subject testing, create a feedback system, and explore research criteria. In the medical area, this technology can fully check a variety of aspects. It certifies the functionality of medical instruments and technologies.[5,9]

The rapid spread of the coronavirus (COVID-19) has resulted in a surge in demand for easily installed healthcare equipment. CRISPR-Cas9-based paper strip, nucleic-acid-based, antigen-Au/Ag nanoparticles-based electrochemical biosensor, optical biosensor, and Surface Plasmon Resonance are among the newly developed biosensors used to detect RNA viruses. In addition, major technical developments in recent years, as well as advances in nanotechnologies, are important drivers of the growing biosensors demand. Biosensor's demand is expected to reach high up to USD 36.7 billion in 10 years, with a market value of $25.5 billion in 2021. Biosensors are transitioning from laboratory-based healthcare to decentralized healthcare. A patient sample is taken and sent to an off-site analytical laboratory in traditional scientific diagnosis. By doing care-based examinations, the patient might expect results within a few hours.[4,6] Rapid pathogen detection allows time to respond.

A sensor converts a process involving medical components into a signal that can be read. Nonbiological compounds, like glucose, and other medical substances, like protein, can be used as biomarkers. An electrochemical transducer or an optical transducer, both typical of the process, can turn into readable signals.[1,18] Biomedical diagnostics and biosensors are becoming increasingly important as the world population ages and grows, as well as the issues created by chronic lifestyle illnesses. New molecular diagnostics would have a significant impact of healthcare system. One of the most important advantages of biomarker screening is that it allows for accurate and individualized treatment as well as early diagnosis. It allows for a diagnosis that is far more likely to be accurate, as it is not reliant on outward clinical symptoms or the patient's molecular knowledge.[3,22] With the growth of portable sensors, various biomarkers may be detected for continuous, real-time monitoring of human health. They look at molecular markers utilizing measurements that are simple to acquire, such as temperature, cardiac-amplitude, concentration, and dehydration. To track human health, the laboratory is developing a fully developed small detecting system for protein biomarkers.[8,14] Doctors should not rely on the patient's recollections or readiness to express their feelings; instead, they should focus on the specifics. Patients and doctors collaborate to create personalized treatment plans.

The data collected by biosensors may be saved for future comparison and form part of the patient's data; it is reducing the need of retesting and supports its infrastructure.[7,15] Treatment becomes more immediate, easing the strain on physicians and hospitals. The world of healthcare is changing, and everyone is aware of it. Nothing could be more fascinating than hearing about cutting-edge biotech applications and technologies that might help us maintain our health in the next few years. Healthcare must focus its efforts on products that make use of cutting-edge biochemical and chemical sensor technologies. Biosensors offer a fantastic user experience and a wealth of possibilities for digital health devices.[12,17] Sensors technology can improve the efficiency, diagnosis, and treatment procedures in today's healthcare applications. To allow accurate multimodal testing, biosensor instruments were also employed in labs and natural settings. Biosensor technology can improve the efficiency, diagnosis, and treatment procedures in today's healthcare applications. Biosensors were also utilized in labs and in the field to enable for accurate multimodal testing. In hospitals, devices for rapid detection and treatment of a variety of illnesses are essential. These sensing devices might be sophisticated and easy to use.

The patient wants to evaluate the device's condition and usefulness, as well as see if the data can be saved. To compare the results to parallel observations, they utilized traditional ethnographic techniques for usability study. Complex bimolecular-based sensors require research, development, and commercialization to meet a variety of healthcare needs. Biosensors that are worn on human skin and may be easily utilized within the body are widely employed for different health applications and everyday health monitoring. It is also critical to develop this technology and discover acceptable materials for connecting and attaching sensors to surfaces[15] Biosensors are used in a variety of sectors, including DNA, intelligent textiles, the automobile industry, and a seemingly endless list of others. As a result, biosensors have a wide range of uses; it is critical to comprehend their capabilities, applications, and technological advancements. The importance of biosensors in the medical profession is discussed in this study. The successful deployment of this technology can cover a wide range of applications.

7.2 TYPE OF MEDICAL SENSORS

Sensors are devices that detect physical, chemical, and biological signals and allow for measurement and recording of such signals. They have been utilized in a variety of disciplines, including research, health, automated production, and environmental monitoring. Mechanical thermal sensors are all categorized according to the primary sensing concept.

Temperature sensors, pressure detectors, flow sensors, acoustic sensors, and gas sensors are all utilized in the medical industry, as are cameras, image sensors, and magnetic field sensors. Optical, X-ray, and ultrasonic image sensors and cameras are utilized in the medical field. The medical industry's usage of sensors is highly dependent on the application. Healthcare sensors can detect and send information about specific biological, chemical, or physical processes. These sensors may also be found in systems that handle clinical samples, such as lab-on-a-chip devices, which are becoming increasingly popular. Healthcare sensors are used to collect information about the patient. Due to the escalating expenses of medical treatments in hospitals and medical care clinics, people are increasingly opting for home healthcare services. As a result, demand for a variety of healthcare equipment is projected to rise in the coming years.

- Physical sensor
- Chemical sensor
- Biopotential electrodes
- Biosensor or bioanalytic

7.2.1 PHYSICAL SENSORS

Physical parameters might be measured with a physical sensor. Physical sensors are utilized in healthcare applications in a variety of ways like an

X-ray and gamma ray-based sensors are used by radiation sensors. Ultrasound and pressure sensors are examples of mechanical sensors. Thermocouple, thermistor, thermopile, optical fiber devices, P–N junction diode, and infrared sensors are examples of thermal sensors, Blood flow monitoring sensors and magnetic resonance imaging devices are examples of magnetic sensors is shown in Figure 7.1.

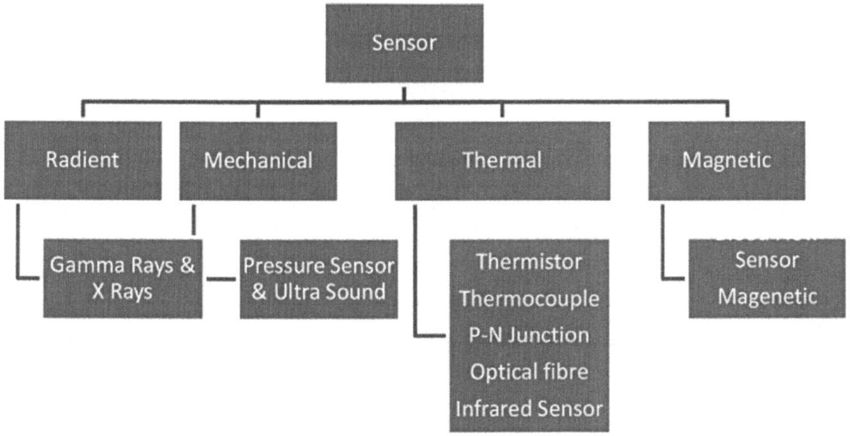

FIGURE 7.1 Type of medical sensor in IOT.

7.2.2 CHEMICAL SENSORS

The component and bodily fluids ratio, such as PH value, glucose concentration, and so on, are detected using chemical sensors; different microbes are all detected by the biosensor.

7.2.3 BIOSENSOR

Biosensors are devices that integrate biological material with an appropriate platform to detect pathogenic organisms, carcinogenic, mutagenic, and/or toxic substances, as well as to report a biological consequence. A huge number of various types of biosensors have been built and developed in recent years for a variety of medical purposes. Healthcare sensors are categorized according to the sensor device they use:

i) The electrochemical sensor works on an enzymatic catalytic process that either takes or produces electrons.

ii) The physical sensors are classified into two categories: piezoelectric sensor and thermometric sensor.
iii) The optical biosensor is a device that works on the principle of optical measurement. Fiber optics and optoelectronic transducers are used.
iv) The optical biosensor is a device that works on the principle of optical measurement. Fiber optics and optoelectronic transducers are used in glucose, BP, the rate of heartbeat, etc.

All these sensors are utilized in a variety of applications in healthcare. Furthermore, nano-biotechnology can be used to improve these sensors.

7.3 SENSORS USED IN MEDICAL EQUIPMENT

7.3.1 VENTILATORS

Sensors used in hospital ventilators are

- Airflow sensors: These sensors monitor a patient's breathing and verify oxygen supply.
- Basic and Aml Switches: These sensors cover panels and doors and are used as switch operator controls.
- Humidity and Temperature Sensors: These sensor switch operator controls and detected are provided by covers, panels, and doors.
- Magnetic Sensors: These sensors are used to control motors of ventilators.
- Oxygen Sensors: These sensors are used to measure and regulate oxygen amount in the air.
- Temperature Sensors: These sensors improve patient's comfort level by monitoring the temperature of the air given to the patient.
- Pressure Sensors: These sensors monitor a patient's respiration cycle.
  - It also checks blocked air and oxygen filters that need to be changed.
  - It controls flow of air and flow of oxygen provided to a patient in ventilators and it is controlled by pressure sensors.

7.3.2 OXYGEN CONCENTRATORS

- Sensors Used in Oxygen Concentrators Airflow Sensors: It monitors a patient's breathing to identify it inhales cycle, it also ensures that oxygen level is delivered to each patient.

- Basic and Aml Switches: It covers panels and doors that are used as a switch to operate and control.
- Magnetic Sensors: It controls motors and controls the speed of the motors and measures.
- Oxygen Sensors: It measures and regulates the oxygen level in the air which is supplied to the patient.
- Pressure Switches: It generates pressure alerts.
- Pressure Sensors: It monitors the pressure in the patient's bed to check that adequate oxygen is produced.

7.3.3 PATIENT-MONITORING SYSTEMS

The practice of maintaining track of a patient's temperature is referred to as "temperature monitoring." Blood pressure can be measured invasively with a pressure transducer implanted in the arm using a blood pressure cuff. It monitors the level that is used to determine the glucose level in the interstitial fluid. Researchers can examine how insulin, exercise, food, and other factors impact blood glucose levels by using continuous monitoring.

- Airflow Sensors: The patient's breathing cycle is monitored by ensuring the flow of air in the respiratory path.
- Humidity and Temperature Sensors: It checks and verifies the ambient temperature and level of humidity in the hospital wards.
- Oxygen Sensors: It ensures the level of oxygen in the air which is given to the patient.
- Pressure Sensors: These sensors monitor a patient's breathing cycle and ensure blood pressure to regularly check the patient's health. It also checks the clogged air and oxygen input filters and is replaced time to time.
- Temperature Sensors: It is used to determine the temperature of the patient.
- SpO2 Sensors: It calculates the proportion of hemoglobin that contains oxygen in the blood compared to the total amount of hemoglobin (oxygenated and nonoxygenated hemoglobin).

7.3.4 INFUSION PUMPS

- Infusion pumps sensors: It verifies the blockages and decides the replacement of the fluid bag that needs to be changed and it also monitors the level of fluids or nutrients to the patient.

- Barcode Scan Engines and Software: It scans the barcodes on the bag and the patient band to ensure the right therapy.
- Basic and Aml Switches: These switches cover panels, doors that are used as switch operator and controls.
- Magneto-Resistive Position Sensor: It checks the tube's location in the pump chamber and pump motor's speed.
- Pressure Sensors: These sensors are used to regulate the pincher rollers.
- Subminiature Load Cells: It monitors the weight of the fluid bag of patients.

7.3.5 LABORATORY EQUIPMENT

These sensors are used in blood and saliva analyzers.

- Barcode Scan Engines and Software: At each level patient's metadata are tracked and validated by these sensors.
- Force Sensors: Force sensor measures the force inside the pump.
- Position Sensor: This sensor checks the location of the magnet in extraction needle or the patient sample.
- Pressure Sensors: It regulates the flow of the fluids.
- Temperature Sensors: These sensors maintain the track of the sample holder's temperature.
- Barcode Scan Engines Sensor: It keeps track and maintains sample metadata and its related information.
- Pressure Sensors: Applying pressure on specimens causes blood cells to condense into smaller and fewer cells.
- Magneto-resistive Position Sensor: This sensor is used to track the sample tube's position.
- Airflow Sensors
- Ventilators and Oxygen Concentrators: In ventilators and oxygen concentrators, airflow sensors are used to check and regulate the airflow to the patient as well as to monitor the patient's inhale cycle.
- Respiratory Monitoring: Airflow sensors keep track of a patient's respiration. The flow of air is measured by using Airflow Sensors. They can be used to ensure that the patient receives the desired combination, as determined by the doctor.
- Gas Chromatography: It is precise and accurate in gas flow monitoring and control. The ceramic flow tube airflow sensor is intended to reduce outgassing while improving accuracy and reliability.

7.4 MAJOR PROPERTIES OF SENSORS

Basically, sensor nodes are tiny wireless devices that can collect different physiological indicators and are put on, around, and in the human body. They generally have the following features[2]:

- Because the sensor nodes are small (less than 1 cm^3), the battery size of the sensors is small, and the amount of energy accessible is typically limited. Furthermore, sensors are expected to function for an extended length of time, and replacing sensor batteries, particularly when they are implanted within the human body, is extremely difficult. As a result, in medical sensors, finding solutions to decrease energy use and capture more energy is constantly necessary.
- Sensor nodes are often diverse, and depending on the sort of data they gather, they demand varying energy resources from the network showing the variability of sensors depending on different data rate requirements.[19,23] The data rate might range from a Kbps to many Mbps, as seen in the table.
- There are no nodes that are redundant. All the nodes have an equal amount of significance and are added according to the application's requirements from a few kilobits per second to many megabits per second.
- To maintain the interference and health issues, nodes have a very less power to transmit.
- Because nodes are often controlled by medical personnel rather than engineers, they should enable self-organization and self-maintenance features. When a node is attached to the human body and turned on, it should automatically join the network and establish connections.

7.5 SENSORS' WIRELESS COMMUNICATION TECHNOLOGIES

The sensors' wireless communication forms three types of networks[16,20]:

- Communication wearable sensor device or implanted sensors in the body and a receiver placed outside the body is known as in-body network communication.
- On-body network communication: utilized to communicate between worn sensors and coordinate with other devices that collect data and send it to a local machine.

- External network communication: Used to communicate between the coordinator and a back-end server on the other side of the world. It lists the various technologies and protocols that are utilized for short- and long-range communication between sensors, coordinator devices, and external back-end servers. It demonstrates that short-range communication protocols such as the Industrial Scientific and Medical (ISM) band and the Medical Implant Communication Service band are utilized to establish in-body and on-body network connections.

7.6 MAJOR ADVANTAGES OF SENSOR IN HEALTHCARE

Improved sensitivity during data gathering, virtually lossless transmission, and continuous, real-time analysis are just a few of the benefits of sensors. Processes are active and executed properly thanks to real-time feedback and data analytics services.

Sensing technology has evolved throughout time, resulting in today's smart and intelligent sensors. Smart sensors, unlike traditional analogue sensors that have no active components, have electrical circuits that allow them to collect measurements and output values as digital data. These sensors have a variety of sensing devices mounted atop a signal converter, as well as embedded CPU units.

Intelligent sensors can perform a variety of inherently intelligent operations, such as self-testing, self-validation, self-adaptation, and self-identification. They are familiar with process needs, can manage a wide range of circumstances, and can identify problems in real time to aid in decision-making. These smart sensors are configured to respond to a variety of process circumstances, allowing executives to get the most out of them.

Companies may enhance productivity, improve energy efficiency, and lower the total cost of ownership of their facilities by switching to smarter sensors and efficiently utilizing data acquired from processes and assets. Process leaders may identify regions of excessive energy usage and take necessary actions to reduce energy waste with continuous monitoring and data retrieval. This strategy will assist businesses in achieving the United Nations' Sustainable Development Goals (SDGs), such as "Affordable Clean Energy," "Industry, Innovation, and Infrastructure," and "Responsible Consumption and Production."

The importance of healthcare software solutions cannot be overstated, as technology promises to improve healthcare services and relieve the strain imposed on healthcare providers. This is especially important given the ageing population and the rise in chronic illness cases.

The main advantages of IoT sensor implementation in healthcare:

1. **Remote monitoring:** In the medical emergency, remote monitoring that linked IoT devices may detect illnesses, cure diseases, and save patients' life.
2. **Prevention:** Sensor nodes assess health conditions, nutrition's choices, and the environment and suggest survival steps to decrease the development of illnesses.
3. **Reduction of healthcare costs:** The Internet of Things decreases the cost of medical visits and hospital stays while also making testing cheaper.
4. **Medical data accessibility:** Medical records with digital storage make it possible for patients to obtain good treatment; they also assist healthcare practitioners in the best medical decisions and discard problems.
5. **Improved treatment management:** IoT devices help with medication administration and treatment response tracking, as well as reducing medical mistakes.
6. **Improved healthcare management:** Healthcare organizations may gain important information on equipment and employee effectiveness and utilize it to recommend changes using IoT devices.
7. **Research:** Due to IoT devices that can gather and analyze huge amounts of data, they have a lot of potential in medical research.

7.7 CHALLENGES AND THREATS

Although the Internet of Things has a lot of potential in healthcare, there are still a lot of obstacles to overcome before it can be fully implemented. The following are the risks and drawbacks of utilizing linked devices in healthcare:

1. **Security and privacy:** Healthcare systems have the potential to be compromised; thus, security and privacy is a key issue discouraging people from utilizing IoT used for medical purposes. Loss of sensitive information regarding health of patients and whereabouts, as well as tampering with sensor data, might have serious repercussions, negating the IoT's benefits.

2. **Risk of failure:** Sensor and other tool performance can be affected by hardware failure, putting healthcare operations at risk. Furthermore, loss of a planned application update might be much riskier.
3. **Integration:** Due to less consensus on IoT protocols, devices may not be compatible. The lack of consistency hinders IoT from being fully integrated, limiting its potential usefulness.
4. **Cost:** While the Internet of Things has sufficient potential to save medical costs, the overall cost of implementing and training of staff is rather significant.

7.8 CONCLUSION

It is hard to judge and diagnose many diseases early on in their course, allowing patients to get effective therapy. To successfully identify illnesses, it is tough to create simple, sensitive, and cost-effective diagnostic systems such as healthcare. Healthcare sensors have a wide range of medical applications that benefit clinicians and patients for a variety of reasons, including disease control, clinical care, preventative therapy, patient health information. Micromaterials have seen a lot of use in the creation of biosensors in recent years. The primary goal of clinical medicine is to classify patients using biosensors that are simple to use. Healthcare sensors enable customised medicine, which provides a fresh approach to medical practise today, according to most experts. This strategy has a huge impact on healthcare, resulting in a plethora of therapeutic and diagnostic options. In this regard, the collaboration of biosensors and the advancement of science and technology can result in a larger range of experience, more complex things, and commodities. This technology will help to solve a variety of medical problems. Healthcare sensors will aid in the development of tailored cancer therapy. Many medical sensor design and development opportunities will arise because of the nanomaterial-based mechanism's application to a variety of difficult medical challenges.

Sensors are crucial in the healthcare business. Different sensors can be used in various healthcare applications. This will make it easier for patients to survive in a constantly changing environment. Different types of sensors may be used to check the most fundamental functions of the human body, which might assist to avert serious issues in healthcare. Diseases can be easily identified at an early stage, allowing for the avoidance of catastrophic consequences. Sensors are inexpensive and dependable equipment. In this study, we successfully investigated sensors and their types, as well as sensors used in healthcare applications.

KEYWORDS

- **biomedical applications**
- **biosensors**
- **digital health monitoring**
- **wearable device**
- **sensor**

REFERENCES

1. Pourasl, A. H.; Ahmadi, M. T.; Rahmani, M.; Chin, H. C.; Lim, C. S.; Ismail, R.; Tan, M. L. Analytical Modeling of Glucose Biosensors Based on Carbon Nanotubes. *Nanoscale Res. Lett.* **2014**, *9* (1), 1–7.
2. Latre, B.; Braem, B.; Moerman, I.; Blondia, C.; Demeester, P. A Survey on Wireless Body Area Networks. *Wirel. Netw.* **2011**, *17* (1), 1–18.
3. Baird, C. L.; Myszka, D. G. Current and Emerging Commercial Optical Biosensors. *J. Mol. Recogn.* **2001**, *14* (5), 261–268.
4. Three, D.; Nazirizadeh, Y.; Gerken, M. Photonic Crystal Biosensors Towards On-Chip Integration. *J. Biophot.* **2012**, *5* (8-9), 601–616.
5. Rodrigues Ribeiro Teles, F. S.; Pires de TavoraTavira, L. A.; Pina da Fonseca, L. J. Biosensors as Rapid Diagnostic Tests for Tropical Diseases. *Crit. Rev. Clin. Lab Sci.* **2010**, *47* (3), 139–169.
6. Dutta, G.; Regoutz, A.; Moschou, D. Enzyme-Assisted Glucose Quantification for a Painless Lab-on-PCB Patch Implementation. *Biosens. Bioelectron.* **2020**, *167*, 112484.
7. Alhadrami, H. A. Biosensors: Classifications, Medical Applications, and Future Prospective, *Biotechnol. Appl. Biochem.* **2018**, *65* (3), 497–508.
8. Yu, J.; Yang, A.; Wang, N.; Ling, H.; Song, J.; Chen, X.; Lian, Y.; Zhang, Z.; Yan, F.; Gu, M., Highly Sensitive Detection of Caspase-3 Activity Based on Peptide-Modified Organic Electrochemical Transistor Biosensors, *Nanoscale* **2021**, *13* (5), 2868–2874.
9. Zhang, J.; Zhang, X.; Wei, X.; Xue, Y.; Wan, H.; Wang, P. Recent Advances in Acoustic Wave Biosensors for the Detection of Disease-Related Biomarkers: A Review, *Anal. Chim. Acta.* **2021**, 338321.
10. Choi, J. R. Development of Point-of-Care Biosensors for COVID-19, *Front. Chem.* **2020**, *8*, 517.
11. Choudhary, M.; Yadav, P.; Singh, A.; Kaur, S.; Ramirez-Vick, J.; Chandra, P.; Arora, K.; Singh, S. P., CD 59 Targeted Ultrasensitive Electrochemical Immunosensor for Fast and Noninvasive Diagnosis of Oral Cancer, *Electroanalysis* **2016**, *28* (10), 2565–2574.

12. Chauhan, N.; Maekawa, T.; Kumar, D. N. Graphene-Based Biosensors—Accelerating Medical Diagnostics to New Dimensions, *J. Mater. Res.* **2017**, *32* (15), 2860–2882.
13. Dutta, N.; Lillehoj, P. B.; Estrela, P.; Dutta, G. Electrochemical Biosensors for Cytokine Profiling: Recent Advancements and Possibilities in the Near Future, *Biosensors* **2021**, *11* (3), 94.
14. Chandra, P.; Koh, W. C.; Noh, H. B.; Shim, Y. B. In Vitro Monitoring of i-NOS Concentrations with An Immunosensor: The Inhibitory Effect of Endocrine Disruptors on i-NOS Release, *Biosens. Bioelectron.* **2012**, *32* (1), 278–282.
15. Gui, Q.; Lawson, T.; Shan, S.; Yan, L.; Liu, Y. The Application of Whole Cell-Based Biosensors for Use in Environmental Analysis and in Medical Diagnostics, *Sensors.* **2017**, 17 (7), 1623.
16. Caytiles, R. D.; Park, S. A Study of the Design of Wireless Medical Sensor Network Based U-Healthcare System. *Int. J. Bio-Sci. BioTechnol.* **2014**, *6* (3), 91–96.
17. Rebelo, R.; Barbosa, A. I.; Caballero, D.; Kwon, I. K.; Oliveira, J. M.; Kundu, S. C.; Reis, R. L.; Correlo, V. M. 3D Biosensors in Advanced Medical Diagnostics of High Mortality Diseases, *Biosens. Bioelectron.* **2019**, *130*, 20–39.
18. Parker, R. N.; Grove, T. Z. Designing Repeat Proteins for Biosensors and Medical Imaging, *Biochem. Soc. Trans.* **2015**, *43* (5).
19. Movassaghi, S.; Abolhasan, M.; Lipman, J.; Smith, D.; Jamalipour, A. Wireless Body Area Networks: A Survey, *IEEE Commun. Surv. Tut.* **2014**, *16* (3), 1658–1686.
20. Hamida, S. T. - B.; Hamida, E. B.; Ahmed, B. A New Mhealth Communication Framework for Use in Wearable Wbans and Mobile Technologies, *Sensors* **2015**, *15* (2), 3379–3408.
21. Zadran, S.; Standley, S.; Wong, K.; Otiniano, E.; Amighi, A.; Baudry, M. Fluorescence Resonance Energy Transfer (FRET)-Based Biosensors: Visualising Cellular Dynamics and Bioenergetics, *Appl. Microbiol. Biotechnol.* **2012**, *96* (4), 895–902.
22. Pollard, T. D.; Ong, J. J.; Goyanes, A.; Orlu, M.; Gaisford, S.; Elbadawi, M.; Basit, A. W. Electrochemical Biosensors: A Nexus for Precision Medicine, *Drug Discov. Today.* **2020**.
23. Lai, X.; Liu, Q.; Wei, X.; Wang, W.; Zhou, G.; Han, G. A Survey of Body Sensor Networks, *Sensors* **2013**, *13* (5), 5406–5447.

CHAPTER 8

IoT-Based Smart Security and Home Automation System

M. SURESH*, SANDIPAN MALLIK*, YASHWARDHAN KUMAR, OBBINTI SANKAR RAO, MADHVI SINGH, DEEPIKA ADHIKARY, SAI SWARUP PATNAIK, and MONIK RAJ SAHU

Department of Electronics Communication Engineering, NIST, Berhampur, Odisha 761008, India

Corresponding author. E-mail: msuresh73@gmail.com, sandi.iitkgp@gmail.com

ABSTRACT

The automation and security concept attribute the application of the Internet of Things (IoT) in a vast advanced topographical territory. IoT refers to the interconnection between numerous sensors and the cloud. It is a network where devices are embedded with sensors, actuators, and software. IoT is implemented in smart home security and automation to improve the environment with the ease of digitalization, promising to make every small task effective while improving personal satisfaction for home occupants. When the talk comes with the house, the concept of making it safer, smarter, and automated comes to play. The huge arrangement of the IoT is empowering smart security and automation techniques all over the world. This chapter concentrates on developing an automation system for smart home and wireless security, which convey the alerts to the homeowner by using web (through ThingSpeak) triggering an alarm, needed according to the range of values set, using Wireless Sensor Network (WSNs); Wi-Fi as a communication protocol and android app for remote controlling of devices. The

Intelligent Technologies for Sensors: Applications, Design, and Optimization for a Smart World.
S. Kannadhasan, R. Nagarajan, & Alagar Karthick (Eds.)

proposed system works on the real-time monitoring with the app. Alert system in case of fire, air saturation level. App-based remotely controlled energy-efficient and data encryption controls system for home appliances like fans, window cartons, energy consumption, tube lights, using various sensors and actuators. This chapter also explores the different pieces of smart home. It includes gas leakage information, temperature, and humidity using various sensors connected to the microprocessor and internet.

8.1 INTRODUCTION

Internet technology connects services, appliances, and instruments to the world wide web by the medium of wireless technologies. Over 9 billion "things" are linked to the internet as of now. "Things" associated with the internet are predicted to cross 20 billion in the forthcoming year. The Internet of Things (IoT)[1] provides us with an environment that is the interconnection of devices such as appliances embedded with software and hardware; thus, devices are able to fetch data over a network and are also able to communicate with each other. Today, the demands of automated systems are increasing rapidly to reduce human intervention.[2] With this increasing demand comes the increase in competition which forces the competitors to come out with more advanced, intelligent, and user-friendly models. The IoT has multiple applications and facilities in various domains, but the two main aspects that are discussed are smart security and home automation system.[3] This technique is incorporated in working places or houses to make the gadgets more convenient and automated. Smart security can be achieved with motion of the sensor. It will sense the movement at the entrance and send a notification if anything goes wrong. Home automation relies on the IoT that can operate the task easily and in a more efficient manner. Safety is the most crucial and effective thing for people in different sectors to protect their essentials. Since the rate of crime is increasing nowadays[4] and to cope with it, a smart security system is structured to secure the place in the absence of an owner. On the other hand, home automation refers to controlling and managing home appliances by using a micro-controller computer technology.[5] This system can be used in many ways such as to control switching lights, fan, door locking security purpose, etc.[6] Atomization or self-regulating is the intelligence of the system to schedule the appliances associated with the internet through the time-scheduling programs.[7,8] The client can access the whole system in real time from anywhere and in any time using internet. This system aims to gain satisfaction combined with simplicity. Smart security and home automation systems have many advantages associated with it. One of them is enormous capability for energy economies.[9]

8.2 IMPLEMENTATION

The flowchart and the proposed home automation system are shown in Figure 8.1 and Figure 8.2 respectively that display the overall structure of IoT-based Smart Security and Home Automation.. This framework operates basically in two stages. At the first stage, the application Blynk takes the input command from the user and passes the required function to the relay module to operate the further process such as if a person demands for switching off fan or turning up lights then it will respond accordingly. In this process the Wi-Fi module ESP32 acts as an interface between that application and relay, providing them with Wi-Fi and Bluetooth services to proceed the working in an efficient manner. At the next stage, sensor parts are arranged in such a way so that if any kind of unwanted activity occurs then the respective sensor will react accordingly and transfer the following data over the internet and will also maintain the record of it in a database format. Along with that it also gives the alert notification through email, displays them on LCD, and will trigger the speaker as a warning sign so that the owner could take the suitable action to avoid any such accidents. This is a loop type process that works continuously to secure the place from danger.

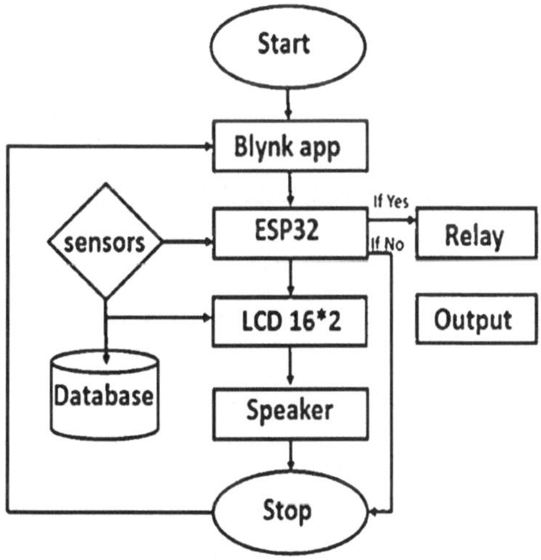

FIGURE 8.1 Flowchart.

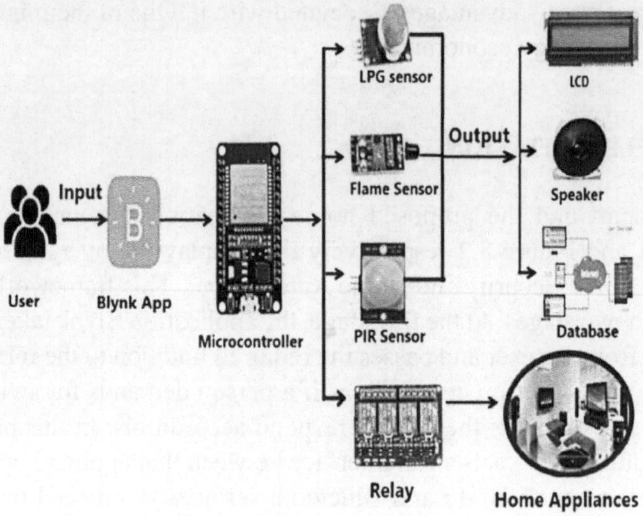

FIGURE 8.2 Proposed home automation system.

8.3 PROPOSED SYSTEM

In this prototype we have proposed real-time monitoring and controlling of different electrical appliances through an android-based mobile application as well as we keep track of the security of homes using different sensors.

8.3.1 SMART HOME SECURITY SYSTEM

A PIR sensor will be installed at the entrance of the building/home to monitor the physical movement of humans. Each day, the body emits thermal energy around the wavelength of 9–10 micrometers. A passive infrared sensor (PIR) uses pyroelectricity to measure temperature. If a human being is in the vicinity, this wavelength can be detected. It is necessary to utilize a simple lens to detect a great distance. In the same way, sensors can be calibrated so that they operate at maximum efficiency to set a higher threshold for sensitivity to domestic pets. Making sure that the room's floor remains out of focus is one way to accomplish this. Any time a person tries to enter premises without the permission of the owner/authority.[19] PIR sensor will detect motion and send the signal to ESP32, which will activate the alert system, that is, microcontroller sends an email/SMS to the owner and sounds a police siren so that the visitor backs off.[20]

8.3.2 SMART HOME AUTOMATION SYSTEM

Home automation may provide an effective method of sharing the data between family members and trusted people for personal security, and may help to reduce energy costs and improve the environment. In this system, all the device (i.e., lights, fans, TV etc.) can be turned ON/OFF with the help of ESP32 board and relay module to control the appliances from anywhere in the home through the mobile phone by using the blink application and LPG gas sensor is installed in the kitchen so that it can recognize if any leakage of gas and alert the person present over there through the speaker or if it happened in the absence of the people then also it will send email/SMS to the owner and the LPG gas sensor will be taking the gas level readings and feed it in the ThingSpeak website where the person can monitor the gas levels all the time.[21] Similarly, a flame sensor was also installed in the home to detect if a fire accident occurred and then it will send a high pulse to the ESP32 which will trigger the speaker and set it to a fire alarm and alert everyone.[23,24] Thus, we can prevent a huge amount of damage caused by an accident. Additionally, it will allow people with disabilities and older adults to remain at home safely and comfortably.

8.4 WORKING OF PROPOSED SYSTEM

ESP32 is used to read the input values from the pin, gas, and flame sensors and here a 16*2 LCD is used for displaying the gas levels surrounding us and all gases detected from the gas sensor will simultaneously send data over the internet to keep a record and to monitor the gas levels frequently.[17] This data can be even useful for future study. As shown in Figure 8.3,[18] Table 8.1 is the specifications of components being used.

8.5 SYSTEM DESIGN

The circuit diagram of home automation and security systems is simple. Starting with power supply, a regulated 5 V power is required for ESP32 and all the components are connected to it; the female power jack is given to connect a 5 V adaptor or battery. For the security system structure, we are using the PIR sensor (i.e. HC-SR501) and the output pin of the sensor is connected to the GPIO pin of the ESP32. When the output of the sensor is high then ESP32 will process it and send the signal to MAX98357 which will

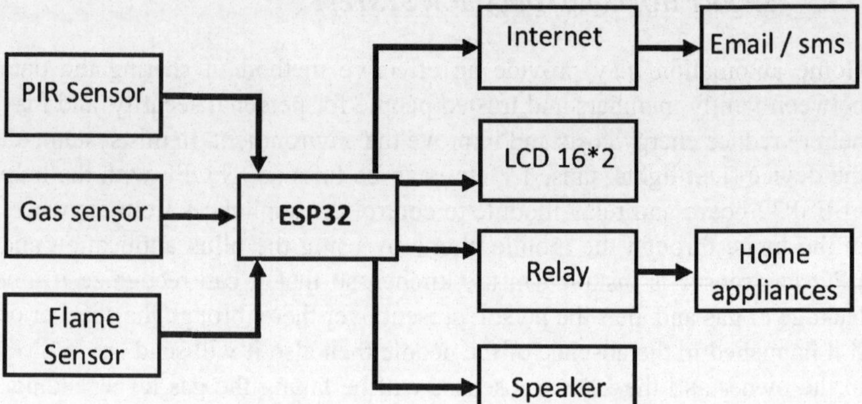

FIGURE 8.3 Block diagram of device.

TABLE 8.1 Technical Specification of Sensors and Device Used.

Sensors	Module Name	Supply Voltage	Output Signal	Operating Range	Use
PIR Sensor	HC-SR501	4.8V - 20V	OV/3V (Output high when motion detected)	Temperature: -20°C - 80°C	Thermal Sensing
Gas Sensor	MQ-5	5V	Analog Signal	300-10000ppm	Measure Hydrocarbon gases
Flame Sensor	YL-38	3.3V - 5V	Digital Signal	760nm-1100nm (Light wavelength)	Hydrogen Station
Wi-Fi Module	ESP32	2.2V - 3.6V	Analog signal	Frequency: 80MHz-240MHz	Provide Wi-Fi and Bluetooth
Display	LCD 16*2	4.7V - 5.3V	—	Temperature: -20°C - 70°C	Output Display
Switch	4-Channel Relay	3.3V - 5V	Digital Signal	Temperature: -40°C - 85°C	Time Delay

transmit to the speakers connected to it. Flame sensor (i.e. YL-38) and LPG gas sensor (i.e. MQ-5 or 6) are connected to ESP32[25] and if the values are above the desired output levels it will send a high pulse to the microcontroller and then it will again enable the speaker and set it to provide tone for the alert system. For the home automation system, we have connected a 4- channel relay module and all the relays are SPDT[26] (i.e. Single Pole and double throw). As all the connections of the relays are the same, it has 5 pins, two relay coil pins, common, normally closed, and normally open. To control these relays we connected to GPIO pins of ESP32. The neutral wire from the 220 V AC supply is directly connected to the neutral point of home appliances, while the phase wire is connected with the home appliances through the relays and finally we have a 16*2 display connected to the microcontroller to fetch the data collected by the sensors and display in it. All the connections are shown in Figure 8.4.[27]

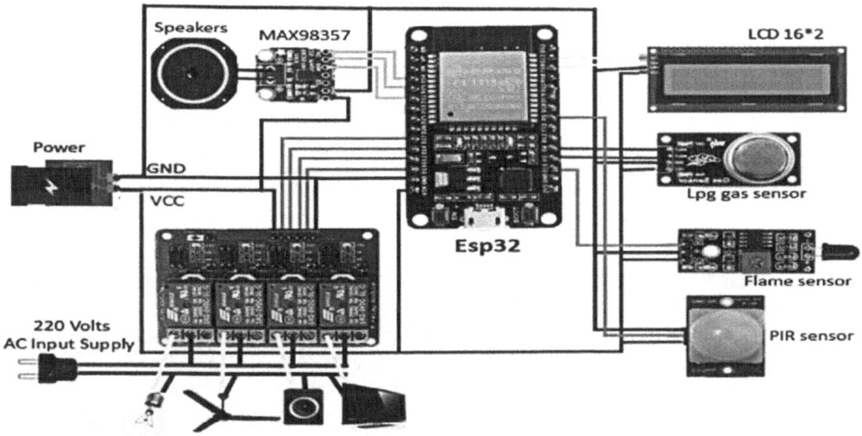

FIGURE 8.4 Circuit diagram of smart home automation and smart security system.

8.6 APPLICATION AND WEBSITE

The application Blynk can control Arduino, Raspberry Pi, and similar devices with Android and iOS devices. This tool lets you create a visual interface for your projects by dragging and dropping widgets. You can start experimenting immediately after completing it in less than five minutes. On the other hand, ThingSpeakTM is an IoT analytics platform that aggregates, visualizes, and analyzes live data streams in the cloud. It provides instant visualizations of the data posted by your devices to ThingSpeak. Anyone who can use it can analyze data from it because the user interface is so easy to understand.

8.6.1 BLYNK APPLICATION

The Internet of Things (IoT) company Blynk offers a platform for the development of mobile (iOS and Android) apps that allow access to the internet and the monitoring and control of electronic devices remotely over the Internet. We can create our own dashboard through which we can arrange buttons, sliders, graphs, and other widgets. We can turn pins ON and OFF and display data from sensors using widgets.[28] Blynk application is ideal for interfacing with IoT projects such as monitoring the temperature or turning on and off home appliances remotely. Using the Blynk library, we can control pins from your phone. So first create a new account in Blynk application and log in into account as shown in Figure 8.5(a). We need to create a new project and choose the device ESP32 Dev Board and connection type wi-fi as shown in Figure 8.5(b).[29] Authentication tokens will establish the connection between our hardware and smartphone. Then we need to add widgets to our projects from the widget box we can select buttons to operate our system as in Figure 8.5(c). Now our application is ready to run as shown in Figure 8.5(d), which lets us interact with hardware and control remotely.[30]

FIGURE 8.5 Application (a) account creation page, (b) create new project, (c) menu page, and (d) working page.

8.6.2 THINGSPEAK

It is an open-source communication system written in Ruby, so users can chat with internet-enabled devices via the internet.[31] API access is provided to social networks and devices, making it easier to access, retrieve, and carry out logging. Its main advantage is the ability to send data to devices via several popular IoT protocols and to visualize data in real time.[32] Furthermore, data from third-party sources can be obtained, aggregated, and analyzed automatically by scheduling or triggering IoT analytics. It is possible to prototype and build IoT systems without requiring servers or web coding. Connecting to ThingSpeak server is easy; we need to create an account and then create channels for individual sensor and your requirements.[33]

It is important that network credentials, channels, and API keys are inserted before uploading code to the ESP32 board and selecting the baud rate. As soon as ESP32 is connected to the ThingSpeak server it will collect the data from the gas sensor at a certain interval of time,[34] it will be going to plot a graph by connecting all the values as shown in Figure 8.6. For LPG gas it is considered immediately dangerous at a level of 2000 parts per million, 10% of the lower explosive limit. Whenever the values are in dangerous ranges, an alarm will be triggered and an alert mail will be sent.[35]

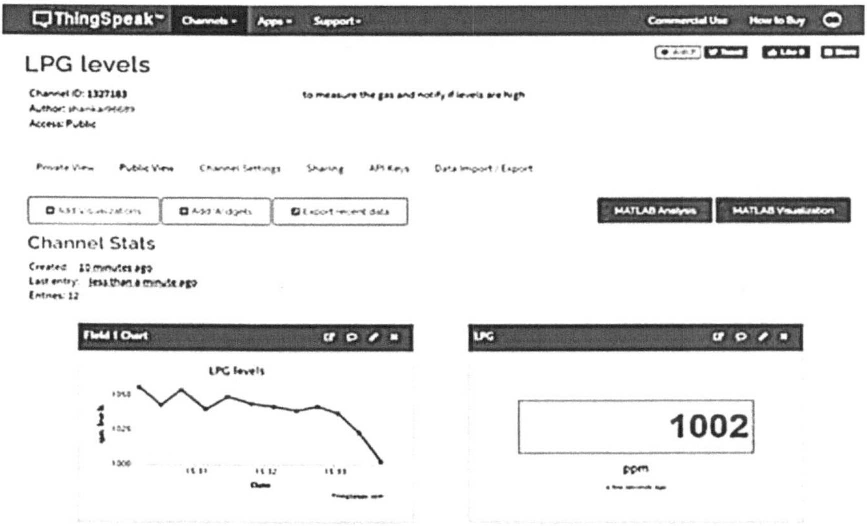

FIGURE 8.6 Gas level values and graph in ThingSpeak.

8.7 RESULT

The proposed smart security system home automation system is being set up and tried-and-true over a prototype room of college in which four 220V devices are connected and all are operated with a blynk application as shown in Figure 8.4(d) along with the speaker, PIR sensor, gas sensor, and flame sensor. All these outcomes were noted and screenshots were taken and results that have come from the above setup can be reflected in the ThingSpeak that has been shown in the graph below in Figure 8.7(a), where we see the values of present and past values also and along with the LCD displaying the value of gas level frequently as shown in Figure 8.7(c). At last we checked our alert system where we send mail to the owner if the gas level is above the threshold value and it is successfully done as we can see in Figure 8.7(b).

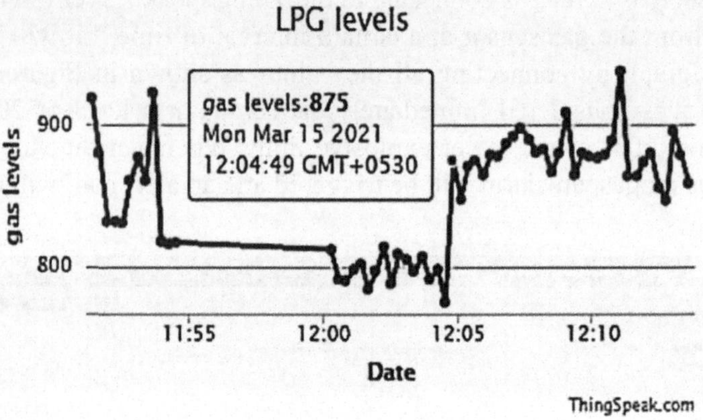

FIGURE 8.7(a) Graph of gas level in ThingSpeak.

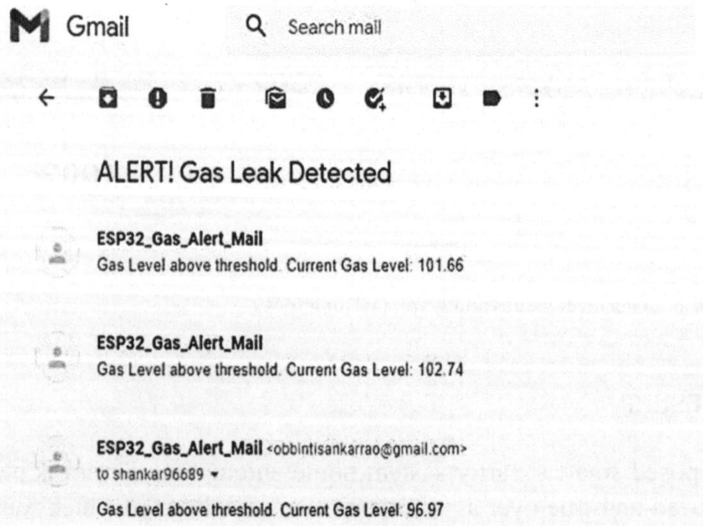

FIGURE 8.7(b) Alert email received with date and time by the owner.

FIGURE 8.7(c) LPG gas levels of a sample having concentration of gas 865 ppm indicating no gas leakage status.

8.8 CONCLUSION

In this paper, smart home security and automation are optimized by using an ESP32 Wi-Fi module. The idea is to automate different electrical appliances without the need for human interference. A smart and intelligent home keeps track and record of various environmental variables and guides the appliances based on the requirements of its occupants. In addition to automating household appliances, it also notifies the user via mail if the gas level rises above the threshold and if it leaks. Taking the above features into consideration, we developed a prototype and tested it.

KEYWORDS

- **wireless sensor network (WSNs)**
- **smart home**
- **automation**
- **security**
- **energy efficient**
- **innovation**

REFERENCES

1. Atzori, L.; Iera, A.; Morabito, G. The Internet of Things: A Survey. *Comput. Netw.* **2010,** *54* (15), 2787–2805.

2. Jurcut, A. D.; Ranaweera, P.; Xu, L. Introduction to IoT Security. *IoT Secur. Adv. Authent.* **2020,** 27–64.
3. Pavithra, D.; Balakrishnan, R. IoT based Monitoring and Control System for Home Automation. In *2015 Global Conference on Communication Technologies (GCCT)*, 2015, pp. 169–173.
4. Tseloni, A.; Thompson, R.; Grove, L.; Tilley, N.; Farrell, G. The Effectiveness of Burglary Security Devices. *Secur. J.* **2017**, *30* (2), 646–664.
5. Asghar, M. H.; Negi A.; N. Mohammadzadeh, Principle Application and Vision in Internet of Things (IoT). In *International Conference on Computing, Communication & Automation*, 2015, pp. 427–431.
6. Vaidya, V. D.; Vishwakarma, P. A Comparative Analysis on Smart Home System to Control, Monitor and Secure Home, Based on Technologies Like GSM, IOT, Bluetooth and Pic Microcontroller with Zigbee Modulation. In *2018 International Conference on Smart City and Emerging Technology (ICSCET)*, 2018, pp. 1–4.
7. Abdulraheem, A. S. *et al.* Home Automation System Based on IoT, 2020.
8. Gupta, P.; Chhabra, J. IoT based Smart Home Design Using Power and Security Management. In *2016 International Conference on Innovation and Challenges in Cyber Security (ICICCS-INBUSH)*, 2016, pp. 6–10.
9. Adriano, D. B.; Budi, W. A. C. Iot-Based Integrated Home Security and Monitoring System. *J. Phys.* **2018**, *1140* (1), 12006.
10. Pravalika, V.; Prasad, R. Internet of Things Based Home Monitoring and Device Control Using Esp32. *Int. J. Recent Technol. Eng.* **2019**, *8* (1S4), 58–62.
11. Moghavvemi, M.; Seng, L. C. Pyroelectric Infrared Sensor For Intruder Detection. In *2004 IEEE Region 10 Conference TENCON 2004*. 2004, *500*, pp. 656–659.
12. Surantha, N.; Wicaksono, W. R. Design of Smart Home Security System Using Object Recognition and PIR Sensor. *Procedia Comput. Sci.* **2018**, *135*, 465–472.
13. Korotcenkov, G. Handbook of Gas Sensor Materials. *Conv. Appr.* **2013**, *1*.
14. Yamazoe, N. Toward Innovations of Gas Sensor Technology. *Sensors Actuators B 108* (1–2), pp. 2–14, 2005.
15. Suma, V.; Shekar, R. R.; Akshay, K. A. Gas Leakage Detection based on IOT. In *2019 3rd International conference on Electronics, Communication and Aerospace Technology (ICECA)*, 2019, pp. 1312–1315.
16. Choi, I.; Shim, H.; Chang, N. Low-Power Color TFT LCD Display for Hand-Held Embedded Systems. In *Proceedings of the 2002 International Symposium on Low Power Electronics and Design*, 2002, pp. 112–117.
17. Kodali, R. K.; Jain, V.; Bose, S.; Boppana, L. IoT based Smart Security and Home Automation System. In *2016 International Conference on Computing, Communication and Automation (ICCCA)*, 2016, pp. 1286–1289.
18. Somani, S.; Solunke, P.; Oke, S.; Medhi, P.; Laturkar, P. P. IoT based Smart Security and Home Automation. In *2018 Fourth International Conference on Computing Communication Control and Automation (ICCUBEA)*, 2018, pp. 1–4.
19. Robles, R. J.; Kim, T.; Cook, D.; Das, S. A Review on Security in Smart Home Development, *Int. J. Adv. Sci. Technol.* **2010**, *15*.
20. Wadhwani, S.; Singh, U.; Singh, P.; Dwivedi, S. Smart Home Automation and Security System Using Arduino and IOT. *Int. Res. J. Eng. Technol.* **2018,** *5* (2), 1357–1359.
21. Harsha, S. L. S. S.; Reddy, S. C.; Mary, S. P. Enhanced Home Automation System Using Internet of Things. In *2017 International Conference on I-SMAC (IoT in Social, Mobile, Analytics and Cloud)(I-SMAC)*, 2017, pp. 89–93.

22. Jacobsson, A.; Boldt, M.; Carlsson, B. A Risk Analysis of a Smart Home Automation System, *Futur. Gener. Comput. Syst.* **2016**, *56*, 719–733.
23. Gunge, V. S.; Yalagi, P. S. Smart Home Automation: A Literature Review, *Int. J. Comput. Appl.* **2016**, *975*, 8887.
24. Assaf, M. H.; Mootoo, R.; Das, S. R.; Petriu, E. M.; Groza, V.; Biswas, S. Sensor based Home Automation and Security System. In *2012 IEEE International Instrumentation and Measurement Technology Conference Proceedings*, 2012, pp. 722–727.
25. Jabbar, W. A. *et al.* Design and Fabrication of Smart Home with Internet of Things Enabled Automation System. *IEEE Access*. **2019**, *7*, 144059–144074.
26. Kousalya, S.; Priya, G. R.; Vasanthi, R.; Venkatesh, B. IOT Based Smart Security and Smart Home Automation. *Int. J. Eng. Res. Technol. IJERT.* **2018**, *7* (04) 181–2278.
27. Singh, H.; Pallagani, V.; Khandelwal, V.; Venkanna, U. IoT based Smart Home Automation System using Sensor Node. In *2018 4th International Conference on Recent Advances in Information Technology (RAIT)*, 2018, pp. 1–5.
28. Durani, H.; Sheth, M.; Vaghasia, M.; Kotech, S. Smart Automated Home Application using IoT with Blynk app. In *2018 Second International Conference on Inventive Communication and Computational Technologies (ICICCT)*, 2018, pp. 393–397.
29. Vikram, N.; Harish, K. S.; Nihaal, M. S.; Umesh, R.; Kumar, S. A. A. A Low Cost Home Automation System using Wi-Fi based Wireless Sensor Network Incorporating Internet of Things (IoT). In *2017 IEEE 7th International Advance Computing Conference (IACC)*, 2017, pp. 174–178.
30. Serikul, P.; Nakpong, N.; Nakjuatong, N. Smart Farm Monitoring via the Blynk IoT Platform: Case Study: Humidity Monitoring and Data Recording. In *2018 16th International Conference on ICT and Knowledge Engineering (ICT&KE)*, 2018, pp. 1–6.
31. Sabancı, K.; Yigit, E.; Üstün, D.; Toktaş, A.; Çelik, Y. ThingSpeak based Monitoring IoT System for Counting People in a Library. In *2018 International Conference on Artificial Intelligence and Data Processing (IDAP)*, 2018, pp. 1–6.
32. Santoso, F. K.; Vun, N. C. H. Securing IoT for Smart Home System. In *2015 International Symposium on Consumer Electronics (ISCE)*, 2015, pp. 1–2.
33. Pasha, S. ThingSpeak based Sensing and Monitoring System for IoT with Matlab Analysis. *Int. J. New Technol. Res.* **2016** *2* (6), 19–23.
34. Razali, M. A. A.; Kassim, M.; Sulaiman, N. A.; Saaidin, S. A ThingSpeak IoT on Real Time Room Condition Monitoring System. In *2020 IEEE International Conference on Automatic Control and Intelligent Systems (I2CACIS)*, 2020, pp. 206–211.
35. Nettikadan, D.; Raj, S. Smart Community Monitoring System using ThingSpeak iot Plaform. *Int. J. Appl. Eng. Res.* **2018**, *13* (17), 13402–13408.

PART II

Design of Intelligent Technologies for Sensors

CHAPTER 9

Design and Development of Web-Based ECG Signal Monitoring and Vital Parameters Measurement

W. S. NIMI*, P. SUBHA HENCY JOSE, and R. JEGAN

Department of Biomedical Engineering, Karunya Institute of Technology and Sciences, Coimbatore 641114, India

**Corresponding author. E-mail: nimiwsrec@gmail.com*

ABSTRACT

Prevalence of chronic disorder and increased aging population leads to continuous monitoring of vital signs for proper diagnosis. This chapter deals with the design of a web-based monitoring system for continuous ECG signal monitoring in real time and feature extraction for estimating the heart rate. Different signal processing techniques are implemented for filtering the signal and extracting the appropriate features to estimate the physiological parameters. The filtering techniques and multiresolution wavelet analysis methods are applied on the real-time signal for quality assessment and vital parameter measurement. Different experiments are carried out for measuring the vital signs from the extracted features. The research proposes a web-based monitoring of real-time ECG signal for continuous monitoring. The design and development of the proposed system is implemented in the graphical programming environment. The collected information is displayed to the physician and the user via a web-based communication for an efficient remote monitoring.

Intelligent Technologies for Sensors: Applications, Design, and Optimization for a Smart World.
S. Kannadhasan, R. Nagarajan, & Alagar Karthick (Eds.)

9.1 INTRODUCTION

Cardio Vascular Disease (CVD) leads to a major cause of mortality globally that constitutes a significant challenge to health care network practitioners. Based on the report provided by WHO, the death rate due to CVD per year is nearly 17.9 million, which makes life-threatening disease to human health globally. The causes for CVD are improper dieting, unhealthy habits of individuals, excess weight, and lack of exercise. The sudden changes in heart and blood vessels are the main reason behind the CVD. There are various diseases included in CVD such as coronary artery diseases, abnormal heart rhythms, rheumatic heart disease, heart failure, hypertensive heart disease, stroke, cardiomyopathy, congenital heart disease, valvular heart disease, peripheral artery disease, thromboembolic disease, and venous thrombosis. Each of these diseases has several symptoms that include changing blood pressure, variation in heart beat, and increased glucose level. In healthcare, the mentioned heart-related problems are easily monitored with the support of healthcare devices and measurement system. Early identification of these symptoms highly reduces the mortality rate which enables the healthcare providers to deliver necessary treatment.[1] Early identification is possible through continuous monitoring of the individual's health status by perceiving the various vital parameters such as heart rate, breathing rate, blood pressure, oxygen saturation, and body temperature.

For continuous monitoring, the utilization of noninvasive methods with more flexible and reliable measurement system for accurate measurement is highly preferred in the healthcare system. Electrocardiogram (ECG) recording is one of the most widely used methods for acquiring the biosignals for estimating the heart rate.[2] Recently, more devices have been developed in continuous monitoring without affecting the daily activities that are used in both home and hospitals. These devices are mainly used for monitoring the individual's health status at home and help in predicting and identifying CVD. The electrocardiogram (ECG) is used for recording electrical activity and states the main direction of electrical impulses throughout the heart. ECG signals are significantly used in a wide range of biomedical applications that lie on bio-signal acquisition for chronic patient surveillance, cardiovascular disease diagnosis, identifying different arrhythmias, sleep apnea detection, predicting sudden cardiac arrest, emotional, and physical activity recognition systems. Chronic heart disease and sudden cardiac mortality can be reduced by early identification and giving proper diagnosis at right time. For diagnosing CVD, the most efficient way is to examine the ECG signal continuously. The rapid change in features of ECG signals represents the change in the physiology of the heart muscle. The abnormal electrical activity of the heart shows the different arrhythmias in ECG. The morphological features of the ECG signal are measured accurately by extracting the fiducial points such as its onset, offset, and peak points. For an effective analysis of ECG, a quality signal acquisition method becomes more important. Most common noises are baseline wander, power line interference (PLI), and electromyogram (EMG) noise. To avoid these noise signals and to minimize the false alarms, there is a need for automatic assessment of ECG signal after real-time signal acquisition. This can be done by removing ECG noises using filtering techniques.

The signals for preprocessing and feature extraction can be acquired from real-time patients and also MIT-BIH Arrhythmia database. Signal processing approaches play a crucial function for extracting the necessary information at specific intermission for accurate measurements. Various learning algorithms and advancement in signal processing approaches lead to an effective early prediction and detection of CVD so that timely treatment can be given to the patient before the disease becomes critical. For the early detection of CVD, continuous monitoring is very essential for providing the accurate diagnosis. During emergency situation, it is necessary to provide the treatment remotely. The data can be wirelessly transmitted to the physician with the support of a web server, so that the physician can monitor the patient's health status remotely. It also provides system flexibility to monitor the patients in isolated place at any time. With the support of web-based communication technology, more than one patient can be monitored simultaneously by the physician and can provide appropriate treatment. The entire section deals with the development of a web-based system, where the ECG sensors are used for acquiring the real-time signal to measure the vital parameters. Filtering techniques are applied to discard the noise signal and the quality of the signal is improved for possible feature extraction. The features such as QRS amplitude, PR interval, and QT interval are extracted from the ECG signals with the support of multiresolution wavelet analysis and the measured vital parameters are displayed to the physician for long-term monitoring by developing a web-based monitoring system. The paper is organized as follows: Section 9.2 of this paper gives the background of the ECG signal which explains the interpretation of ECG signal and survey on the related works, Section 9.3 provides various materials and methods used with system architecture for vital parameters measurement and monitoring, Section 9.4 shows the experimental results and their discussions, Section 9.5 concludes the paper with future scope of the system.

9.2 BACKGROUND

9.2.1 INTERPRETATION OF ECG

ECG contributes a major part in identifying the problems related to cardio vascular system. ECG records the electrical activities of the heart. These electrical activities can be measured by placing the electrode above the skin at the particular location of the body. For the diagnosis of the heart diseases, different wave patterns are obtained from human heart activity represented by P, Q, R, S, and T waves. The diseases related to heart can be easily identified by visualizing the quality ECG signal. The obtained wave pattern includes the peaks and valleys of the P, Q, R, S, T, and U waves. Table 9.1 explains various representations of the ECG wave pattern. Normal human ECG signal has a frequency range of 0.05–120 Hz and is of a very small range in millivolts (mV). The various features are obtained mainly on the P wave amplitude, duration of QRS complex, and RR interval. Figure 9.1 shows the wave pattern generated in ECG signal.

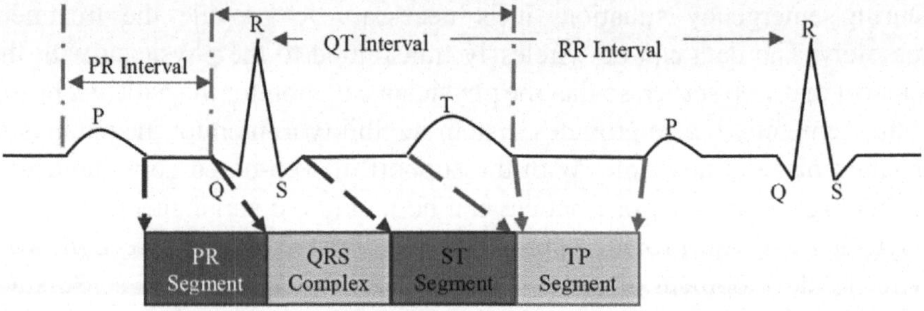

FIGURE 9.1 A schematic structure of ECG wave pattern.

One of the most important features obtained from ECG for heart rate measurement is R peak. The location of other wave patterns can be identified using R peak as a reference point. Hence, the detection of R peak becomes more important during ECG signal processing. QRS detection has the most significant part in disease diagnosis. The QRS detection system is classified into two categories.[3] One of these categories needs extra computational support through digital signal processors (DSP) like wavelet transform and the second category merges DSP and analog-to-digital converters (ADC) that processes data in real time. Apart from this, the information needed for diagnosis can be obtained by detecting the intervals and amplitudes generated by the characteristic wave peaks and boundaries. From the various literature, it was observed that the methods chosen for detecting R peak from real-time ECG signal should be very effective. The improper signal acquisition procedure and methods lead to an interference of various noises like motion artifact, baseline wandering, and the powerline interference.[4] The occurrence of motion artifact is due to the movement of patient. Baseline wandering is a low frequency noise signal caused due to the respiration and changes in electrode impedance. Powerline interference is caused by the stray effect of the alternating current fields because of the loops generated in the cables used by patients. Therefore, it is highly recommended to preprocess the signal that contains different filtering techniques for maintaining the quality of the signal.

TABLE 9.1 Representation of ECG Wave Patterns.

Wave pattern	Representation	Duration (s)	Amplitude (mV)
QRS complex	Ventricular depolarization	0.08–0.10	0.005–0.030
P wave	Atrial depolarization	≤ 0.11	0.05–0.25 mV
Q wave	First downward wave of the QRS complex	≤0.03	<0.2 mV
R wave	Initial positive deflection	Right ventricle <0.035 Left ventricle<0.045	Variable
S wave	Negative deflection following the R wave		
T wave	Ventricular repolarization	0.10–0.25	0.1–0.5 mV
U wave	General repolarization of the Purkinje fibers	–	0.1–0.33 mV
RR interval	Time elapsed between two successive R-waves of the QRS signal	0.6–1.2	–
PR interval	Duration from the initial atrial depolarization to the initial ventricular depolarization and the atrioventricular nodal delay	0.12–0.20	–
QT interval	Measure of both the ventricular depolarization and repolarization	0.20–0.40	–

9.2.2 RELATED WORK

For the past few years, various developments have been sustained for recording the ECG signals. Different researches were carried out in eliminating the noise signals interfered along with the original signal. Different methods were implemented to identify the heart rate, heart rate variability, respiration rate, oxygen saturation (SpO2), and blood pressure. One of the easiest ways for eliminating baseline wandering in ECG is using a high-pass filter (HPF) which helps in blocking all the drifts and allows the important ECG components within the specified wave band. Baseline wander may affect the ST-segment which is of very low-frequency components.[5] Also, the frequency spectrum of baseline wander gets altered during the occurrence of ectopic beats during various diseases. Hence, an efficient method is needed for a broad range of applications such that the removal of these baseline drifts should not affect the original ECG waves.

Various linear phase filters such as finite impulse response (FIR) and infinite impulse response (IIR) are used for preventing the issues generated by phase distortion.[6] The flexibility of the cut-off frequency used in different applications can be increased by using various time-variant filters.[7] Many researchers use multirate system wavelet transform for removing baseline wandering.[8] Empirical mode decomposition (EMD) is also used as there is no need for any a-priori known basis to represent the signal.[9] The different techniques have been investigated by many researchers and tested with the MIT-BIH database and also tested for various methods by removing this noise using different methods such as stationary wavelet transform,[10] combination of dual threshold approach and band pass filter,[11] combined approach of Hilbert transform and band pass filter,[12] and dual median filter[13] where the baseline drift is corrected. Table 9.2 shows different methods used by various researchers for removing the artifacts generated in the original ECG signal. In healthcare network, the ECG features are extracted for disease identification using different methods. R peaks can be detected by suppressing this noise signals. Signal envelop filtering weakens the amplitude of instantaneous peaks occurred during the physical activities.[14] The QRS complex is further improved by consolidating Savitzky-Golay filter with the Shannon energy envelop. Based on the occurrence of R wave interval the heartbeat interval can be measured.[15] A phase-space reconstruction with the help of bandpass filter is proposed for R peak detection from noise signal. Table 9.3 gives different techniques used by different researchers for extracting the features in ECG signal for different applications.

9.3 MATERIALS AND METHODS

9.3.1 ECG SYSTEM ARCHITECTURE

The sensing module of the proposed system is made up of ECG sensor for acquiring the required quantity of physiological signal. The heart rate, QRS

TABLE 9.2 Different Noise Removal Techniques Used in ECG.

Reference	Methods used in ECG	Operation	Pros/cons
Vargas et al. (2020)	Genetic algorithm- minimization of a new noise variation estimate	Eliminates Gaussian noise, power line interference and muscle artifact	The entire noise is not removed but is minimized.
Jin Z et al. (2019)	Iterative majorization-minimization algorithm	Utilizes the convexity of a function to find its maxima or minima and suppress noise	The source impedance substantially affects filtering properties
Wang et al. (2019)	Convex optimization method, which combines linear time-invariant filtering with sparsity	Correction of baseline wandering and denoising of ECG signals	Mode-mixing (IMF has components of different frequencies)
Hsin-Tienet et al. (2019)	Fully convolutional network-based denoising autoencoders	ECG signal denoising	Generates fake samples close to real data
Chen et al. (2017)	Peak detection and moving average method	Removes motion artifacts based on QRS peak detection	It alters the morphology of the ECG signal and fails to remove low-frequency motion artifact
Shuto et al. (2017)	Stationary wavelet transform (haar wavelet)	Removes QRS complexes, P and T wave from motion affected noise signal	Often fails to detect the QRS peak of ECG signal, so correlation factor is less
Waltenegus (2016)	Adaptive filter (normalized least mean square)	Eliminates motion artifacts using accelerometer and gyroscope sensor output	Some random movements were not removed and it needs extra sensors for motion artifact removal
F. R. Hashim et al. (2013)	Discrete wavelet transform DWT (Daubechies)	DWT with various thresholds is applied to getter better SNR	Artifacts having a spectral overlap with the ECG signal are not removed

TABLE 9.3 Methods for Feature Extraction and Identified Parameters.

Reference	Methods used	Identified parameters	Application	Pros/cons
Maheswari A et al. (2019)	db8 wavelet	R peaks, RR intervals	Arrhythmia identification	Computational complexity due to multiple decomposition levels
Rajesh et al. (2019)	Fourier–Bessel series expansion-based empirical wavelet transform	Amplitude, beat variability kurtosis, skewness, entropy	Detects myocardial Infarction	Variations in the T-wave morphology and inter-beat duration
Wenliang et al. (2019)	Pan Tompkins	QRS complex, R peaks, P and T waves and its features	Arrhythmia recognition and classification	Robust in noise sensitivity, less computational load, and higher accuracy
Antoni (2019)	QRS candidate search and QRS optimization based on a sliding window	QRS complex, R peaks	Arrhythmia detection	Results are almost indistinguishable except for large peaks and valleys
S Raj et al. (2017)	Discrete cosine transform-based discrete orthogonal stockwell transform	Pre-RR, post-RR, local-RR, and average-RR interval.	Arrhythmia detection	Feature appears to be similar for each cardiac event
Hongqiang et al. (2016)	Kernel-independent component analysis, discrete wavelet transform	Frequency domain feature and nonlinear features	Automatic recognition of arrhythmias	Computational complexity due to multiple decomposition levels
Changyue et al. (2016)	Hidden Markov models, Baum–Welch algorithm	RR intervals and ECG-derived respiratory signals	Obstructive sleep apnea detection	For short time interval the signal obtained is inappropriate
Bayasi et al. (2016)	Adaptive search windows along with adaptive thresholds	QRS complex, T-wave delineation, and P-wave delineation	Predicts ventricular arrhythmia	During T- and P-wave delineation, false indication due to instantaneous change in sign
X. Tan et al. (2014)	Empirical Mode decomposition	QRS-peak detection, Q-wave onset and offset	Electrocardiogram delineation	If Q-wave or S-wave are absent the onset and offset are distorted

amplitude, and different wave intervals are measured from acquired signal using the ECG sensor. ECG architecture consists of different layers wherein the first layer offers the mechanism for signal acquisition. The desired level of input quantity of the acquired signal is achieved by using a low-power amplifier that is further transmitted in the LabVIEW environment. The foremost section of the measurement system is the analog front-end structure that comprises data acquisition and the processing unit. The next layer is for preprocessing and processing the signal. Here the various artifacts or noise signal interfered with the acquired ECG signal are filtered, and QRS detection and ECG wave delineation operations are performed. For estimating the heart rate, processing of physiological signal and extracting the features play a significant role.[33] In this section, myDAQ analog device, filtering techniques, and wavelet-based decomposition model for vital sign recording, processing the signal, monitoring and extracting the necessary features are proposed. Figure 9.2 shows the overall workflow of the proposed system.

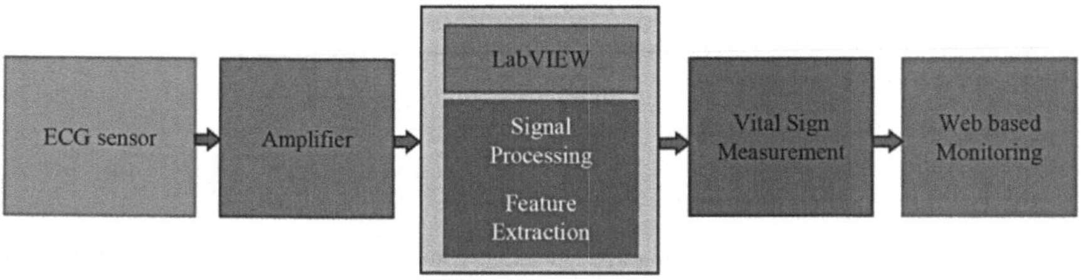

FIGURE 9.2 Overview of proposed system work flow.

9.3.2 *SIGNAL ACQUISITION AND PROCESSING*

The recording of ECG signal with ECG sensor is an essential segment of biomedical field for measuring patient's physiological parameters.[34] The extensive collection of research article in this paper consolidate the use of ECG sensor for quality acquisition of signal. For investigations, ECG signal is obtained through a single-lead ECG sensor and the measurement is carried out at the wrist. Figure 9.3 shows the typical ECG sensor and their output ECG signal representation.

9.3.3 *PREPROCESSING UNIT*

The advanced signal processing methods are applied on the acquired ECG signal. Real-time ECG is corrupted by noises that are taken in different environmental conditions. The original ECG can be interfered with several noise signals and artifacts whose frequency band is the same to that of ECG signal. The wide range of preprocessing methods plays a vital role in removing various ECG noises. Signal preprocessing with filtering techniques provides the removal of unwanted noise signal and delivers the quality signal for proper diagnosis. Preprocessing of the ECG signal is carried out by using the band-pass filtering method. Band pass filtering enhances the signal-to-noise ratio and allows the signal within a certain range of frequencies and blocks the signal outside that frequency range. BPF consists of two cut-off frequencies, upper cut-off frequency for providing the higher frequency limit and lower cut-off frequency for providing the lower frequency limit. Selecting the exact bandwidth for filtering the noise level is a major challenge as the ECG signal has very low amplitude.

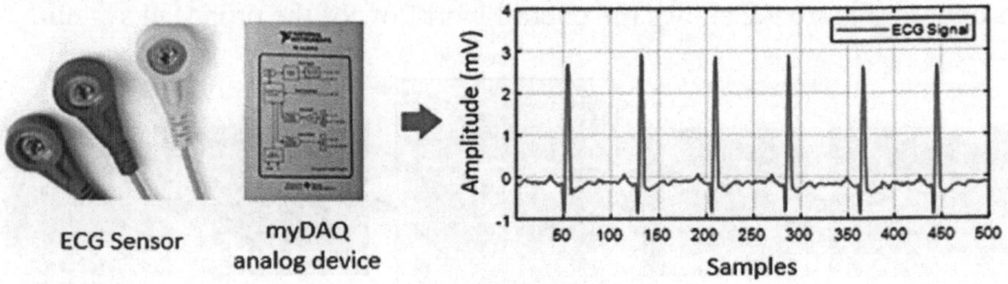

FIGURE 9.3 A schematic representation of ECG signal acquisition with ECG sensor and its output.

9.3.4 *FEATURE EXTRACTION AND VITAL PARAMETERS MEASUREMENT*

The most important stage carried out during ECG signal processing is feature extraction. It is done after the removal of noise signals from the ECG signals. Advanced signal processing techniques are implemented for extracting the required features from the obtained ECG signal. Among the different ECG time domains, frequency domain, and statistical features, R peaks detection becomes more effective for estimating heart rate and also other features are extracted for better diagnosis. Wavelet-based multiresolution analysis is proposed for feature extraction. The wavelet-based analysis method is applied on the signal for removing signal trending. For estimating the heart rate, a large number of signals from the different subjects have been acquired and the quality of the signal for vital parameter measurement is improved using the wavelet analysis approach.

9.3.5 WAVELET ANALYSIS

The advantage of using wavelet-based signal analysis approach is to provide time domain and frequency domain information of the real-time signal. Wavelet analysis is widely preferred for obtaining the nonstationary signals. It can be used for a low range of frequencies to a high range of frequencies due to its varying window sizes. The important concern in wavelet transform is related to the signal decomposition and the signal reconstruction. This is carried out by downsampling and upsampling respectively. According to Haddadi[35] discrete wavelet transform uses the scale value (s) and position value (τ) based on the power of two and can be represented as

$$\psi_{s,\tau}(t) = \frac{1}{\sqrt{2^j}}\psi\left(\frac{t-k*2^j}{2^j}\right) \tag{9.1}$$

where ψψ the fundamental mother wavelet, $s = 2^j$, $\tau = k * 2^j$ and $(j, k) \in Z^2$.

The attributes of wavelet functions in both time and the frequency localization are called wavelet packet. Based on the research focused by Rostaghi,[36] wavelet packet function $\psi^i_{j,k}$ can be defined using three indices that are oscillation parameter (i), scale parameter (j), and translation parameter (k). Mathematically, it can be represented as

$$\psi^i_{j,k}(t) = 2^{j/2}\psi^i\left(2^j t - k\right), \quad i = 1, 2, 3, \ldots \tag{9.2}$$

Here the recursive equations are used to obtain the wavelet function ψ^i. The recursive equations can be written as

$$\psi^{2i}(t) = \sqrt{2}\sum\nolimits_{k=-\infty}^{\infty} h(k)\psi^i(2t-k) \tag{9.3}$$

$$\psi^{2i+1}(t) = \sqrt{2}\sum\nolimits_{k=-\infty}^{\infty} g(k)\psi^i(2t-k) \tag{9.4}$$

where $h(k)$ and $g(k)$ are the discrete filters. These are the quadrature mirror filters analogous with the scaling function and the mother wavelet function.

$$C^i_{j,k} = \int_{-\infty}^{\infty} f(t)\psi^i_{j,k}(t)\,dt \tag{9.5}$$

Using eq 9.5 the wavelet packet coefficient $C^i_{j,k}$ can be determined. For regenerating the original signal each of the wavelet packets should be reconstructed. The signal reconstructions are done with the help of inverse wavelet transform. In this work the wavelet packets are reconstructed independently. For the given wavelet packet (j, i) the signal can be reconstructed as f^i_j using the equation.[6]

$$f_j^i(t) = \sum_{k=-\infty}^{\infty} C_{j,k}^i \psi_{j,k}^i(t)\,dt \tag{9.6}$$

The reconstructed signals are added using the wavelet packets of the *j*th decomposition level to regenerate the original signal as given in equation.[7]

$$f(t) = \sum_{i=0}^{2^j-1} f_j^i(t)\,dt \tag{9.7}$$

The independent wavelet packet's frequency band with sampling frequency F_s is given as equation.[8]

$$F_j = \frac{F_s}{2^{j+1}} \tag{9.8}$$

To avoid the redundancies, downsampling is done by following every decomposition level. Eight levels of decomposition are done for reconstructing the original signal and peak detection. Choosing the correct mother wavelet plays an important role in the wavelet-based analysis method. The mother wavelet is chosen based on the application used and the type of the signal that is used for analysis. The improper selection of mother wavelet can influence the outcome of the signal. One of the merits of the discrete wavelet transform is its diversity in mother wavelet. The mother wavelet in this work is chosen based on the maximum energy to the Shannon entropy criterion. The mother wavelet with peak vitality to the Shannon entropy ratio is considered to be a perfect wavelet. Figure 9.4 shows the db04 wavelet functioning and the scaling function.

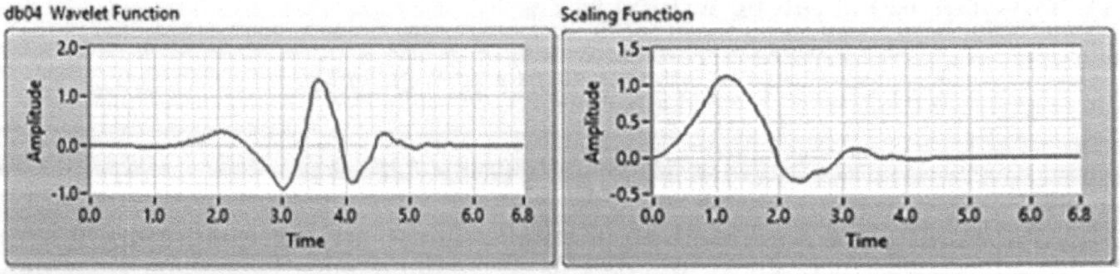

FIGURE 9.4 db04 wavelet function and scaling function for processing an ECG signal.

9.3.6 WEB-BASED VITAL PARAMETERS MONITORING

Continuous monitoring of patient signal and its vital parameters becomes timely important for analyzing the condition of patient health.[37] This will provide an early detection of many diseases and also enable the physician to provide an effective treatment.[38] This can be done with the help of wireless communication technology. This paper proposed a creation of web page to transmit the patient data and their signal to the web page. The advantage of creating web page provides continuous online monitoring of patient signal effectively.

9.4 MEASUREMENT RESULTS AND DISCUSSION

The MIT-BIH database is one of the most commonly available standard datasets for analyzing different arrhythmias. It was used for the basic researches carried out in cardiac disorders all over the world from the 1980s.[39] The real-time signals are used in this paper for testing the performance of the proposed system. The proposed system with the wavelet analysis method is used to extract the time domain and frequency domain features from the ECG signal. The following features are estimated using the graphical programming environment. The ECG sensor is attached to the wrist of the human body and the signals are acquired from different patients and processed to get noise removed signal as shown in Figure 9.5. The different morphological features are identified and extracted with the help of the advanced biomedical signal processing tool kit and recorded in Table 9.4.

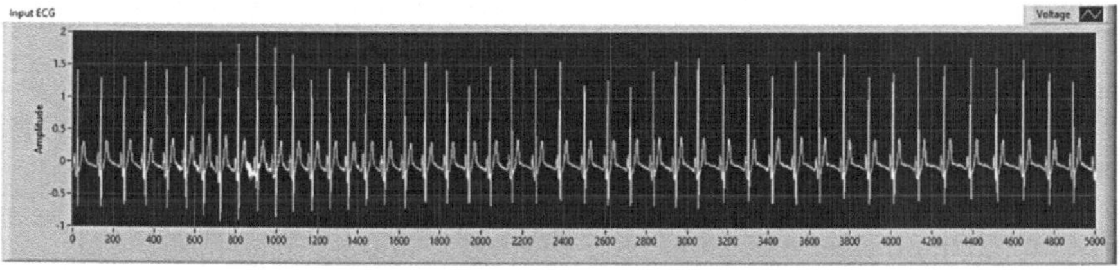

FIGURE 9.5 Real-time ECG signal acquisition with ECG sensor.

The proper identification of the ECG pattern helps to predict the nature of heart diseases. The dynamic change of heart activity causes an irregular heart rhythm called arrhythmias. It is an abnormal signal generated from the heart activity that may result in heart fast and slow functioning. Based on heartbeat, there are many arrhythmias that can be found from the pattern of ECG signal. The fast heart beat caused by ventricles of the heart is called Ventricular tachycardia. The very low range of P-wave amplitude and elevated high PR interval in the ECG signal is categorized into Hyperkalemia arrhythmia. The decreased T-wave amplitude in the ECG wave pattern is called hypokalemia arrhythmia. The signal with low QT interval is a kind of arrhythmia called hypercalcemia. The fast beat in atria causes atrial tachycardia. Figure 9.6 shows various representations of abnormalities in ECG signal.

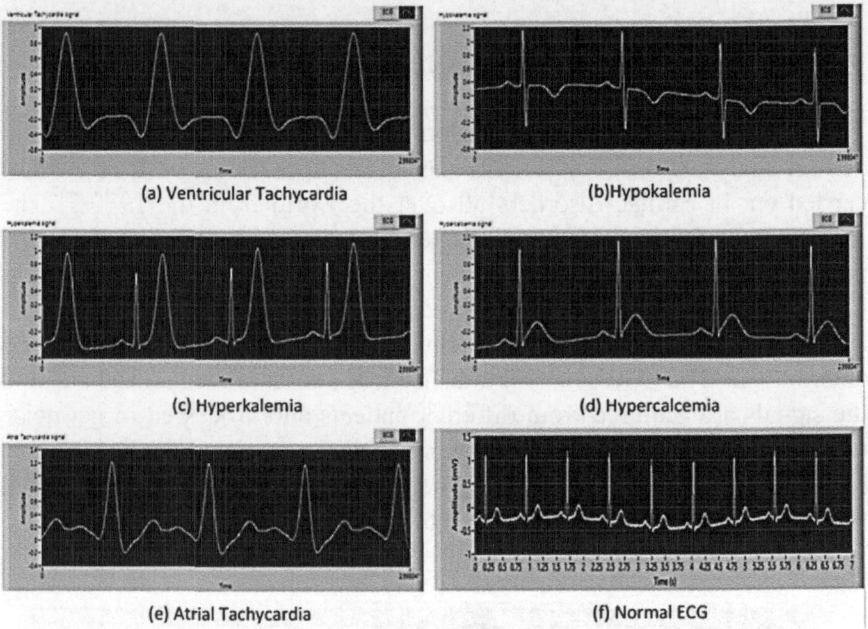

FIGURE 9.6 Representation of various ECG arrhythmias.

The advanced signal processing techniques play a vital role in processing the input signal for extracting the desired features. The multiresolution wavelet analysis method brings a solution to analyze the signal effectively with the help of mother wavelet. The most widely used wavelet techniques such as Harr wavelet, Biorthogonal wavelet, Coiflet wavelet, Discrete Meyer wavelets, Symmlet wavelet, Daubechies wavelets are used for ECG signal processing. In this paper, Daubechies wavelet is used as it has the shapes similar to that of the QRS complex obtained in ECG signal and the energy spectrum is centered on low frequencies.[40] ECG signal is reconstructed from the 8th level db4 decomposition for detecting the QRS amplitude. The original signal is down sampled by sampling the signal to a very low frequency compared to the original signal. This will lead to eliminating the low-frequency noises and the QRS complex is enhanced.

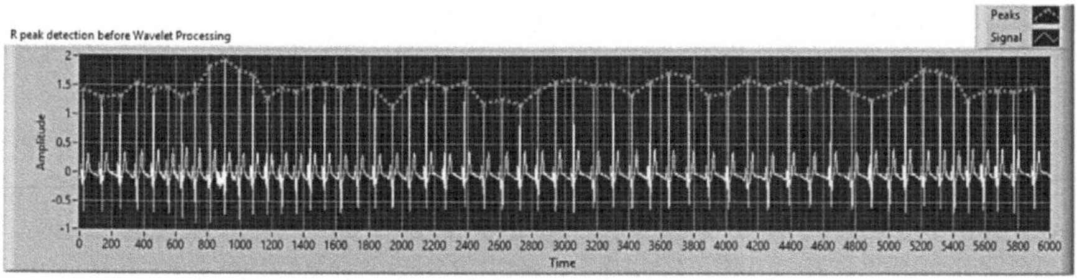

FIGURE 9.7 R peak detection before wavelet processing.

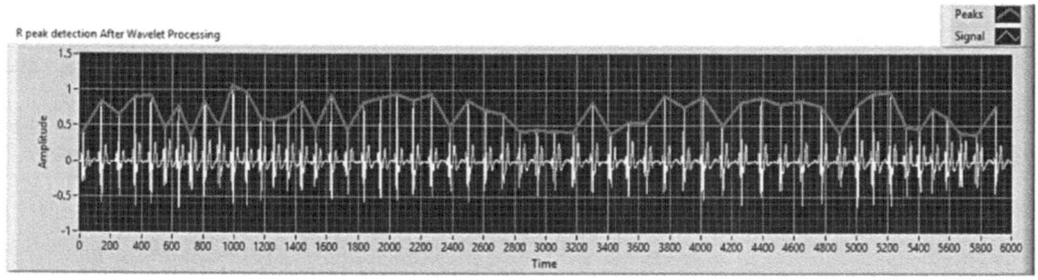

FIGURE 9.8 R peak detection after wavelet processing.

The soft threshold peak detector is applied on the input ECG signal to detect the original R peak as shown in Figure 9.7. The results indicate the R peak detection from ECG signal before the wavelet analysis method was employed. The multiresolution wavelet analysis method is implemented with discrete wavelet transform for reconstructing the original signal to enhance the frequency component of original signal. Figure 9.8 shows the reconstructed ECG signal by the wavelet analysis method with peak detection. The number of R peaks is counted with the programming environment for 60 s to estimate the heart rate of the particular signal. Finally, the proposed measurement system is configured with web publishing technology to transmit the ECG measurement to the cloud for continuous monitoring as shown in Figure 9.9. Table 9.5 indicates the performance assessment of the proposed method for estimating heart rate with comparison of heart rate measured with standard device.

9.5 CONCLUSIONS

This research provides web-based continuous monitoring of electrocardiogram signal and vital sign measurement. The developed ECG front end system supports the measurement and analysis of the obtained ECG signal.

The filtering techniques and multiresolution wavelet analysis method have been used to process the signal for extracting the desired features. The results indicated that the proposed method estimates the heart rate from an ECG signal with an error percentage of 1.31%. The multiresolution wavelet analysis with Daubechies wavelets provides an efficient reconstruction of original signal by presenting approximation and detail coefficients and is more suitable for ECG signal processing. The proposed web-based communication methodology provides long-term continuous monitoring and allows the user to visualize their vital information at home.

TABLE 9.4 Summary of Extracted ECG Features From Normal, Tachycardia, and Hyperkalaemia.

ECG features	Normal	Tachycardia	Hyperkalemia
Heart rate mean (bpm)	78	120	84
Heart rate S.D (bpm)	0.86	0.98	0.85
QRS amplitude mean (mV)	0.863	1.184	0.636
QRS amplitude S.D (mV)	0.021	0.011	0.018
QRS time mean (s)	0.062	0.143	0.058
QRS time S.D (s)	0.002	0.008	0.002
PR interval mean (s)	0.15	0.126	0.141
PR interval S.D (s)	0.009	0.023	0.012
QT interval mean (s)	0.338	0.337	0.339
QT interval S.D (s)	0.009	0.06	0.008
ST level mean (mV)	−0.036	−0.211	0.108
ST level S.D (mV)	0.029	0.039	0.021

TABLE 9.5 Performance Assessment of Heart Rate Measurement.

Subjects	Measured (bpm)	Estimated (bpm)	Error (%)
Subjects 1	78	75	3.84
Subjects 2	82	78	4.87
Subjects 3	65	61	6.15
Subjects 4	87	84	3.44
Subjects 5	74	72	2.70
Subjects 6	76	75	1.31
Subjects 7	79	77	2.53

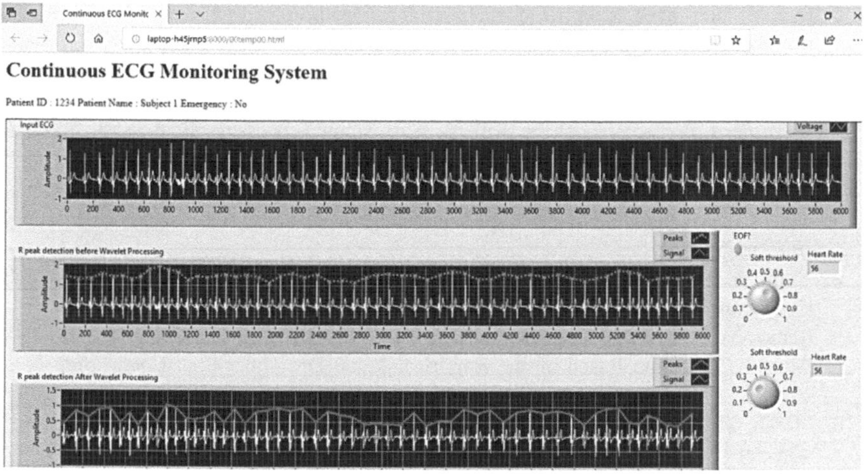

FIGURE 9.9 Web-based continuous monitoring of ECG signal and R peak detection.

KEYWORDS

- **ECG signal**
- **web monitoring**
- **heart rate**
- **vital parameters**
- **feature extraction**

REFERENCES

1. Karolin, A.; Jegan, R. Embedded Based Patient's Vital Sign Monitoring System using Wireless Methods. *Int. J. Eng. Res. Technol.* **2014,** *3*(4), 835–837.
2. Jegan, R.; Anusuya, K. V. Bio Sensor based Feature Extraction and Physiological Parameters Measurement For Bio Medical Applications. *Int. J. Biomed. Eng. Technol*, **2018,** *28*(1), 67–80.
3. Tang, X.; Hu, Q.; Tang, W. A Real-Time QRS Detection System with PR/RT Interval and ST Segment Measurements for Wearable ECG Sensors Using Parallel Delta Modulators. *IEEE Trans. Biomed. Circuits Syst.* **2018,** *12*(4), 751–761.
4. Jegan, R.; Nimi, W. S. In *Low Cost and Improved Performance Measures on Filtering Techniques for ECG Signal Processing and TCP/IP Based Monitoring Using LabVIEW,* 4th International Conference on Advanced Computing and Communication Systems (ICACCS), Coimbatore, 2017, pp 1–7.

5. Luo, Y.; Hargraves, R. H.; et al. A Hierarchical Method for Removal of Baseline Drift from Biomedical Signals: Application in ECG Analysis. *Bioinformat. Biomed. Informat.* **2013,** 2013, 896056.
6. Harting, L. P.; Fedotov, N. M.; Slump, C. H. In *On Baseline Drift Suppressing in ECG-Recordings*. Proceedings of the IEEE Benelux Signal Processing Symposium, 2004, pp 133–136.
7. Chouhan, V. S.; Mehta, S. S. In *Total Removal of Baseline Drift from ECG Signal*. Proceedings of International Conference on Computing: Theory and Applications (ICCTA'07), 2007, pp 512–515.
8. Nimi, W. S.; Jose, P. S. H.; Jegan, R. Review on Reliable and Quality Wearable Healthcare Device (WHD). *IJRQEH* **2021,** *10*(4), Article 1.
9. Blanco-Velasco, M.; Weng, B.; Barner, K. E. ECG Signal Denoising and Baseline Wander Correction Based on the Empirical Mode Decomposition. *Comp. Biol. Med.* **2008,** *38*(1), 1–13.
10. Merah, M.; Abdelmalik, T. A.; Larbi, B. H. R-Peaks Detection Based on Stationary Wavelet Transform. *Comput. Methods Programs Biomed.* **2015,** *121*(3), 149–160.
11. Sangketkit, P.; Wongsa, S. In *Robustness Evaluation of a Dual-Threshold QRS Detection Method for Wearable ECG Recorders*. 2017 14th International Conference on Electrical Engineering/Electronics, Computer, Telecommunications and Information Technology (ECTI-CON), Phuket, Thailand, 2017, pp 230–233.
12. Benitez, D. S.; Gaydecki, P. A.; Zaidi, A.; Fitzpatrick, A. P. In *A New QRS Detection Algorithm Based on the Hilbert Transform,* Proceeding Computers in Cardiology; Cambridge, 2020, 27, pp 379–382.
13. Dohare, A. K.; Kumar, V.; Kumar, R. An Efficient New Method for the Detection of QRS in Electrocardiogram. *Comput. Elect. Eng.***2014,** *40*(5), 1714–1730.
14. Lee, M.; Park, D.; et al. A Novel R Peak Detection Method for Mobile Environments. *IEEE Access*, **2018,** *6*, 51227–51237.
15. Jegan, R.; Anusuya, K. V. High-performance ECG Signal Acquisition for Heart Rate Measurement. *Int. J. Biomed. Eng. Technol*, **2013,** *12*(4), 371–381.
16. Vargas, R. N.; Veiga, A. C. P. Electrocardiogram Signal Denoising by a New Noise Variation Estimate. *Res. Biomed. Eng.* **2020,** *36*(1), 13–20.
17. Jin, Z.; Dong, A.; Shu, M.; Wang, Y. Sparse ECG Denoising with Generalized Minimax Concave Penalty. *Sensors (Basel)*, **2019,** *19*(7),1718.
18. Wang, X.; Zhou, Y.; Shu, M.; Wang, Y.; Dong, A. ECG Baseline Wander Correction and Denoising Based on Sparsity. *IEEE Access*, **2019,** *7*, 31573–31585.
19. Chiang, H-T; Hsieh, Y.; Fu, S.; Hung, K.; Tsao, Y.; Chien, S. Noise Reduction in ECG Signals using Fully Convolutional Denoising Autoencoders. *IEEE Access* **2019,** *7*, 60806–60813.
20. Zou, C.; Qin, Y.; Sun, C.; Li, W.; Chen, W. Motion Artifact Removal Based on Periodical Property for ECG Monitoring with Wearable Systems. *Pervasive Mobile Comput. **2017,** 40*, 267–278.
21. Nagai, S.; Anzai, D.; Wang, J. Motion Artefact Removals for Wearable ECG using Stationary Wavelet Transform. *Healthcare Technol. Lett.* **2017,** *4*(4), 138–141.
22. Dargie, W. In *A Stochastic Fusion Technique for Removing Motion Artefacts from the Measurements Of A Wireless ECG.* IEEE International Conference on Multisensor Fusion and Integration for Intelligent Systems (MFI), Baden-Baden, 2016, 449–454.
23. Hashim, F. R.; Safie, S. I. Wavelet Based Motion Artifact Removal for ECG Signals. *Proc. IEEE-EMBS Conf. Biomed. Eng. Sci.*, **2013,** 339–342.

24. Maheswari, A.; Sangaiah, A. K. Arrhythmia Identification and Classification using Wavelet Centered Methodology in ECG signals. *Concurrency Comput. Pract. Exp.* **2019,** *32*(17).
25. Tripathy, R. K.; Bhattacharyya, A.; Pachori, R. B. A Novel Approach for Detection of Myocardial Infarction From ECG Signals of Multiple Electrodes. *IEEE Sensors J.* **2019,** *19*(12), 4509–4517.
26. Zhu, W.; Chen, X.; Wang, Y.; Wang, L. Arrhythmia Recognition and Classification Using ECG Morphology and Segment Feature Analysis. *IEEE/ACM Trans. Comput. Biol. Bioinform.* **2019,** *16*(1), 131–138.
27. Antoni, B. Fast QRS Detection and ECG Compression Based on Signal Structural Analysis. *IEEE J. Biomed. Health Informat.* **2019,** *23*, 123–131.
28. Raj, S.; Ray, K. ECG Signal Analysis using DCT-Based DOST and PSO Optimized SVM. *IEEE Trans. Instrum. Meas.* **2017**, 1–9.
29. Li, H.; Yuan, D.; Wang, Y.; Cui, D.; Cao, L. Arrhythmia Classification Based on Multi-Domain Feature Extraction for an ECG Recognition System. *Sensors* **2016,** *16*(10).
30. Song, C.; Liu, K.; Zhang, X.; Chen, L.; Xian, Z. An Obstructive Sleep Apnea Detection Approach using a Discriminative Hidden Markov Model From ECG Signals. *IEEE Trans. Biomed. Eng.* **2016,** *63*(7), 1532–1542.
31. Bayasi, N.; Saleh, H.; Mohammad, B.; Mohammed, I. In *65-nm ASIC Implementation of QRS Detector Based on Pan and Tompkins Algorithm.* 10th International Conference on Innovations in Information Technology (IIT), 2014, pp 84–87.
32. Tan, X.; et al. EMD-Based Electrocardiogram Delineation for a Wearable Low-Power ECG Monitoring Device. *Canadian J. Electri. Comp Eng.* **2014,** *37*(4), 212–221.
33. Jegan, R.; Anusuya, K. V. Real-Time ECG Peak Detection for Heart Rate Measurement Using Wavelet Packet transform. *Int. J. Biomed. Eng. Technol.* **2015,** *19*(3), 244–254.
34. Jegan, R.; Nimi, W. S. In *Sensor Based Biomedical Frame Work for Monitoring Patient Vital Parameters.* IEEE International Conference on Circuits and Systems, 2020, pp 96–100.
35. Haddadi, R.; Abdelmounim, E.; Hanine, M. El.; Belaguid, A. In *Discrete Wavelet Transform Based Algorithm for Recognition of QRS Complexes*. International Conference on Multimedia Computing and Systems (ICMCS), Morocco, 2014, pp 375–379.
36. Rostaghi, M.; Mehrdad, N. K. Comparison of Feature Extraction from Wavelet Packet based on Reconstructed Signals Versus Wavelet Packet Coefficients For Fault Diagnosis Of Rotating Machinery. *J. Vibroeng*, **2016,** *18*(1), 165–174.
37. Jegan, R.; Nimi, W. S. Sensor Based Smart Real Time Monitoring of Patients Conditions Using Wireless Protocol, *Int J EHealth Med. Commun.* **2018,** *9*(3), 79–99.
38. Jegan, R.; Anusuya, K. V. Heart Signal Monitoring For Remote Patient Assistant Based On Optical Remote Sensing. *Optoelectron. Adv. Mater. Rapid Commun.* **2015**, *9*(5), 570–574.
39. Moody, G. B.; Mark, R. G. The Impact of the MIT-BIH Arrhythmia Database. *IEEE Eng. Med. Biol. Mag*, **2001,** *20*(3), 45–50.
40. Mahmoodabadi, S. Z.; Ahmadian, A.; Abolhasani, M. In *ECG Feature Extraction using Daubechies Wavelets.* Conference: Proceeding of the Fifth IASTED International Conference Visualization, 2005.

24. [illegible] Samanta, A. [illegible] Based [illegible] and Classification [illegible]. [illegible] 2019 [illegible].

25. [illegible] Signal [illegible]. [illegible] 2019 [illegible].

26. [illegible] Decomposition and [illegible]. [illegible] 2019, [illegible].

27. [illegible] Based on [illegible] Signal [illegible] Analysis. [illegible] 2019, [illegible].

28. [illegible] Signal Analysis using [illegible] and [illegible]. [illegible].

29. [illegible].

30. [illegible].

31. Daniel, [illegible].

32. [illegible] 2014, [illegible].

33. [illegible].

[illegible]

[illegible]

[illegible]

CHAPTER 10

Efficient Internet of Things Enabled Smart Healthcare Monitoring System Using RFID Security Scheme

ANITHA G[1*], G. RAMKUMAR[1], R. THANDAIAH PRABU[1], S. RAMESH[1], V. MOHANAVEL[2], and ALAGAR KARTHICK[3]

[1]*Department of Electronics and Communication Engineering, Saveetha School of Engineering, Saveetha Institute of Medical and Technical Sciences (SIMATS), Chennai 602105, Tamil Nadu, India*

[2]*Centre for Materials Engineering and Regenerative Medicine, Bharath Institute of Higher Education and Research, Chennai 600073, Tamil Nadu, India*

[3]*Department of Electrical and Electronics Engineering, KPR Institute of Engineering and Technology, Coimbatore 641024, Tamil Nadu, India*

ABSTRACT

The concept of Internet of Things (IoT) and its associated developments are utilized on several latest smart applications such as agriculture, healthcare industries, security services and so on. Specifically, the logic of applying the Internet of Things in healthcare field is so important and it is considered to be the boom to healthcare industry. Because many patient life is involved in such industry so, that the adaptation of latest technologies with such field provides a lot of impact and support to physicians to provide best treatments in proper time. The IoT device is operating like a bridge to interconnect the locally available environment to the remote cloud server. In hospitals, it's really a complex task to maintain the patient details in the local system and

Intelligent Technologies for Sensors: Applications, Design, and Optimization for a Smart World.
S. Kannadhasan, R. Nagarajan, & Alagar Karthick (Eds.)

preserve it for a longer period. Due to the failure cases, it is hard to recover the backup from such local machines as well as it is hard to maintain the backups for periodical days. In order to avoid such issues a remote cloud server-based health record monitoring system is introduced over this article.

10.1 INTRODUCTION

The concept of Internet of Things (IoT) and its associated developments are utilized on several latest smart applications such as agriculture, healthcare industries, and security services. Specifically, the logic of applying the Internet of Things in healthcare field is so important and it is considered the boom to the healthcare industry. Because many patients' life is involved in such industry, so the adaptation of latest technologies with such a field provides a lot of impact and support to physicians to provide best treatments in proper time. The IoT device is operating like a bridge to interconnect the locally available environment to the remote cloud server. In hospitals, it is really a complex task to maintain the patient's details in the local system and preserve it for a longer period. Due to the failure cases, it is hard to recover the backup from such local machines as well as it is hard to maintain the backups for periodical days. To avoid such issues a remote cloud server based health record monitoring system is introduced over this article. This book provides a smart health care monitoring system with the adaptation of latest technologies such as Internet of Things, Radio Frequency Identification (RFID)-based security management, Smart Sensor associations, and Disease Prediction Logic. All these benefits are accumulated in a single pack called Enhanced Learning-based Healthcare Monitoring System (ELHMS). The logic of RFID security is utilized in this approach to provide a security to the medical professionals such as physicians, nurses, and so on who can access the smart health care device to monitor the details. In addition, this proposed health care system provides a provision to the caretakers to monitor the required details instantly from their smart mobile phone/tablet/PC/laptop without any delay and interventions. This is possible with only the appliance of Internet of Things as well as once the data is pushed into the server it is considered completely safe and secured due to the efficiency of the server management system. The server management system efficiently encrypts the health records by using a cipher mechanism called Advanced Encryption Standard (AES), so that the records maintained into the server are totally safe. The sensor unit contains multiple health-monitoring sensors such as

Heart Rate Monitoring Sensor, Temperature Monitoring Sensor, Position Identification Sensor, and Global Positioning System (GPS) to identify the present location of the respective patient and Global System for Mobile Communications (GSM) for sending an appropriate alert to the caretakers and hospital representatives regarding any emergency. For the entire health care monitoring system it is perfect to provide an efficient support to the hospitals during this complex situation with proper security and intelligence.

10.1.1 MAJOR OBJECTIVES OF THIS STUDY

The major objective of this smart health monitoring system is to provide a sufficient health care details management scheme without any flaws over server handling to the users. This provision allows several patients to avoid keeping the physical records everywhere as well as this scheme allows the respective guardian to view the details of the patients from anywhere at any time. The adaptation of Internet of Things (IoT) allows the Smart HealthCare Monitoring Kit (SHCMK) on the patient end to connect with the server end and store the health details instantly without any delay. The security concerns are the major concentration over this study as well as the respective security perceptions are carried out in two ways such as: Access Control and the Health Data Security. The Access Control logics are handled in association with the adaptation of RFID and the data security is achieved by means of the cipher mechanism called Advanced Encryption Standard (AES). A supervised learning scheme is utilized to provide a good healthcare-assisted prediction to the patient and doctors with respect to the analysis of existing health records. The existing patient health records are maintained as a dataset and preserved separately over the server unit for processing. Those records are processed according to the machine learning principles and created a model based on the training samples. Based on the created model the health prediction for the present patient is analyzed as well as this testing record is also appended to the dataset for further reference. With respect to all these specialized sensors, latest technologies, supervised learning abilities, and the security principles, a novel smart healthcare monitoring system is introduced with intellectual skill set. The proposed approach is termed Enhanced Learning-based Healthcare Monitoring System (ELHMS), in which it associates all the specified techniques in a proper way to provide an efficient healthcare surveillance scheme to the people. The following summaries will illustrate all these details in a clear manner with graphical representations.

10.1.2 INTERNET OF THINGS (IOT) IN ASSOCIATION WITH HEALTHCARE

The logic of Internet of Things is associated with most of the present smart applications to operate and control the entities through online with the help of internet services. Most of the companies adapt such technology to carry the organizational records into the server end and maintain over there for security reasons. The IoT acts as a bridge between the client and the server end with respect to the nature of internet-enabled services. The proposed healthcare scheme utilizes the logic of Internet of Things to transfer the local available health related sensor data from the SHCMK to the remote cloud server. The data maintained into the remote cloud server is considered to be safe due to the robust security norms of the cloud environment. Usually, the cloud-based servers preserve the data into the storage unit by means of cryptographic principles. Similarly, in this proposed scheme, a cryptographic logic is followed to preserve the data into the server with proper security, in which it is called Advanced Encryption Standard. This cipher logic processes the plain text and converts it into cipher text with respect to the frequency of 512-bits. The enhancement of the healthcare industry through the use of various Internet of Things enabled capabilities, including remote access and consistent monitoring of patient medical status, long term evaluation of healthcare data, efficiency gains and the advancement of technologies for patient centric healthcare rather than hospital oriented treatment.

The Internet of Things technology allows the use of such a diverse number of different applications, sensors, portable smart gadgets, electronic hand-held devices, and wearable gadgets that have either been connected or disconnected (wireless). These devices measure as well as transmit data to adjacent networks or access points for some more future evaluation, allowing for more rapid response to patients. The proliferation of modern communications technology and the association of IoT provide us with potential to aid the health care business through the provision of advanced ideas as well as to assist patients with severe disease and to meet the needs of those with learning disabilities in the context of medical environments. In the past, patients and respective caretakers could only communicate with physicians via medical visits and phone conversations. The physician had very little contact with the patient and it appeared as though able to keep up would be unattainable. However, with the advent of Internet of Things, global and instant patient monitoring of patients has now become conceivable, unleashing the opportunity to push individuals healthier and assist physicians in providing prompt recommendations or medication. The advantages and challenges of the specified things are depicted in Figure 10.1. The further description details the several advantages and research issues confronting the health care environment.[1,2]

The following summary illustrates the advantages of the IoT adaptation over the healthcare industry in a clear manner.

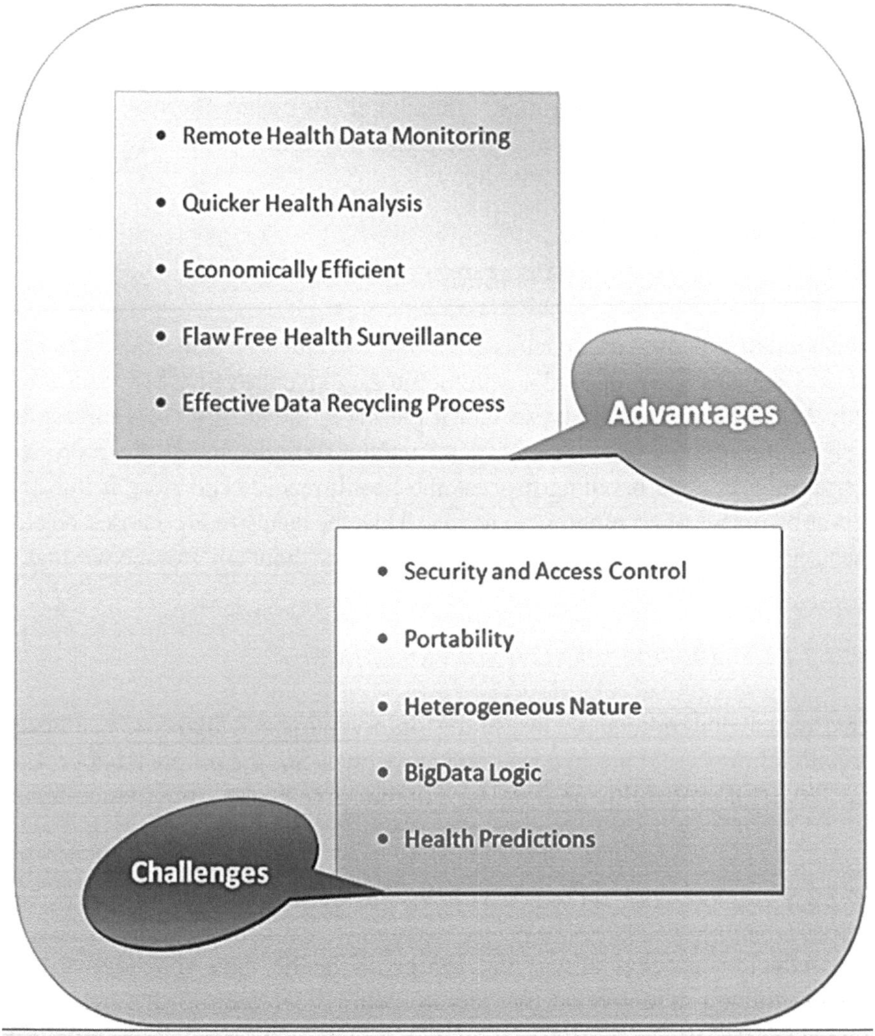

FIGURE 10.1 Advantages and challenges of adapting IoT in healthcare field.

10.1.2.1 REMOTE HEALTH DATA MONITORING

With the inclusion of Internet of Things over the healthcare field, the caretakers and doctors can easily monitor the patient's health records immediately from remote areas as well as the patients are not required to carry their health reports every time.

10.1.2.2 QUICKER HEALTH ANALYSIS

The proposed learning approach ELHMS assists the physicians to analyze the health conditions of the patient immediately by means of cross-validating it with the already trained model. It helps the physicians to analyze the health conditions instantly without any delay.

10.1.2.3 ECONOMICALLY EFFICIENT

This kind of proposed approach is efficient in wealth wise with respect to two aspects, such as: the proposed health monitoring kit called SHCMK consumes only 12 V DC power supply, so that it preserves the energy consumption as well as the second one is called data maintenance, by using the proposed learning based data handling process the health records and store it into the server by means of cryptographic format. Thus the data size is compressed and the users can store multiple records in that space without any storage wastages.

10.1.2.4 FLAW FREE HEALTH SURVEILLANCE

This system does not require any human intervention and all the surveillance-oriented things are done in a systematic manner. So, the proposed approach has less error probability and provides human-free health surveillance logic.

10.1.2.5 EFFECTIVE DATA RECYCLING PROCESS

The recycling process allows the testing of health data appended to the already trained model in an intellectual manner for future references. The data appended to the dataset is available in clear view for manipulating the further records in an efficient manner.

The following summary illustrates the challenges of the IoT adaptation over the healthcare industry in a detailed manner.

10.1.2.6 SECURITY AND ACCESS CONTROL

The data maintained into the local environment is complex to manage during disaster periods but the data maintained into the remote cloud server is considered to be safe. However, the data is not maintained into the client place, it is in the place of a remote entity and other service providers. So, security is a serious concern over this approach. The Access Control logics are the other crucial issue needed to consider as well as the access control logics need to be improved by using some recent technologies apart from the classical credential-based authentication scheme.

10.1.2.7 PORTABILITY

In literature[3, 4] lots of health care monitoring kits are designed and implemented in real-time exposures. But all are having certain portability issues such as the hospital entity needing to customize their products for adapting such latest provision or else the smart health monitoring device needs to be customized according to the convenience of the hospital. The logic of portability is again a great challenge in the proposed approach design.

10.1.2.8 HETEROGENEOUS NATURE

The health care monitoring application needs some centralized data processing unit for manipulating the data from multiple smart devices connected with the patient end as well as all the connected smart devices in single hospital area need to share the same internet service for accumulating the health records of the patient.

10.1.2.9 BIG DATA LOGIC

The data maintenance over the remote cloud server unit is complex in dealing with health records, in which the records are continuously collected from the smart device end (SHCMK) and passed to the server end for manipulation and storage. In this case, a separate data-handling logic such as Big Data is required to manage these continuous records in an intellectual manner.

10.1.2.10 HEALTH PREDICTIONS

Along with all the specifications mentioned earlier, proper prediction logic is required to evaluate the testing (present/live) health records of the patient with the server end dataset model and provides proper risk factor results in a clear manner. Thus, the physicians can easily analyze the health condition of the patient in a time-efficient manner.

Table 10.1 illustrates the traditional cloud-based data-handling system by using Internet of Things interfacing scheme.

TABLE 10.1 Analysis of Conventional Cloud-based Data-Handling System.

Controller used	Connectivity	Remote server storage
Smart sensor association unit[5]	Global positioning system (GPS)-based location details obtaining and netbook simulation software.	Available
Arduino controller[6]	WiFi-enabled connectivity is obtained and programmed used Arduino IDE.	Available
Arduino YUN controller[7]	Short range connecting mediums such as Bluetooth/ZigBee. Programmed using Arduino IDE.	Available
Arduino UNO controller and IDE[8]	Global system for mobile communications (GSM) module is used to establish the connectivity between client and storage unit.	Unavailable
Arduino UNO controller and IDE[9]	Short range connecting mediums such as Bluetooth are used and for identifying the location details GPS is used.	Available
Arduino UNO controller and IDE[10]	Short range connecting mediums such as bluetooth are used.	Unavailable
Raspberry-PI[11]	WiFi-enabled connectivity or else the connection is established by means of Ethernet.	Available
ESP8266 with think device[12]	WiFi-enabled connectivity.	Available
Arduino UNO R3 controller[13]	WiFi-enabled connectivity.	Available
Raspberry-PI[14]	WiFi and Bluetooth-assisted connectivity.	Available
Arduino mega controller[15]	WiFi/ZigBee-enabled connectivity.	Available

10.2 SMART HEALTHCARE MONITORING KIT (SHCMK)

This study describes the detail on proposed approach smart health care monitoring kit called SHCMK, in which it is associated with number of smart sensors to accumulate the patient's health summary in a clear manner. A novel WiFi-enabled ESP32 model microcontroller is utilized in this approach to design a smart healthcare kit, in which the sensor unit is associated with it in proper pin configurations. The respective sensors used in this kit are heart rate estimation sensor, body temperature monitoring sensor, patient position identification sensor, RFID reader, GPS, and GSM modules. All these sensors are associated together to collect the appropriate data from the real-time environments and send those details to the remote cloud server end using Internet of Things services. The details of all sensors are described clearly in detail over the following summary.

10.2.1 ESP32 WI-FI-ENABLED MICROCONTROLLER

All the electronic applications need to be controlled according to the principles of respective microcontroller connected into it. The microcontroller is generally termed controller, in which all the associated sensors are handled with respect to the logic programmed into the controller. Based on such a program, the controllers can acquire and analyze the sensor readings and operate accordingly. In this Smart HealthCare Monitoring Kit (SHCMK), a novel microcontroller called ESP32 is utilized, in which it may operate independently or even as a device driver to a client microcontroller unit, in which it minimizes the connection protocol complexity over its primary system processor. The ESP32 WiFi-enabled microcontroller can communicate with several other devices via its Standard Programming Interface, Standard Input/output Logics, UART, and I2C protocols to enable WiFi as well as Bluetooth capability. The following are the features associated with the ESP32 WiFi-enabled module.

i. It consists of 32 bit Xtensa Dual Core or Single Core processor, which can operate under the frequency ranges of 160 or else 240 megahertz.
ii. It consists of minimal power coprocessing unit called Ultra-Low-Power (ULP) Co-Processor unit.
iii. It contains the memory capacity of 520 Kilobits Static-Random Access-Memory (S-RAM) and 448 Kilobits Read-Only-Memory (ROM).

iv. ESP32 module inbilt consists of Bluetooth and WiFi with the specifications of BLE support, in which the WiFi can operate under the frequency standard of 802.11 b/g/n and the 4.2 version Bluetooth model.
v. It consists of 34 programmable general-purpose input and output pins.
vi. Flash encryption and cryptographic security support association is available in this device. That is the reason the association of AES is included into this approach.
vii. Low power consumption of less than 5 V DC.

Figure 10.2 illustrates the perception of ESP32 Wi-Fi-enabled microcontroller module in a clear manner.

FIGURE 10.2 ESP32 controller.

10.2.2 HEART RATE MONITORING SENSOR

The heart beat analyzing module is used to estimate the pulse ratio of the respective patient, in which it gives the digital output as either LOW or HIGH. The digital outputs are accumulated into an array for a particular period of time (Ex. 1 min) and analyze the pulse level of the patient in an accurate manner. Usually, all the heart beat analyzing sensors consist of LED to indicate the beat level via flashing it. The beat calculations are mentioned as Beats Per Minute (BPM) ratio, in which the following formulation is used to estimate the heart beat level for a minute.

$$H \leftarrow (1+TL)^{N}.\frac{N}{T} \tag{10.1}$$

where H is the heart beat measurement variable, TL indicates the time limit taken for measuring the beat, N indicates the overall beat with respect to TL, and T indicates the specific time period for heart beat level measurement. According to the statistics of National Health Statistics Reports (NHSR), heart beat ranges are estimated based on the BPM level and listed in Tables 10.2 and 10.3, in which Table 10.2 illustrates the perception of heart beat ratio for Men and Table 10.3 represents the perception of Women heart beat ratio.[16]

TABLE 10.2 Heart Beat Ratio for Men.[16]

Age range	Average beats per minute (BPM)	Total ratio (BPM)
Below 1	118–137	104–158
1	110–125	95–139
2–3	98–114	88–125
4–5	87–104	76–117
6–8	79–94	69–106
9–11	76– 91	66–103
12–15	70–87	60–99
16–19	69–85	58–99
20–39	66–82	57–95
40–59	64–79	56–92
60–79	64–78	56–92
80 and above	64–77	56–93

TABLE 10.3 Heart Beat Ratio for Women.[16]

Age range	Average beats per minute (BPM)	Total ratio (BPM)
Below 1	115–137	102–155
1	107–122	95–137
2–3	96–112	85–124
4–5	84–100	74–112
6–8	76–92	66–105
9–11	70–86	61–97
12–15	70–87	60–99
16–19	66–83	57–97
20–39	61–78	52–92
40–59	61–77	52–89
60–79	60–75	50–91
80 and above	61–78	51–94

Figure 10.3 illustrates the perception of heartbeat sensor module and the patient just needs to hold the finger to the respective sensor for measuring the heart beat level. The sensor consists of three pins, in which the pins are portrayed with the respective labels such as VCC for power, GND for ground, and the data. All these specifications are represented in Figure 10.3 in a proper manner.

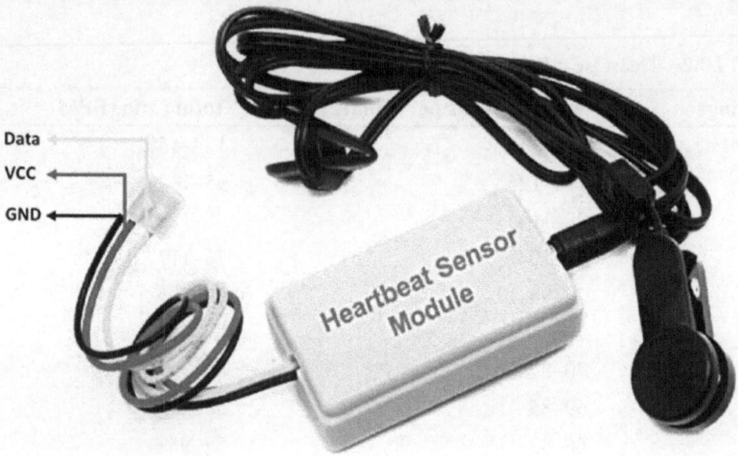

FIGURE 10.3 Heartbeat sensor.

10.2.3 TEMPERATURE-MONITORING SENSOR

This system adapts temperature sensor to measure the body temperature, in which the sensor called DS18B20 temperature measurement sensor is used in this smart health care kit. This DS18B20 sensor is a one-wire programmable sensor and it is globally used to evaluate the temperature of human body as well as some complex environments. This sensor can identify the measurements from −55°C to +125°C, in which the accuracy level is variating on ±5°C. Figure 10.4 illustrates the perception of DS18B20 sensor and the patient just needs to touch the sensor for measuring the body temperature.

10.2.4 POSITION IDENTIFICATION SENSOR

The Position Identification Sensor is used to estimate the position of the patient, in which it provides the value of X, Y, and Z ranges to identify the patient is in normal bed position or felt down. The X, Y, and Z values are in negative state while the patient's position is abnormal and the values in positive range indicate the patient is in normal position. Figure 10.5 illustrates the perception of the position identification sensor, in which the sensor called Micro-Electro-Mechanical-Systems (MEMS) is used to identify the patient's position in a clear manner. This sensor is mainly used to estimate the angle of the patient with respect to specific range determinations. The MEMS sensor acceleration range of X, Y, and Z will be ±100 g, ±200 g, and ±400 g.

FIGURE 10.4 DS18B20 temperature sensor.

FIGURE 10.5 MEMS sensor.

10.2.5 GPS AND GSM MODULES

The GPS module is used to identify the location details of the respective patient, in which it measures the exact location of the patient by means of latitude and longitude specifications. Based on those coordinates the location can easily be monitored by the controller and stored accurately into the server end with specific date and time constraints. The GPS device consists of miniaturized processor units and associated antennas, in which it can acquire information directly from satellites via specific radio-frequency bands. Then it will gather timestamps and certain other information from every accessible satellite. In this application, a novel Neo-6M antenna-enabled GPS module is utilized and it acquires the location details properly. Figure 10.6 will portray the perception of the GPS module with the associated antenna in a graphical manner.

FIGURE 10.6 GPS module.

The Global System for Mobile Communications (GSM) module is used in this smart health care kit to send the respective emergency alerts to the respective caretakers and physicians to take an appropriate action to save the patient's life in a quicker manner. Generally, the GSM modules are used to establish a communication between microcontrollers and the General-Packet-Radio-Service (GPSR). The GSM follows Time-Division-Multiple-Access (TDMS) strategy to send wireless signals over the network environment. The GSM module is adapted over this proposed smart kit to attain the following features such as enhanced efficiency of spectrum levels, global roaming facility to send alerts across countries, improved data quality, scalability and provides good performance with lower frequency bands. In this application, the SIMCOM900 GSM module is utilized to attain higher efficiency in results. Figure 10.7 illustrates the clear perception of the SIM900 GSM module in a graphical manner.

FIGURE 10.7 GSM module.

10.2.6 RFID READER AND CARDS

In this smart health care kit, a specialized access control module called RFID reader is connected, in which it cross-validates the respective user's identity by means of associated RFID cards. The staff working in a hospital environment must carry their smart identity cards to verify their authentication with the system. The system authorization process is handled by means of the administrator end and the staff including doctors need to prove their identity to the system to access the smart device called SHCMK. Once the respective staff is authenticated into the device, the sensor readings are marked with respect to the identity of the staff member. Thus the caretaker can easily monitor the patient's health details according to the associated staff handling the patient. A RFID reader is a network-interfacing gadget, in which it can be compact or substantially affixed. It communicates with the respective smart RFID cards using radio signals. While engaged, the RFID card transmits a signal to the transmitter, which converts it into information. The RFID smart card itself contains the transceiver. In this application, a novel RC522 RFID reader module is utilized to provide higher efficiency in outcome. The RC522-based RFID reader operates according to the frequency range of 13.56 Megahertz electro-magnetic field. The associated RFID reader can communicate with an ESP32 controller at the maximum data transmission rate of 10 Mega bits per second. Figure 10.8 illustrates the clear perception of the proposed RC522 RFID reader module in a graphical representation.

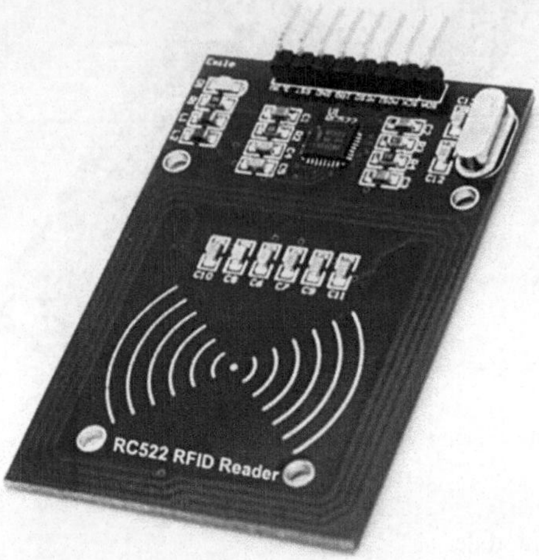

FIGURE 10.8 RFID reader.

10.3 SHCMK DESIGN

This study clearly describes the purpose and efficiency of the proposed smart health care monitoring system with the adaptation of latest technologies over the previous summaries. This illustration provides a logical overview and the proposed SHCMK system design with respect to clear block diagram and the associated working nature in a clear manner. Figure 10.9(a) illustrates the perception of the proposed SHCMK block diagram; Figure 10.9(b) illustrates the receiver unit block diagram with clear sensor specification.

10.4 CONCLUSION AND FUTURE DIRECTIONS

This article clearly describes the efficiency of the proposed healthcare monitoring scheme with the adaptation of several latest technologies to improve the conventional systems. The association of Internet of Things over the proposed approach improves the performance of the smart healthcare system at a drastic level as well as the caretakers attain gains from such

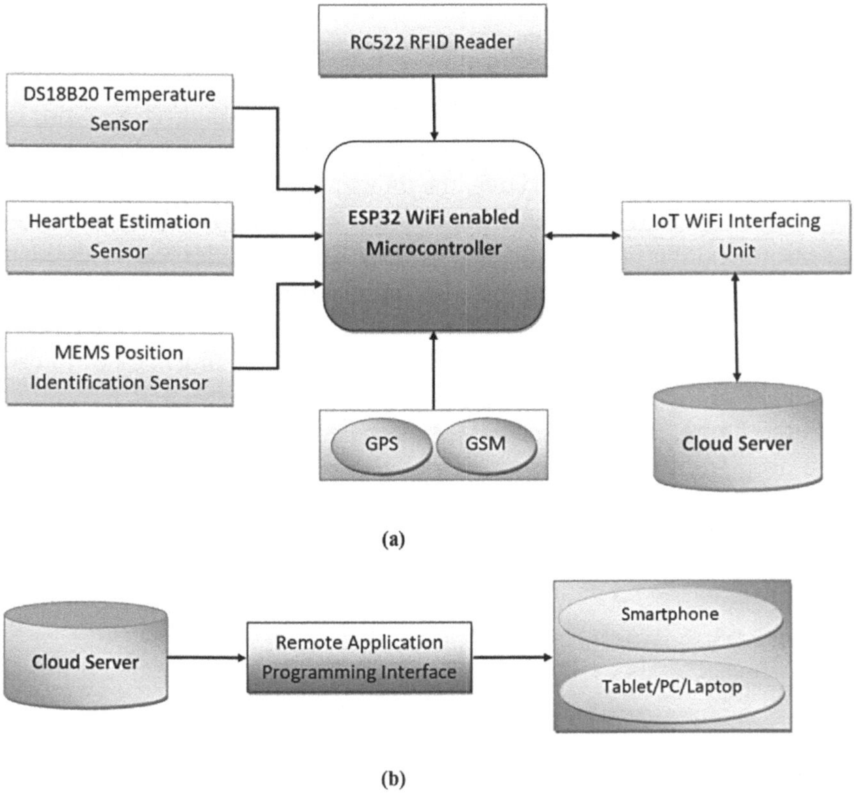

FIGURE 10.9 (a) SHCMK (transmitter) and (b) receiver end block diagram.

development. The proposed Smart HealthCare Monitoring Kit-based health surveillance provides higher efficiency, low implementation cost, power efficiency, reliability, security, and accuracy in the proper manner. The association of RFID-based access control strategy provides more security in terms of avoiding the unknown device operations and it is easy to identify the respective physicians to provide treatments to patients. By using this kind of innovative approach, caretakers can easily monitor the patient's health-related details from anywhere at any time as well as with respect to this logic patients need not to keep the health records physically at all times. This article provides an in-depth examination of established Internet of Things assisted healthcare-monitoring services, including their various communication protocols as well as the opportunities and limitations associated with remote cloud servers. This study has demonstrated the importance of incorporating IoT in residential as well as medical environments. Additionally, this research analyzes Artificial Intelligence models that could be utilized in conjunction with IoT-enabled systems. The system makes use of multiple sensors to monitor the user's healthcare outcomes using remote cloud platforms to gather and analyze the health records. The analytics findings will also be used to classify and forecast data in order to recognize potential diseases based on risk factors analyzed over the trained model. The controller used in this application provides global IoT-enabling support with the specification of sensor handling in an intellectual manner. The controller has an ability to provide crypto analysis using the AES, in which it has a predefined library to process the plain text based on the cipher keys with 512 bit frequency. All the proposed solutions provide a sufficient tool to monitor the health care constraints of the patient in a proper way.

In the future, the proposed model can be enhanced by means of adding some deep learning algorithms such as Deep Neural Networks (DNN) to reduce the time complexity for processing the huge health records as well as it will improve the processing accuracy in a fine manner.

KEYWORDS

- **IoT**
- **SHCMK**
- **healthcare**
- **RFID**
- **GPS**
- **sensor**

REFERENCES

1. Darwish, A.; Hassanien, A. E.; Elhoseny, M.; Sangaiah, A. K.; Muhammad, K. The Impact of the Hybrid Platform of Internet of Things and Cloud Computing on Healthcare Systems: Opportunities, Challenges, and Open Problems. *J. Ambient Intell. Humanized Comput.* **2019**, *10*(10), 4151–4166.
2. Raja, T. Y. M. V. Internet of Things: Benefits and Risk of Smart Healthcare Application. *Innovation* **2019,** *10*(3), 37–42.
3. Qi, J.; Yang, P.; Min, G.; Amft, O.; Dong, F.; Xu, L. Advanced Internet of Things for Personalised Healthcare Systems: A Survey. *Pervasive Mobile Comput.* **2017,** *41*, 132–149.

4. Islam, S. R.; Kwak, D.; Kabir, M. H.; Hossain, M.; Kwak, K. S. The Internet of Things for Health Care: A Comprehensive Survey. *IEEE Access* **2015,** *3*, 678–708.
5. Rodrigues, J. G.; Kaiseler, M.; Aguiar, A.; Cunha, J. P. S.; Barros, J. A Mobile Sensing Approach to Stress Detection and Memory Activation for Public Bus Drivers. *IEEE Trans. Intell. Transp. Syst.* **2015**, *16*(6), 3294–3303.
6. Wasnik, P.; Jeyakumar, A. In *Monitoring Stress Level Parameters of Frequent Computer Users*. 2016 International Conference on Communication and Signal Processing (ICCSP), Apr 2016, pp 1753–1757: IEEE.
7. Mulfari, D.; Celesti, A.; Fazio, M.; Villari, M. In *Human-Computer Interface Based on IoT Embedded Systems for Users With Disabilities*. International Internet of Things Summit, Springer; Cham, Oct 2014, pp 376–383.
8. Nowshin, N.; Rashid, M. M.; Akhtar, T.; Akhtar, N. In *Infrared Sensor Controlled Wheelchair for Physically Disabled People*. Proceedings of the Future Technologies Conference, Nov 2018. Springer: Cham, pp 847–855.
9. Bhatnagar, V.; Chandra, R.; Jain, V. In *IoT Based Alert System for Visually Impaired Persons*. International Conference on Emerging Technologies in Computer Engineering. Springer: Singapore, Feb 2019, pp 216–223.
10. Biswas, S.; Misra, S. In *Designing of a Prototype of e-health Monitoring System*. 2015 IEEE International Conference on Research in Computational Intelligence and Communication Networks (ICRCICN), Nov 2015, pp 267–272; IEEE.
11. Garbhapu, V. V.; Gopalan, S. IoT Based Low Cost Single Sensor Node Remote Health Monitoring System. *Procedia Comp. Sci.* **2017,** *113*, 408–415.
12. Marques, G.; Pitarma, R. IAQ Evaluation Using an IoT CO2 Monitoring System for Enhanced Living Environments. In *Trends and Advances in Information Systems and Technologies*, 2018.
13. Ahmad, S.; Hasan, M.; Shahabuddin, M.; Tabassum, T.; Allvi, M. W. IoT Based Pill Reminder and Monitoring System. *Int. J. Comp. Sci. Netw. Secu.* **2020,** *20*(7), 152–158.
14. Abdelgawad, A.; Yelamarthi, K.; Khattab, A. In *IoT-Based Health Monitoring System for Active and Assisted Living*. International Conference on Smart Objects and Technologies for Social Good; Springer: Cham, 2016, pp 11–20.
15. Gondalia, A.; Dixit, D.; Parashar, S.; Raghava, V.; Sengupta, A.; Sarobin, V. R. IoT-Based Healthcare Monitoring System for War Soldiers Using Machine Learning. *Procedia Comp. Sci.* **2018,** *133*, 1005–1013.
16. https://www.samsung.com/us/heartratesensor/

CHAPTER 11

Modeling of an Active Voltage Doubler: Resonant DC–DC Converter for Wide Range DC Drive Applications

R. BANUPRIYA[1*], J. MANJUSHREE KUMARI[2], and L. SUGANYA[1]

[1]*Department of Electrical and Electronics Engineering, PGP College of Engineering & Technology, Namakkal, Tamil Nadu, India*

[2]*Department of Electrical and Electronics Engineering, Gnanamani College of Technology, Namakkal, Tamil Nadu, India*

[*]*Corresponding author. E-mail: priyarangasamy85@gmail.com*

ABSTRACT

Solar power is utilized as a power-producing and feeding unit to the dc-driving system in this study. For the driving application, a high-efficiency phase shift full bridge resonant converter is utilized. The suggested converter boosts the input power to a greater range while minimizing losses. For the ease of design, the converter is activated using fixed pulse width modulation. The whole system is mathematically modeled in MATLAB Simulink, and a prototype closed-loop system model has been created. The solar equivalent circuit is used to create the proposed converter and mathematical modeling for the solar panel. With a PI controller regulating the motor speed, the motor load is driven at maximum speed. This controller is simple to use and reasonably priced. In both simulation and hardware, the PIC controller is utilized, and the output of the proposed system is assessed and compared.

Intelligent Technologies for Sensors: Applications, Design, and Optimization for a Smart World.
S. Kannadhasan, R. Nagarajan, & Alagar Karthick (Eds.)

11.1 INTRODUCTION

Solar energy, rain, waves, and geothermal heat are examples of renewable energy sources that are naturally regenerated on a human timeframe. Electricity production, air, water heating, water cooling, transportation, and rural energy services are all common uses for renewable energy. Traditional biomass accounts for 8.9% of total energy consumption, with heat (modern biomass, geothermal, and solar heat) accounting for 4.2%, hydropower accounting for 3.9%, and electricity from wind, geothermal, biomass, and solar accounting for 2.2%. In 2015, global renewable technology investments totaled over $286 billion, with China and the United States spending significantly in wind, hydro, solar, and biofuels. With 7.7 million people employed worldwide, solar photovoltaic employs the most individuals of any renewable energy industry. In 2015, renewable energy accounted for almost half of all new power capacity added globally. Unlike fossil fuels that are concentrated in a small number of nations, renewable energy sources are widely available. Renewable energy and energy efficiency technology has resulted in substantial improvements in energy security, climate change mitigation, and economic gains. According to the findings of a recent literature analysis, when greenhouse gas (GHG) emitters are held responsible for damages caused by GHG emissions that cause climate change, a high value for liability mitigation would offer substantial incentives for renewable energy deployment. In public opinion polls throughout the world, renewable energy sources increase like solar and wind power has broad support.

11.2 SOLAR ENERGY

Solar energy is gathered in a variety of ways, including solar heating, photovoltaic, thermal energy, architecture, power plants, and artificial photosynthesis. It is a significant source of renewable energy, and its technologies are classified into passive or active solar based on how they gather and distribute solar energy or convert it into solar power. Photovoltaics, concentrated solar energy, and solar water heating are all examples of active solar methods. Orienting a structure to the Sun, choosing materials with favorable thermal mass or light-dispersing characteristics, and designing areas with natural air circulation are all examples of passive solar methods. Photovoltaic (PV), often known as solar PV, has developed from a small-scale specialized industry to a mainstream energy source in the past two decades. A solar cell is a device that uses the photoelectric effect to turn light directly into electricity. Charles

Fritts invented the first solar cell in the 1880s. Dr. Bruno Lange, a German engineer, developed a photo cell in 1931 that used silver serenade instead of copper oxide. The crystalline silicon solar cell was invented by Gerald Pearson, Calvin Fuller, and Daryl Chapin in 1954, building on Russell Ohl's work from the 1940s. With a 4.5–6% efficiency, these early solar cells cost $286 per watt. By 2012, available efficiencies had surpassed 20%, with research photovoltaics exceeding 40% is shown in Figure 11.1.

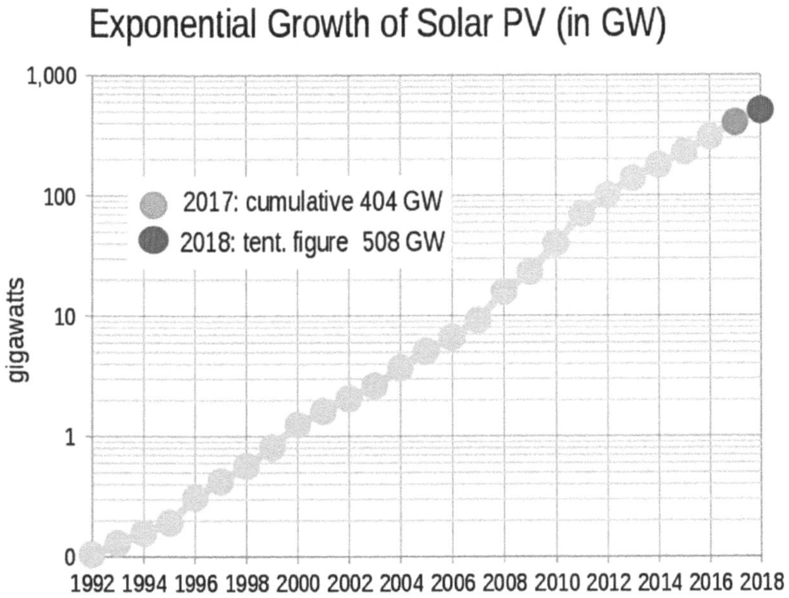

FIGURE 11.1 Worldwide growth of solar PV.

11.3 DC–DC CONVERTER

Prior to the advent of power semiconductors and related technologies, one method of increasing the voltage of a DC source for low-power applications is to convert it into AC through a vibrator, to step up a transformer and a rectifier. An electric motor was utilized to operate a generator with the required voltage for greater power (sometimes combined into a single "dynamotor" unit, a motor and generator combined into one unit, with one winding driving the motor and the other generating the output voltage). When there was no other option, such as powering a vehicle radio, these inefficient and costly methods were used just a few times. Methods such as converting a DC power supply to high-frequency AC, changing the voltage using a small, light, and inexpensive transformer due to the high frequency, and rectifying back to DC became economically viable thanks to the introduction of power semiconductors and integrated circuits. Despite the fact that dynamotors and transistorized vibrator supplies for high voltage mobile transceivers are readily available.

11.4 ELECTRICAL DRIVES

Both the electrical and mechanical sectors today use modern power electronics and drives. Power converters are used in electrical motor drives to generate DC or AC outputs and to operate from either a DC (battery) or a traditional AC source. We will go through the most essential elements that apply to all kinds of drive converters here. Despite the fact that there are many distinct kinds of converters, all of them, with the exception of the extremely low-power ones, use some sort of electronic switching. The importance of using a switching approach is highlighted in the Wrist example, which delves into the implications in great detail. We will see that switching is necessary for high-power conversion, but the waveforms that follow are never optimal from the motor's perspective. The thyristor DC drive is still a popular speed-controlled industrial drive, particularly when the greater maintenance cost of the DC motor brushes (compared with an induction motor) is acceptable. The motor armature receives a low impedance DC voltage adjustable from the regulated rectifier, allowing for speed adjustment.

The kind of converter utilized has a big influence on the drive's efficiency. In this research, a high efficiency phase shift full bridge resonant converter is used for the DC driving application. The performance of the proposed approach is assessed using MATLAB Simulation and hardware. As a power supply for the system, MATLAB is used to simulate the solar panel mathematically. A soft-switching active voltage divider DC–DC converter with minimal power loss has been developed. For small-scale manufacturing, this suggested method is utilized to operate a DC motor.

The suggested research focuses on a voltage doubler rectifier DC/DC converter circuit with a broad input range. This approach may be utilized in large-scale businesses that need a lot of electricity. Low-power applications such as DC drives utilized in small companies, on the other hand, have lacked such research.. For low-power applications, selective control techniques and power modulators will be utilized. The suggested work in this article is on a step-down DC–DC converter with intrinsic balancing. When the inherent balancing method is utilized, this task is adequate for extremely low-power applications. However, the MPPT methods utilized in solar power production are not compatible with this DC–DC converter. The driving mechanism also requires a greater current to operate. To solve the issue, a balanced step-up converter of comparable design may be used.

This article uses a bidirectional resonant DC–DC converter for a high-voltage fain application. The cost of a resonant converter with bidirectional switches is high, and the control and loss reduction methods are complex. A design employing a voltage gain technique with a simple design may be utilized to overcome the cost at all levels. This article describes a full-bridge LLC resonant converter for CC and CV battery charges. The LLC filter's design is very complex, and it comes at a high price. For a small-scale application, the requirement filter must be simple and low-cost. The use of zvs/zcs in a simple design may assist to reduce the cost and size of the system. The aim of this research is to develop and build a brushless DC motor drive using a modified zeta converter as a power source with power control. To get better power quality at AC mains for a broad range of speed control, the suggested BLDC motor drive is constructed and its performance is tested in MATLAB/Simulink environment. It has been suggested as a low-cost option for low-power applications. The performance of the proposed drive was evaluated across a wide range of speed control and supply voltage changes. Furthermore, test results on a constructed hardware prototype were utilized to validate the performance of the proposed drive.

The current system's block diagram shows a parallel connection of a thermal electric generator feeding the DC–DC converter. This combination's combined impact provides a high level of power for the required applications. The different loads linked in the grid connection are driven by a voltage source inverter. The block diagram schematic of the current system is shown in Figure 11.2.

11.5 CIRCUIT DIAGRAM

In Figure 11.3, an active voltage-doubler rectifier is utilized to produce a hybrid resonant dc/dc converter. When the input voltage is higher than the nominal input voltage, the converter switches to the PSFB-SRC mode, which increases power conversion to switching the primary side switches, the converter switches to resonant that boosts the resonant inductor current using the active voltage-doubler rectifier on the secondary. The converter's steady-state waveform looks like a hybrid resonant converter, but it doubles the boost ratio without requiring secondary diodes. As a result, the suggested design lowers the transformer turn ratio, transformer size, and the number of semiconductor devices, resulting in a smaller and less expensive converter. Although the architecture is the same, the operating characteristics are significantly different. Secondary side switches are used in this article to create a closed circuit that increases the resonant inductor current, and secondary side switches are also used to balance the secondary voltage without the need of an extra balancing circuit. The concept was tested using a 600-W trial prototype.

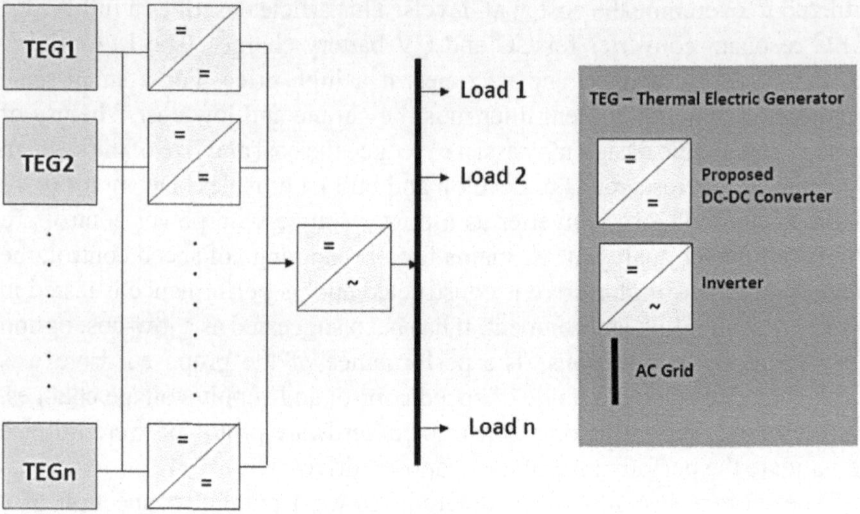

FIGURE 11.2 Block diagram of the existing system.

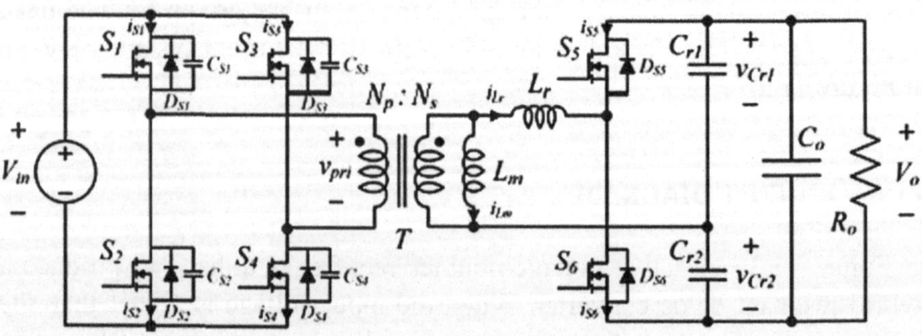

FIGURE 11.3 Circuit diagram of the existing method.

A detailed comparison of various power modulators is shown in Table 11.1. This detailed study suggested the purpose of using a full bridge boost resonant converter with the active voltage doubler output.

TABLE 11.1 Comparison of DC–DC Converters.

Topology		**Full bridge + voltage-doubler**	**Full bridge + active clamp circuit**	**Full bridge + bidirectional GaN AC switches**	**Full bridge + bridgeless boost resonant converter**	**Full bridge + boost resonant converter with active voltage-doubler**
Operation Modes		Buck mode	Buck mode + boost mode	Buck mode + boost mode	Buck mode + boost mode	Buck mode + boost mode
Switches		4	4	6	6	6
Diodes		2	3	2	2	0
Resonant Components		3	3	3	2	3
Inductor current in boost mode	Mode 1	–	Resonant	Linear	Resonant	Resonant
	Mode 2	–	Resonant	Resonant	Resonant	Resonant
Conduction loss (switches and diodes in the path)		–	Boost on: 3	Boost on: 4	Boost on: 4	Boost on: 3
		–	Boost on: 4	Boost off: 3	Boost off: 4	Boost on: 3
Voltage Gain	Buck-mode	2n	2n	2n	n	2n
	Boost-mode $(\gamma_1, \gamma_2, \gamma_3 \geq 1)$	–	$\frac{n}{1-D}$	$2n^*\gamma_1$	$n^*\gamma_2$	$2n^*\gamma_3$
CEC Efficiency	Buck-mode	95.5 %	96 %	97.5%	97.6%	98.1 %
	Boost-mode	–	N/A	≈ 96.1 %	≈ 96.8 %	96.2 %
Rated power		5 kW	1 kW	300 W	300 W	600 W
Cost		Low	Medium	High	High	Medium
Circuit design		Simple	Medium	Medium	Medium	Simple

The proposed block diagram consists of the solar PV panels connected to the proposed DC–DC converter. The motor loads are connected to the grid to operate parallel.

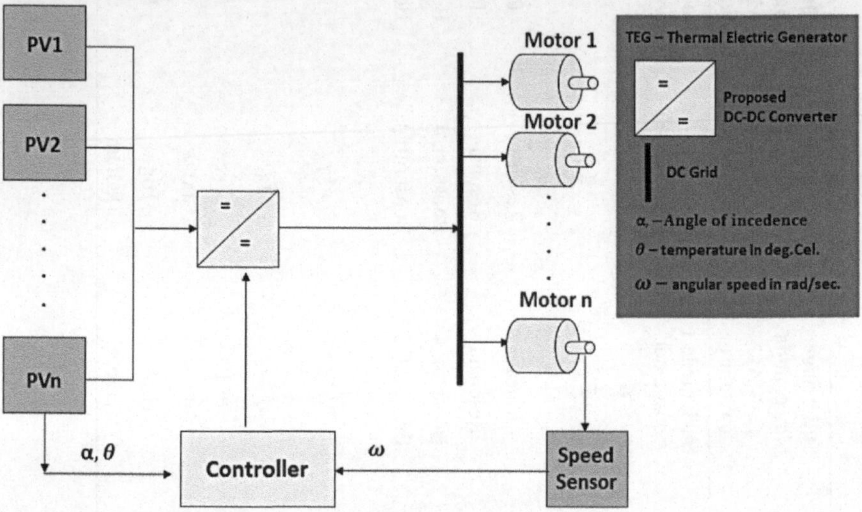

FIGURE 11.4 Block diagram of the proposed system.

Maximum power point tracking technology (MPPT) is used to provide a consistent input power to the PV system from solar power supplied by the PV system. As a result, the power modulator receives solar PV input and produces a high-power output with low ripple. The DC drive for the intended application is verified to be fed by this output. Because of the disturbances, the motor's speed may vary or fluctuate when operating under loaded circumstances. Such disturbances cannot be resolved at the source, but they must be addressed before the load or any other parallel loads are affected. This may be accomplished by using a closed-loop controller to set the motor's speed to the desired value. The suggested system is developed utilizing MATALB for a single input and single load system; however, multiple loads in a closed-loop structure would have the same effect.

11.6 PROPOSED CIRCUIT DIAGRAM

The proposed circuit diagram is shown in Figure 11.5. In the proposed system, the solar PV panels are connected to the recommended DC–DC converter. As a consequence, the power modulator accepts solar PV input and outputs a high-power, low-rising output. This output has been confirmed to feed the DC drive for the specified application. When running under loaded conditions, the motor's speed may change or fluctuate due to the disturbances. Such disturbances are impossible to fix at the source, but they must be handled before the load or any other parallel loads are impacted.

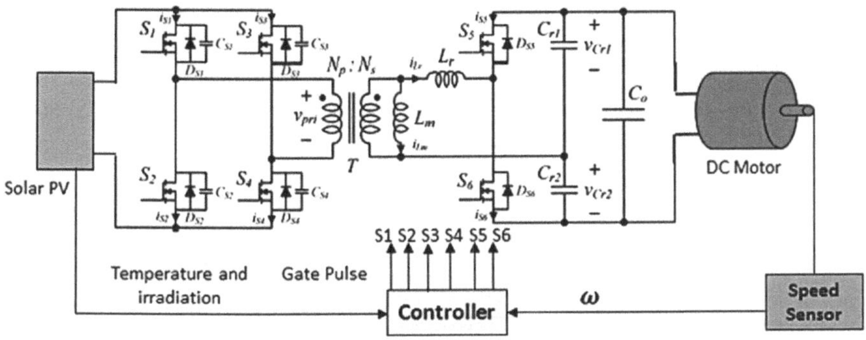

FIGURE 11.5 Circuit diagram of the proposed method.

11.7 CONVERTER SWITCHING TECHNIQUES

The steady-state operation of the converter must be investigated to determine its behavior for any given set of specifications (line-to-line input voltage Vll, rms, output voltage Vo, output current Io, and switching frequency fsw) and component values (inductors La = Lb = Lc = Lin, duty ratio D, and switching frequency fsw) to develop a procedure for designing the proposed converter. Important converter characteristics may be discovered and used to construct the design process once the research is completed. The most essential parameter to compute for the converter's design is the dc bus capacitor voltage Vbus, since it is only after that other parameters such as input current may be determined. In contrast to a conventional two-stage converter, a single-stage converter is not solely controlled by the AC–DC boost PFC stage and cannot be kept constant on purpose. This voltage may be determined by noting that the storage capacitor must be in energy equilibrium while the converter is in steady-state operation.

The energy pumped into the capacitor from the input section must equal the energy supplied to the output section throughout a half-line cycle, resulting in a net dc current flowing in and out of zero. This cannot be computed using an equation with a closed-form solution due to the many possible combinations of input and output modes of operation; instead, it must be solved using a computer programme. For an operating point with specified input voltage Vin, output voltage VL, switching frequency fsw, input inductor Lin, output inductor Lo, transformer turns ratio N = Npri/Nsec, and output voltage Vin, the output current Io may be computed as follows: Select the collection of criteria and component values to be considered. Assume a duty cycle D as a starting "guess" (i.e., D = 0.5) to begin the process of determining a matching dc bus capacitor voltage Vbus. Assume that the output current is constant before calculating Vbus. Check that the output current is continuous with this Vbus value by comparing the peak output current ripple to the average current Io. If this connection is satisfied, Vbus equals the value computed. If this is not the case, the output current is discontinuous, and Vbus must be computed using the formula for discontinuous current mode (DCM). To calculate the average current that flows out of capacitors during a half-line cycle with Vbus known, use either (16) for continuous current mode (CCM) or DCM. The calculated Vbus value is accurate. If this is the case, the operating point to be discovered is invalid, and the procedure must be repeated with a new D value. Figure 11.6 depicts the steady-state analysis technique as a flowchart that may be implemented in a computer application.

11.8 PROPOSED SIMULATION CIRCUIT

The suggested simulation circuit was created using the Simpower system library tools in MATLAB Simulation. The continuous solar output, represented by the DC input supply, is called the input. Four thyristor switches link the first stage of the DC supply to the AC rectification. For optimal functioning of the high boost voltage output, the output is linked to the concept constant power transformer. The transformer's output is fed into single-stage capacitor-coupled high-power converter switches. The drive circuit is made up of a DC drive that is loaded. The simulation circuit created in MATLAB is shown in Figure 11.7.

11.9 SIMULATION RESULTS AND DISCUSSIONS

In the first stage of AC rectification, four thysistor switches are used, and two switches are used in the DC rectification. The thyristor switch G1 and G4 gate pulses are activated with a short delay of 100 and a 40% duty cycle.

The thyristor switches G2 and G3 are programmed to function in the opposite direction of G1 and G4 and are activated with a short delay of 100 with a 40% duty cycle. The pulse produced for the proposed power modulator is shown in Figure 11.8.

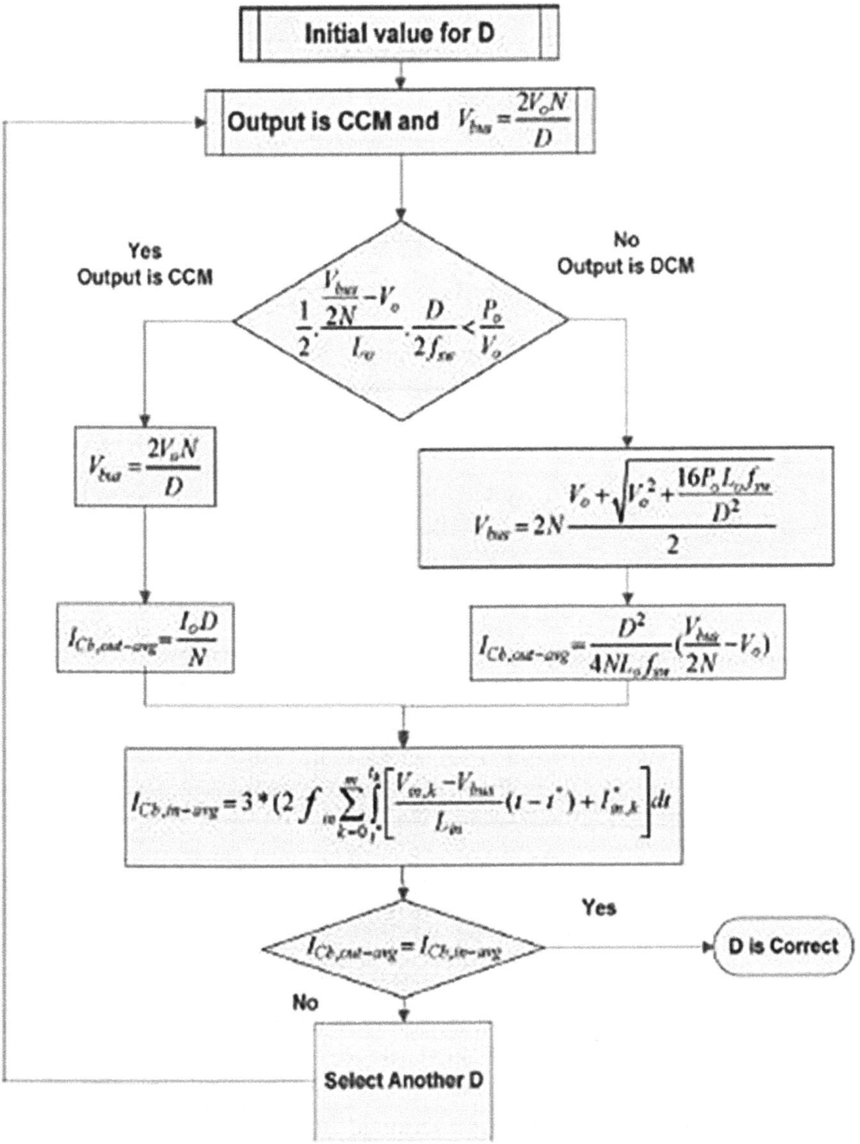

FIGURE 11.6 Control logic of the proposed system.

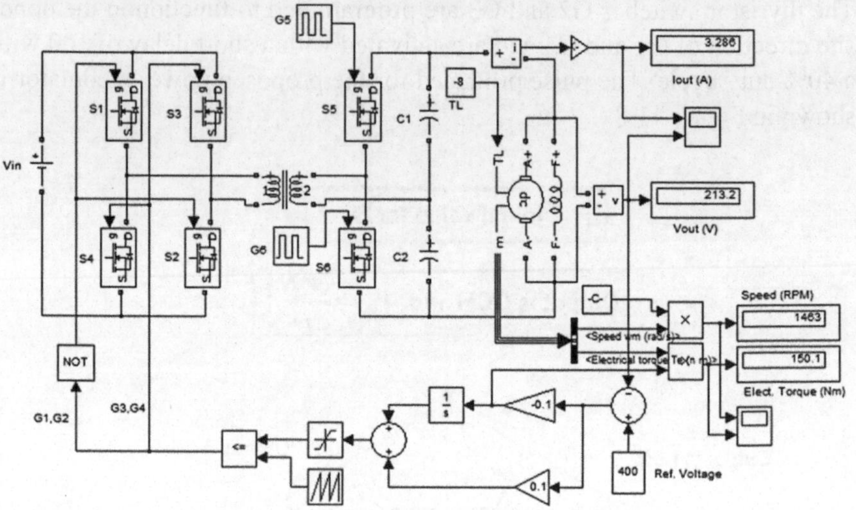

FIGURE 11.7 MATLAB simulation circuit diagram.

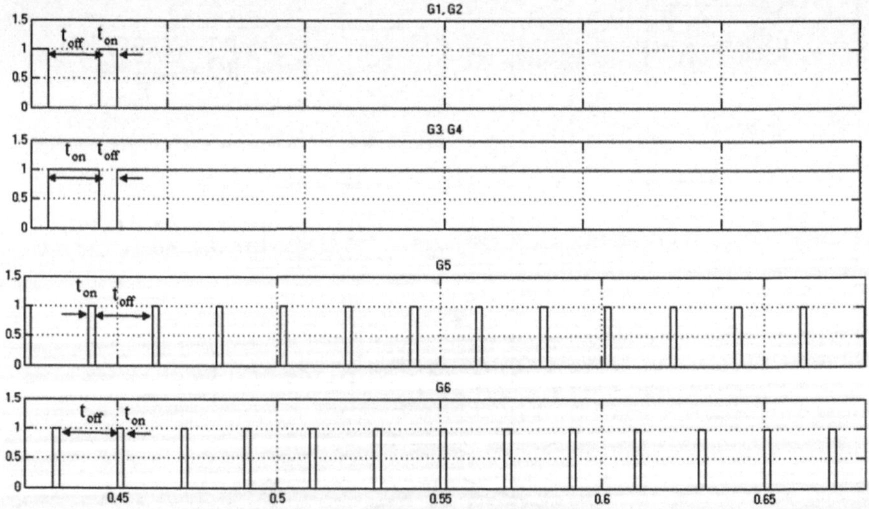

FIGURE 11.8 Gate pulse for the proposed power modulator.

11.10 OUTPUT VOLTAGE AND CURRENT

The voltage and current at the motor terminal are simulated, and the result is the graph shown in Figure 11.9. A 50-VDC input is used to power the proposed system. With a voltage of 215 VDC and a current of 3.2 A, the high-voltage gain with boost output is accomplished, resulting in a power output of 688 W. The voltage generated in the loaded condition is adequate to operate the motor load. With 688-W input power, the motor is rated for 700 W, which is adequate to run the motor load.

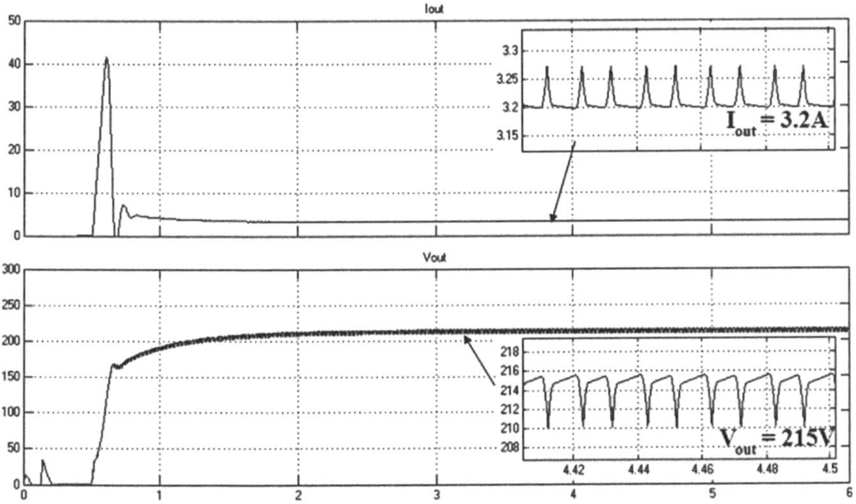

FIGURE 11.9 Motor terminal voltage and current of the proposed converter.

11.11 MOTOR SPEED AND TORQUE

At full load, the power output from the proposed power modulator is supplied to the 700-W DC motor load. The motor is rated for a rated current of 3.0 A and a rated speed of 1500 rpm. A load torque of 8.0 Nm is applied to the motor. The speed and torque waveforms are shown in Figure 20.3. The motor has a closed loop speed of 1462 rpm and a torque of 148 Nm when utilizing the PI converter.

The closed-loop model is much more efficient than the open-loop situation, which has erratic speed and torque circumstances. As a result, the tight-loop PI controller eliminates speed and torque ripple and provides a smooth drive operation, assisting small-scale industrial applications.

Figure 11.10 illustrates the suggested prototype model. The hardware prototype includes a solar panel as well as input electricity as a battery source. The circuit's power comes from a 12-V battery and a 6-V solar panel. The DC–DC rectifier device regulates the input voltage to the motor load using the input from the battery/solar panel. The driver circuit, which is controlled by the PIC controller, connects and operates the DC motor. The motor speed is read by the speed sensor and sent back to the controller as feedback. Closed-loop control is used to obtain the required motor speed.

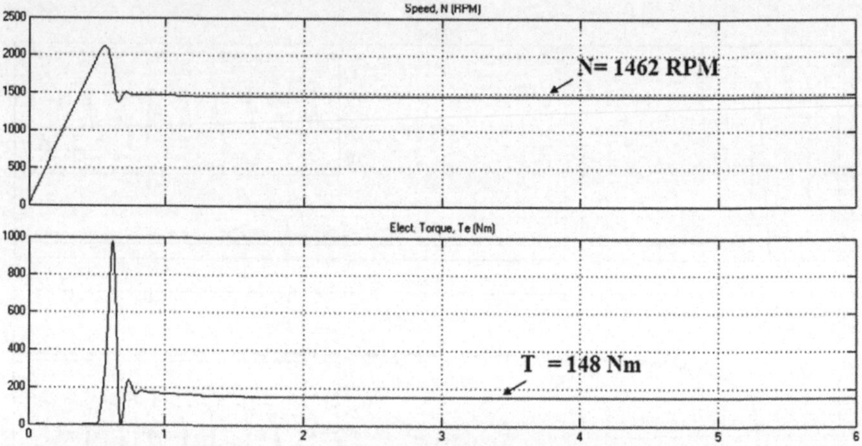

FIGURE 11.10 Speed and torque curve of the proposed motor load drive.

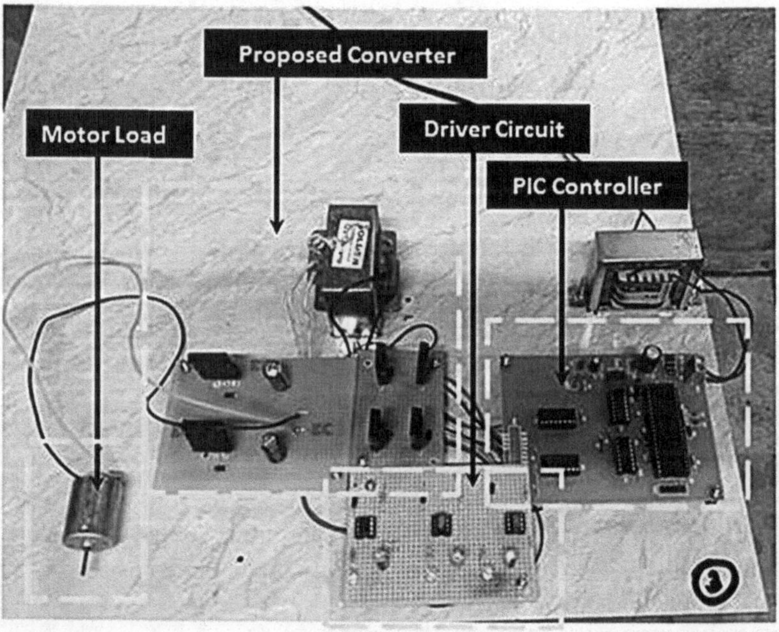

FIGURE 11.11 Hardware prototype model.

11.12 RESULTS AND DISCUSSIONS

For different circumstances, the simulation and hardware results are simulated and observed. The result comparison of the open and closed loops of the motor load in the mathematical modeling using the simulation tool is shown in Table 11.2. The graphical examination of the motor speed is shown in Figure 11.12.

TABLE 11.2 Simulation Comparison on Motor Speed.

Voltage(V)	Speed, N (rpm)		
	Reference	Open loop	Closed loop
215.00	1500.00	1390.00	1462.00

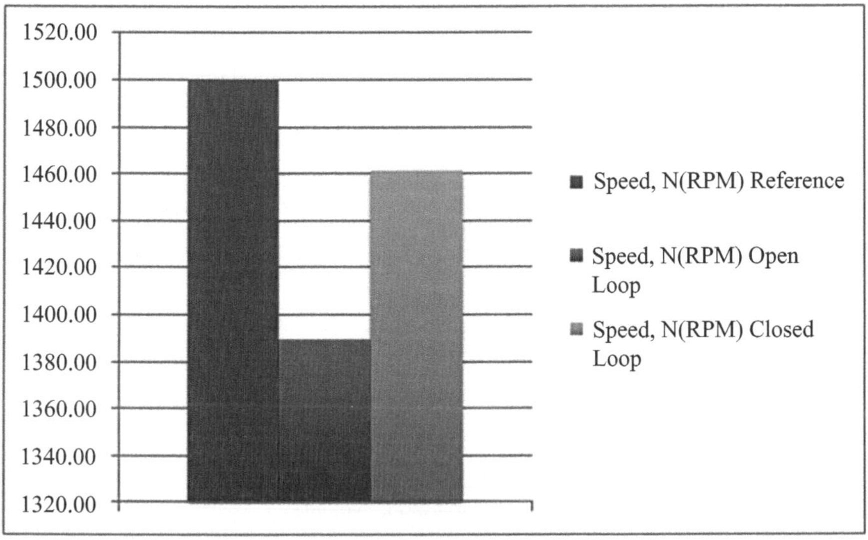

FIGURE 11.12 Motor speed comparison of simulation.

The motor terminal voltages are monitored and visually displayed using the hardware prototype model, which is configured for different speed values. The measured values are shown in Table 11.3. The graphical analysis is shown in Figure 11.13. The change in the motor terminal voltage when the motor speed is adjusted is shown in this thorough study. The motor voltage is not more than 12 V, which is the rated value.

TABLE 11.3 Motor Speed Versus Motor Terminal Voltage.

Set speed, N (rpm)	Motor terminal voltage (V)
480.00	4.88
500.00	5.16
540.00	5.46
580.00	6.20
600.00	6.93
680.00	7.56
700.00	7.84
800.00	10.24

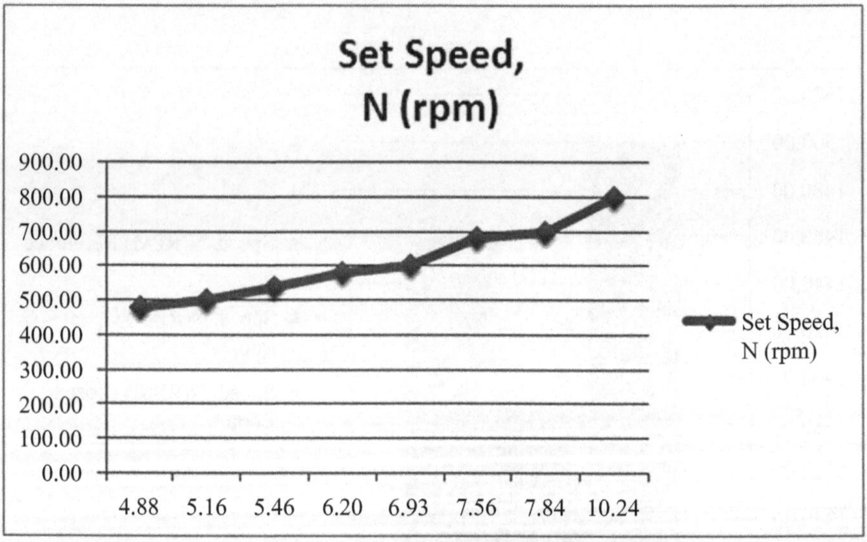

FIGURE 11.13 Motor speed versus motor terminal voltage.

11.13 CONCLUSION

The suggested project effort is based on a review of scholarly publications. This goal is accomplished in the proposed study by mathematically modeling the DC–DC converter and the DC motor drive in both open- and closed-loop modes. The results indicate that DC motor driving in small-scale enterprises is simple when high boost DC–DC converters are used. The speed of closed-loop control has improved, according to the findings of the open- and closed-loop simulation experiments. To get to the required speed, the hardware utilizes a simple control technique.

11.14 FUTURE SCOPE

When higher-end controllers are developed for driving applications, the suggested system may have a greater responsiveness. The prototype's closed loop seems to have a sluggish reaction that might be enhanced with a better control system, such as artificial intelligence.

KEYWORDS

- **converter**
- **DC**
- **voltage**
- **double**
- **speed**
- **control**

REFERENCES

1. Vu, H.-N.; Choi, W. A Novel Dual Full-bridge LLC Resonant Converter for CC and CV Charges of Batteries for Electric Vehicles. *IEEE Trans. Ind. Electron.* **2018,** *65* (3), 2212–2225.
2. Shen, Y.; Wang, H.; Blaabjerg, F.; Durra, A. A.; Sun, X. A Fixed-Frequency Bidirectional Resonant DC–DC Converter Suitable for Wide Voltage Range. In *2017 IEEE Applied Power Electronics Conference and Exposition (APEC)*, 2017; pp 2084–2090. doi: 10.1109/APEC.2017.7930986.
3. Zhang, X.; Green, T. C.; Junyent-Ferré, A. A New Resonant Modular Multilevel Step-down DC–DC Converter with Inherent-balancing. *IEEE Trans. Power Electron.* **2015,** *30* (1), 78–88. doi: 10.1109/TPEL.2014.2301974.
4. Murdoch, C. S.; Reynoso, S. N. Design and Implementation of a MPPT Circuit for a Solar UAV. *IEEE Trans. Ind. Latin Am.* **2013,** *11* (1), 108–111.
5. Al Nabulsi, A.; Dhaouadi R. Efficiency Optimization of DSP-based Standalone PV System Using Fuzzy Logic and Dual-MPPT Control. *IEEE Trans. Ind. Inf.* **2012,** *8* (3), 573–584.

6. Mohd Zainuri, M. A. A.; Mohd Radzi, M. A.; Soh, A. C.; Rahim, N. A. Development of Adaptive Perturb and Observe-Fuzzy Control Maximum Power Point Tracking for Photovoltaic Boost DC–DC Converter. *IET Renew. Power Gener*. **2014,** *8* (2), 183–194.
7. Savio, M.; Vasantharaj, S.; Sasikumar, M. Implementation of Stand-Alone Hybrid System Using SVPWM for Impedance Source Inverter. *Wulfenia J.* **2012,** *19* (11), 404–417.
8. Lin, J. L.; Yao, W. K.; Yang S. P. Analysis and Design for a Novel Single-Stage High Power Factor Correction Diagonal Half-Bridge Forward AC–DC Converter. *IEEE Trans. Power Electron.* **2006,** *53* (10), 2274–2286.

CHAPTER 12

Future Prospects of Electronic Skin

B. LEELAMANI[1*] and V. V. R. RAMAN[2]

[1]Department of Biotechnology, Aurora's Degree and PG College, Chikkadpally, Hyderabad, Telangana, India

[2]Department of Computer Science, Aurora's Degree and PG College, Chikkadpally, Hyderabad, Telangana, India

[*]Corresponding author. E-mail: b.leelamanib@adc.edu.in

ABSTRACT

The skin is one of the most essential organs in the physical body, since it protects internal organs and detects cutaneous impulses, among other things. Artificial, or better still, electronic skin (e-skin) may be a difficult objective to achieve due to its complexity, requiring several different and complementary study fields. Nonetheless, there are many and very important application areas: Humanoids and industrial robots, artificial prosthetics, and biomedical instruments. Despite the fact that it does a tremendous amount of labor, the skin is an often ignored organ. Human skin is sensitive and open to a variety of stimuli. At the same time, it is tough enough to withstand bruising and wounds while still being able to recover in a short amount of time. The difficulty with so-called electronic skin, or e-skin, is to include all of those characteristics. A comprehensive strategy is often used to successfully combat the growth of electronic skin. Starting with the definition of the system specification, the mechanical arrangement of the skin must be designed and fabricated alongside the electronic embedded system to move toward aspects such as tactile processing algorithms and, as a result, the channel interface. The e-skin must be flexible and stretchy, for example to

Intelligent Technologies for Sensors: Applications, Design, and Optimization for a Smart World.
S. Kannadhasan, R. Nagarajan, & Alagar Karthick (Eds.)

accommodate joint motions, and tactile input must be processed in real time inside the system control loop. With sensors that monitor pressure, temperature, humidity, and air movement, this artificial skin may one day resemble actual skin in some ways, but in others, it may be even superior.

12.1 INTRODUCTION

The skin is the biggest organ in the body, and it can sense pressure, temperature, and other complex environmental signals. The skin is connected to an integrated network of elastic sensors that provide data about tactile and temperature stimuli to the brain's CNS, enabling the states to process the organs in our surroundings securely and effectively. Nonliving skin is based on live skin and is made using artificial ways for autonomous intelligent robots and medical biological measurements. Electronic networks made up of adjustable, elastic nursing devices that are unit appropriate for large-area deployment and connected with diverse functions might be a sign of progress in the creation of an electronic skin natural skin. In terms of abstraction resolution and heat sensitivity, electronic skins surpass their biological counterparts. Chemical and biological sensing, as well as critical elements like natural degradation and self-powering, may be rapidly implemented. Continued quick progress in this field suggests that a fully integrated e-skin will be available in the not-too-distant future. The use of physics to mimic natural skin's sensory qualities might be a fertile ground for innovative research in AI and human–machine interfaces. Flexible and elastic pressure detector arrays with a broad variety of transduction techniques and morphological types were devised to replicate tactile sensation with electronic skins. The scientific community is emphasizing the need of more research in this field.

12.2 BACKGROUND

Many researchers looked at the possibility of using tactile-sensing simulation as early as the 1970s, and had unquestionable bound exalting bit sensors, such as an artificial hand with perceptible result and a personal computer with a touchscreen, but with low resolution and hard materials.[8] Until the 1990s, exceptional improvements in everglade flexile and stretchy physical science for a range of applications were produced.[6] Since 2010, tangible sensors with improved performance area units have been created to support a variety of physical transduction processes, including piezo resistivity, energy storage devices, and the piezo effect.[1] These artificial intelligent e-skins will be able to discern between a range of physical inputs from the outside world, such as stretch, bend, heat, and humidity.[5] Self-healing and self-powered devices are particularly beneficial for next-generation flexible electronic skins.[2] Extremely compact, elastic, and useful sensors with high-density integrated circuits and diverse chemical or physical sensors are presently being created for applications in artificial brilliant, biological observation, and medicinal implant services.[7]

12.2.1 FEATURES

The electronic skin idea was first developed for artificial intelligence applications. Artificial men might be given pressure sense, enabling them to hold objects firmly without damaging them (the collect an egg). These e-skins are made up of pressure-sensing polymers and electrical components for measuring pressure, and they might help artificial equipment like prosthetic legs and arms feel more natural. The transistors that amplify weak signals must be flexible enough to mimic human skin, which is a challenge in the development of these devices. The energy needed by transistors to improve signal strength and gain is determined by the quality of the charge carriers in the semiconductor underneath the main layer. Single crystalline unnatural compound wafers are used in most computer chips. Alternative materials are also investigated since the layers are brittle. A greater proportion of inferior carrier quality is found in certain prospective flexible semi-conductors, such as promoting polymers. These components are inappropriate for use as transistors in electronic skin that is in direct contact with the victim's skin due to their higher voltages. Using particular procedures, it can mold fragile semiconductors into a variety of forms. In terms of adaptability, chemical element 32 and atomic number 32 are typically on par with nanowires. Furthermore, although their barrier properties are superior than conducting polymers, they arc still inferior to a doped artificial material. It is hard to imagine the performance required to use these materials to amplify extremely tiny signals that are not inherited from human skin. The e-skin is constructed of a single crystal artificial compound that has outstanding flexibility and a property comparable to that of personal flexible electronics. An oxide unleash layer is protected in this instance by an extremely fine artificial chemical layer. For the chemical element layer, a micrometer-sized lattice is created, and a transfer stamp layer is then attached to the split chemical element's very top. The transfer layer is then manufactured and placed to a flexible substrate after that. Electronic skin is more durable than genuine skin for robots. Human skin has smooth, delicate, and sensitive touch functions. To avoid rejection for prolonged time taking, the electronic skin that may be used for physiological observation must be equipped with a network layer that displays mechanical characteristics that are similar to those of actual skin. The e-skin should have a fitting touch, intimate integration, and adequate adhesion to the natural skin, and it should not be too stiff, inflexible, hard, or noticeable. Special materials, produced appropriately by thorough modeling, were required to understand their features. The layer of the e-nursing skin is constructed of elastomeric polyester with mechanical properties comparable to human skin. The electronic skin is unique in that it contains two protective layers sandwiched between multifunctional middle layers, unlike other electronic skins. Because the saving layers are of similar thickness, they create opposing forces that cancel out, resulting in relatively little stress on the central circuit layer, independent of the device's slant. The intermediate layer, which is included within the curving form that generates an elastic web, contains the metallic, semiconductor, and material components required for sensors, power supply, and light reflecting elements. The net's curved geometries allow it to decay quickly while having little impact on its occurrence. This breakthrough style combines all required elements in an ultrathin layer in terms of hair thickness. As if it were a jaybird, the e-skin is simply a bandage that is mounted onto or removed from natural skin. A high-resolution approach, similar to that used with large electrodes, was employed to acquire physiological data from the heart, brain, and skeletal muscles. Because they may utilize components with a range of highly specialized capabilities, the electronic skin can instantly provide a variety of physiologically specific data. Throughout this illustration, the transfer printing manufacturing process has been shown to be both practical and affordable, allowing for smart medical applications of the electronic skin. The quality of transferable chemical substances connected to the skin has increased as a result of the wireless communication network that may be employed in recent demonstrations of electronic skin devices. In the future, several kinds of electronic skins with applications such as body heat collecting and wearable radios may be used in fascinating new ways.

12.2.2 DESIRABLE PROPERTIES FOR E-SKIN

In terms of the ability to sense physical pressures, electronic skin may replicate actual skin's qualities, and it can also be augmented with natural skin's capabilities by integrating chemical and biological sensing properties.

12.2.2.1 BIOLOGICAL COMPATIBILITY AND BIOLOGICAL DEGRADABILITY

Because e-skin applications rely on tight connections with bio-interfaces, biological compatibility is a must-have feature for devices. Electronic skin should ideally be constructed from nontoxic materials.

12.2.2.1.1 Self-healing

While human skin has the potential to mend itself after being mechanically harmed, this function in e-skin has yet to be discovered. The energy necessary to create changes in both mechanical and electrical damage would be highly advantageous in practical applications.

Incorporating self-healing qualities into substances may be done in two ways:

1. Using materials that already have healing compounds in them.
2. The use of materials having reversible dynamic bonding. The usage of capsules containing healing ingredients was first shown in auto-healing, nonconductive polymers. As a consequence, for electronic skin implications, an electrically active system is required.

12.2.2.2 TEMPERATURE SENSITIVITY

Because most physical sensors are heat sensitive by nature, they must be calibrated using a heat sensor. Temperature sensing may be a fundamental function of human skin that aids in injury prevention and gives information about overall environmental changes; however, most tangible sensors are heat sensitive by nature, and their response must be calibrated with a heat sensor. Many companies have focused on piezoelectric and pyroelectric sensors that can differentiate between temperature and pressure take-ups. Resistive heat detectors (RHD) are intriguing for e-skin applications because of their elastic, natural device form and compatibility with electrical readout methods. RHDs use a fabric composed of metals such as Au and platinum to relate a change in temperature to a change in resistance. To investigate the impact of tangible sensations from temperature sensors, meandering pieces of platinum were validated as temperature-sensing devices. Such devices have a resolution of 0.03°C across a large temperature range. Stretchable temperature sensors, as well as precisely linked Au lines, are also exhibited. While the sensor's linear behavior was impressive, the resistance change was very small, requiring a sensitive reading approach.

12.2.2.3 SELF-POWERING

Providing a constant supply of energy to mobile gadgets might be a long-term issue. The skin, the largest human organ, offers a large storage space for P.E. Due to the body's interface with the outside world, electronic skins may also provide the potential to eliminate energy from environmental sources such as light and machine pressure. Solar cells, energy harvesters, super-capacitors, batteries, and antennas are just a few of the possible technologies for power generation, transmission, and storage in elastic systems that have recently been shown. Light is a commonly available power source that is best captured using large-surface-area devices. Flexible and elastic solar cells are supported by hard GaAs device islands, which are joined by freestanding metal connections. When exposed to a strain of up to 20%, these solar cells have a capacity of 13%. Despite these advantages, GaAs' broad range makes large-scale e-skins challenging to implement. The business uses ordinary materials and processes to make OPVs on ultrathin substrates. By shifting the devices to a pre-elastic substrate, they were able to achieve stretchability of up to 400% with a 4% ability. Mechanical power applications include dielectric elastomer generators and piezoelectric generators. The flexibility of electric dielectric elastomer goods, which consist of an elastomeric dielectric coated by two very compliant electrodes, makes them perfect for electronic skin. The electrodes are charged by exposing the voltage in a compact state. The power of the elastomer grows as it relaxes, allowing for more energy charges to be gathered. Although dielectric elastomer generators may achieve exceptionally high efficiency, the complexity and weight of the dual electronics have historically limited their use. According to current research, employing self-specific systems may minimize circuit complexity. Elastic models based on buckling active materials have been created, and machine energy-producing systems based on nanostructured piezoelectric materials have been enhanced. Machine-compatible energy storage devices have recently improved, owing to advancements in wear and tear materials. At the electrode–electrolyte interface, super capacitors store energy in the form of several layers of charged species, allowing for very high power densities.

12.3 OPTOELECTRONICS

Flexible optoelectronics has a tough time with semi-transparent displays, clinical imaging, and semi-spherical photo-detector (PD) arrays. Traditional planar PD arrays need complicated optics and image correction software to provide an accurate image. Bent PDs that are similar to the natural retina have been developed, which might minimize the amount of optics required, lowering demand and resulting in a weightless detector. Furthermore, by offering dynamically tuneable magnification and focussing, detectors with varying forms may be able to mimic the complicated picture process of insect eyes.

12.3.1 THE FUNCTIONS OF E-SKIN

1. To prevent harm to the intrinsic electronic system as a result of interactions with the outside world.
2. To properly provide machine signals to the dispersed sensor arrays below.
3. To get and preprocess sensor signals in a practical manner.
4. To extract the relevant and necessary information for the job at hand in an efficient and trustworthy manner.
5. To send the data to the system's next higher level of ICT infrastructure.

Each of these operations may be broken down into many tasks that work together to complete the extrinsic tangible system. To accommodate joint movements, the electronic skin may be flexible and elastic in nature, and processing must be done in real time to enable the control system to utilize proper data.

12.4 CHARACTERISTIC PROPERTIES AND DIVERSE FUNCTIONS OR APPLICATIONS OF RECENTLY DEVELOPED DEVICES FOR E-SKINS

12.4.1 RECENT DEVELOPMENTS IN NOVEL E-SKIN

In addition to basic pressure-sensing capacity, E-skins are increasingly being fitted with a growing variety of other features for practical and promising applications. The auto-powered feature is mostly utilized in e-skins for artificial intelligence systems that must function continuously for lengthy periods of time. Many functional integrations, such as auto-healing, multi-stimulus discrimination, and simultaneous heat and pressure sensing, are designed to emulate the sensory capacities of real skin. Additionally, data and energy are sent through wireless technology in combination with electronic skin. In our study, we look at the following recent developments in the industries.

12.4.1.1 SELF-POWERED E-SKIN

The energy consumption of pressure sensors is a key impediment to the development of integrated and practical e-skin systems. Various developments, such as the design of appropriate resistance and operating voltage for piezoresistive sensors and the use of thin dielectric materials with high dielectric constants in capacitive sensors, are regarded crucial methods to reduce power consumption. Fabrication of self-powered sensors is an alternative and ultimate answer to the e-power skin's consumption issues. As a consequence, research on stretchable solar cells, piezoelectric nanomakers, and triboelectric generators or sensors[16] has recently attracted a lot of attention. Tribo-electric generators are quickly progressing among these topics due to their simplicity of design, self-powered capabilities, low cost, and many applications.[9] Fan et al. revealed the first flexible triboelectric generator capable of converting random mechanical energy into electrical signals while using no power. The device's basic functioning mechanism is described below. Charges transfer between the contact surfaces of two stacked polymer films due to friction under mechanical stimuli resulting from the tribo-electric effect, resulting in an electric potential difference between the top and bottom electrodes deposited on the outer surfaces of the films due to electrostatic induction; thus, electrical signals are identified with the top and bottom electrodes deposition on the outer surfaces of the films. By creating ministructures on the contact surfaces of flexible triboelectric generators, their pressure sensitivity was increased, enabling them to operate as self-powered pressure sensors.[7] An array of discrete triboelectric sensor units was gathered to allow mapping of applied pressure using the output voltage from all units. This integration approach is not fixed for the development of large-scale high-resolution arrays with many pixels since each pixel in this array is autonomous. The development of single-electrode turboelectric sensors makes it easier to create a more suitable related approach. An auto-powered motile-tracking tribo-electric sensor was driven by arrayed aluminum electrodes beneath a piece of polytetrafluoroethylene (PTFE) film. The maximum voltage is analyzed using an Al electrode when a sensed object with triboelectric charges moves to the point over the center of the electrode due to the electrostatic-induced potential difference between the electrode and the electrical ground, thus providing details about the moving velocity and acceleration data, revealing the possibility for large-scale integration of triboelectric charges. These studies have shown the effectiveness of triboelectric devices in usage, such as self-powered wearable electronics,[10] healthcare monitoring, and self-powered e-skin systems.[14]

12.4.1.2 MULTIFUNCTIONAL E-SKIN

The ability to duplicate the various qualities and functions of actual skin is critical in the modification of artificial electronic skin. Other features, such as as auto-healing and temperature monitoring, are sought to better adapt to environmental situations, but they also pose significant challenges for electronic skin implementation. In the hunt for a property that matches human skin's ability to self-heal after a rupture, the auto-healing feature has aroused interest for robotic and prosthetic applications.[20] Auto-healing materials were successfully developed in this work[22] by mixing healing agents or dynamic reversible-bonds, such as a healable thermoplastic elastomer material. Furthermore, providing superior electrical conductivity and repeatable pressure sensitivity of the self-healing material is the most challenging aspect in meeting the critical objectives of these e-skin applications. The researchers reported how they used nickel nanostructured particles embedded in a hydrogen-bonded organic supramolecular polymer matrix to produce an elastic and pressure-sensitive device with auto-healing properties for mechanical and electrical activities. The variable range of composite resistance grows with decreasing Ni content due to the bigger changeable space between particles when pressure is applied to the instrument. The healing duration and temperature influence how much of these characteristics are preserved. For the demanding qualities and applications, e-skin conductivity may be maintained with around 90% ability after 15 s. The development of a multi-ability force-sensitive sensor that meets the requirements of high sensitivity and flexibility is currently being considered.[10] Under a variety of machine stimuli, such as pressure, shear, and so on, the difference in electrical resistance created by the change in conducting route between the two arrays was easily recognized, and has a broad range of applications in strong signal monitoring in the ultra-low-pressure sector. Javey and colleagues[15] recently demonstrated the use of exceptionally sensitive and highly flexible electronic whiskers for sensing the 3D distribution of an air channel. Two opposing electrodes of each whisker were mounted on the top and bottom surfaces of an elastic high-aspect-ratio fiber using different amounts of Ag nanoparticles contained in a carbon nanotube paste. These electrodes demonstrated differential variations in resistance to detect changing curvature degrees at various angles. Choi and colleagues then created a pressure- and vibration-sensitive sensor based on nanoscale crack junctions created on the stretched surface of a metal sheet by emulating spider legs' slit organ structure. The electrical transmitter of the sensor varies with only a slight difference in the gaps due to the change in contact resistance of these breaks, enabling the gadget to recognize sound waves for pressure mapping and speech recognition. Temperature sensing is another important feature of natural skin, as it aids in maintaining a thermal equilibrium between the body and its surroundings and represents a significant advancement in the application of temperature-responsive e-skin in human–machine interactions.[12] The main concept of skin electronics was proposed by Rogers and coworkers, who claimed that multifunctional devices can tightly attach to the uneven surface of the skin using Vander Waals forces and have excellent mechanical and electrical properties for all active components, such as heat and strain detectors, microscale circuit elements, and other auxiliary functional electronics. The most challenging challenge is to ensure that after several cycles of deformation, no separation events occur. Low effective moduli and tiny film thicknesses, as well as the design of thin filamentary serpentine interconnectors, have all been employed to reduce the driving force for interface separation. The geographical distribution of epidermal temperature has been detected constantly and accurately using temperature sensor arrays based on temperature coefficient of resistance (TCR) materials or silicon PIN diodes. These skin electronics were more temperature and thermal conductivity sensitive, allowing for the monitoring of a number of human physiological parameters such as skin hydration, tissue thermal conductivity, blood flow status, and wound healing.[18,26] Utilizing a PDMS micro-hair structure, researchers constructed a stretchy sensor that can conformably contact uneven skin and provide different capacitive results for clinical monitoring of heat and electrocardiogram information using wireless transmission technology. A series of multifunctional transistor arrays with pressure and temperature sensitivity have also been created by a number of companies. Researchers used organic transistors to create a pressure and temperature active matrix that reads out the distributions of the external environment at the same time. Along with the transistor film, a PSR layer and organic diodes were built to achieve these processes. Park also created a real-time multi-stimuli-responsive sensor array by directly integrating piezo pyroelectric and piezo thermoresistive materials into the transistor as a gate dielectric and organic semiconductor channel, which could distinguish between temperature and pressure signals using AC to analyze the change in the amplitude and offset values of the drain current. Huge systems with modular components based on a unique sensing mechanism to monitor the intricate external environment have been constructed to attempt to properly duplicate every outstanding attribute of real skin. Kim developed an intelligent prosthetic e-skin that can sense heat, humidity, and a variety of strains all at once, as well as a heater to control body temperature.

The scientific and precise design of these different sensing element arrays enabled the device's fine mechanical reliability and high spatial-temporal sensitivity to the variable ambient environment, describing that these devices have been well received in a wide range of applications in artificial e-skin, man–machine interfacing, and modern robotics. Natural skin functions such as strain, pressure, heat, and humidity sensing are also simulated, in addition to the activities discussed in this article. More qualities of extraordinary electronic skin should be used in the future, in addition to the core functions of real skin. As a result, the high level of integration of the e-skin system is a hot topic of research, implying that flexible electronic-integrated circuits and other chemical or physical sensors must keep up with the e-skin system to meet the demands of robotics, man–machine interactions, health monitoring, and medical implant services. Rogers and his colleagues created advanced stretchable, foldable, and high-quality integrated circuits, including as single-crystalline silicon-based MOSFETs, complementary metal-oxide semiconductor (CMOS) logic gates, and different amplifiers. Inorganic semiconductors and elastomeric substrates are used in these devices, which offer outstanding electrical and mechanical properties. Certain types of modern electronics have been progressively developed with the main goal of using weightless devices. Based on the phenomenon of magnetoresistance, an extremely tiny and lightweight magnetic field sensor was successfully created to detect both static and dynamic magnetic fields. Other specialized electronics have been designed to detect diverse biological signals in human tissue, such as implanted electronics and sticky sensors.

12.5 WIRELESS TECHNOLOGY FOR E-SKIN

We use wireless communication the most in our everyday lives. This type of communication overcomes the limits of unorganized wires and cables, allowing data to be transmitted between disparate devices. To leverage wireless technology, research industries have employed e-skin sensors for wireless sensor arrays, energy transfer, and data transmission. A pressure-dependent capacitive element and an initiative antenna make up the resonant circuit of a flexible wireless sensor array that can monitor and map pressure variations in real time, with the resonant frequency decreasing with applied pressure due to an increase in impressive coupling capacitance. This wireless analysis approach, which uses a lower frequency than natural systems, offers a wider variety of applications in human–machine interactions and biological research. Recent study has produced a transplantable and biodegradable electronics-based device based on a Mg wireless heater that would entirely disintegrate in water, allowing for remote-control treatment within the body, such as heat disinfection near an operating site and medication release triggering. Using wireless data transmission, a flexible and exceptionally sensitive temperature sensor for remotely monitoring the temperature of the human body was created. The device is made up of a resistance-responsive composite film comprised of Ni particles packed within a polyethylene matrix polymer, as well as an indirect RF identification antenna. A polymer bandage based on MEMS with different sensors and a wireless signal transmission material was studied to fulfil the drug delivery function of the wearable and elastic device. Other elastic and wearable systems with transistors and ECG sensors have been presented to monitor human health and to provide the framework for current research on large-scale integrated wearable devices in man–machine interactions through wireless communication.

12.6 ARCHITECTURE

E-skins for robotics, sensors, and actuators are tightly coupled to microelectronic circuits, giving electronics a whole new level of flexibility. Shaped electronics may be subjected to high deformation strains. A hemispheric detector array is formed when the surface area of a disc detector array is doubled. When folded over elbow-like joints, the skin may show elasticity and may be stretched by 15%. In semiconductor integrated circuits and MEMS technology, hard materials that are not employed as flexible structures and thin active substances that shatter at a critical strain of 1% are used. Free-standing thin metal sheets may be shattered by tensile pressures on the scale of 1%.

Subcircuit cells, which are made up of a transducer and an electronic circuit, may be placed on mechanically separated islands made up of a special material that absorbs all of the strain, resulting in elastic and flexible skin. The depiction of such an island with an electronic surface. The islands are sufficiently toughened to prevent them from breaking when the circuit is deformed. Electrically, flexible metal wires connect the subcircuits.

12.7 CONCLUSIONS

As a consequence of e-skin development research, the availability of innovative materials and methodologies has risen in recent years. As a result of this advancement, e-capabilities skins are gradually converging.

1) enable highly evolved interactive and flexible robots capable of performing complex tasks in less regimented situations has piqued curiosity.
2) make sure you have the right screens and optics.
3) provide biometric prosthetics, regular health monitoring technologies, and other clinical research advancements.

Attempts to construct exceedingly flexible and stretchy sensors, as well as efforts to offer high-performance electronic skin, were attempted initially. Recent advances in the use of functional materials and device design optimization are excellent reasons for replicating human skin's distinctive properties. The use of microstructured gate dielectrics in transistors, which starts the active matrix to limit signal crosstalk between pixels and promotes quick addressing and low power consumption, has a substantial influence on sensor pressure sensitivity. In addition, the use of oriented piezoelectric NWs and NBs with greater inherent piezoelectricity and mechanical stability supports the development of high-resolution sensor arrays that go beyond human sensing capabilities. Fabricating multifunctional electronic skin is also a vital goal for addressing industrial needs for a wide range of applications. A lot of research have focused on integrated electronics for the detection of multiple stimuli. Thanks to quickly expanding technologies like wireless technology and the display of small sensors, these initiatives have gotten a lot of attention in the field of clinical monitoring and medical implant services. In terms of actual application, however, there are still certain challenges to solve. New materials and a unique transduction mechanism should be investigated further to obtain a variable pressure measurement range. The pressure measurement range should be expanded to fulfil the demands of various applications. Apart from that, because the energy problem is currently one of our society's most critical challenges, the creation of low-power devices remains an essential subject in this study. In addition, several types of artificial smart e-skins are increasingly required in modern healthcare and clinical medical research. Future e-skins will respond to changes in the environment using new information transmission technologies, comparable to natural human skin, and will be able to adjust and provide feedback in real time in response to a variety of environmental stimuli.

12.8 FUTURE PROSPECTS

The skin possesses a huge number of touch-sensitive receptors, including smooth skin in the palms and soles of the feet, as well as nonsmooth skin or mucosal skin that covers the bulk of the body, according to early research. Deep mechanosensitive extreme organs are found in most load-bearing smooth tissues, notably muscles, and convey sensory information to the brain in the form of palpable sensations that follow the features of transient and persistent mechanical stimuli to propagate across extended distances. Sensory inputs from stationary receptors in the skin provide information regarding direct mechanical contact, such as the motions created by smooth skin as an object is moved over hairy skin. At initially, artificial skin breakthroughs in robotics were limited. Furthermore, the earliest use of e-skin goes back to the 1970s, when an artificial hand covered with skin was tested to detect grip strength, slip, and other grip-related qualities such as texture and stiffness.[27] In the late 1980s, infrared sensors were initially employed to prevent contact in a robotic arm.[24] Robots are increasingly expected and designed to be capable of doing tasks that require physical touch with the environment and humans. Artificial skins with tactile capabilities[11,21] and safe physical contact are becoming increasingly popular as a result. In wearable systems, cutaneous skin-like devices have also been employed as a second skin to monitor health status by detecting several physiological signs.[42] Sensor distribution, readout, and an effective integration technique are all key considerations in any of these scenarios. Electronic skin technology is being developed with the goal of duplicating some qualities of genuine skin, taking into account factors such as sensor types and density. Electrophysiology studies have shown a number of different types of responses to mechanical stimulation. A little indentation of the smooth skin with a sharp object activates the receptors at the contact spot. Surprisingly, the response is generally temporary, meaning it is stronger during the ramping stages of the indentation at the beginning and conclusion of the stimulus, suggesting that the activated receptors are strain rate sensitive. Although some degree of temporal adaptation is always present, the response is less commonly connected to the intensity of the indentation. A number of anatomically chosen end-organs have been related to these responses. The first and most common are Meissner corpuscles nested in cavities visible on the epidermis' internal face. There are around 30 such receptors per centimeter square in the typical skin of finger tips. "Neurite complexes" are the second kind of end-organ. This term describes a condition in which a nerve fiber divides into many branches. Merkel cell–neurite complex is an end-organ found at the dermis–epidermis junction (territory range of 20–2000 m). It corresponds to a slowly adapting response, according to electrophysiology studies. The Pacinian corpuscle is a large end-organ that may be found in a variety of soft tissues all throughout the body. They have a high sensitivity to oscillating stimuli and a phase-locking physiological response. A population of end-organs known as Ruffini endings is thought to be responsible for the most prevalent kind of reaction observed in the human hand, in which broad expanses of smooth skin are pushed apart in a continual response in a relatively small number of nerve fibers. These organs in the hand have never been observed in modern times, which is surprising. Hairy skin varies from smooth skin in histology because it lacks the reticulated structure and Meissner corpuscles seen in smooth or nonhairy skin. There are additional Merkel cell complexes organized in 1000-mclusters called touch domes that are separated by several millimeters.[4,28] It would be inadequate to discuss the presence of so-called C-fibers in all skin types without discussing them. These non-myelinated nerve fibers quietly convey nerve impulses to the brain from the whole skin surface. They are associated with pain, but in recent years, they have also been shown to participate in light touch, in addition to the fast system with the other receptor types outlined above. Tactile sensation in humans has been explored using a unique approach known as microneurography, which involves inserting a small-diameter needle electrode into a peripheral nerve and monitoring the activity of single-nerve fibers.[19] Researchers were able to correlate peripheral nerve activity to experiences using this strategy, which is not employed in people for other sensory modalities. However, a thorough understanding of touch remains difficult since current knowledge of specific receptors is inadequate to develop reliable integrative models capable of predicting the behavioral aspects of tactile function. Tactile sensors for electronic skin have been produced as a result of research and understanding of the properties of various mechanical receptors in actual skin. The focus of e-skin development at the moment is stiff body robotics applications. In addition, robotics is gaining popularity these days, and contemporary robots are designed to be softer and more obedient. Soft robots still lack sensory input, necessitating precise control during manipulating movements and interactions. The challenges are founded on the premise that the soft electronic skin contains transduction sites with different bandwidths, dynamic ranges, resolutions, sensitivity, and mechanical properties, and that these qualities may be taken into consideration using materials and neural designs. Because power consumption, space and time resolution, and compatibility with elastic and smooth materials make circuit design complex, circuits linking large numbers of scattered sensors are often required to be addressed.

KEYWORDS

- **artificial prosthetics**
- **biomedical instrumentation**
- **electronic devices**
- **electronic skin**
- **multifunctional middle layer**
- **triboelectric generators**

REFERENCES

1. (a) Crone, B.; Dodabalapur, A.; Lin, Y. Y.; Filas, R. W.; Bao, Z.; LaDuca, A.; Sarpeshkar, R.; Katz, H. E.; Li, W. *Nature* **2000,** *403*, 521; (b) Lumelsky, V. J.; Shur, M. S.; Wagner, S. *IEEE Sens. J.* **2001,** *1*, 41; (c) Darlinski, G.; Bottger, U.; Waser, R.; Klauk, H.; Halik, M.; Zschieschang, U.; Schmid, G.; Dehm, C. *J. Appl. Phys.* **2005,** *97*, 093708; (d) Cheng, I.-C.; Wagner, S. In *Flexible Electronics: Materials and Applications*; Wong, W. S., Salleo, A., Eds.; Springer Science & Business Media, LLC: New York, 2009; Vol. 11, Ch. 1; (e) Park, J.; Lee, Y.; Hong, J.; Lee, Y.; Ha, M.; Jung, Y.; Lim, H.; Kim, S. Y.; Ko, H. *ACS Nano* **2014,** *8*, 12020.
2. (a) Tee, B. C. K.; Wang, C.; Allen, R.; Bao, Z. *Nat. Nanotechnol.* **2012,** *7*, 825; (b) Wang, S.; Lin, L.; Wang, Z. L. *Nano Energy* **2015,** *11*, 436.
3. (a) Pan, C. F.; Fang, Y.; Wu, H.; Ahmad, M.; Luo, Z. X.; Li, Q. A.; Xie, J. B.; Yan, X. X.; Wu, L. H.; Wang, Z. L.; Zhu, J. *Adv. Mater.* **2010,** *22*, 5388; (b) Pan, C. F.; Guo, W. X.; Dong, L.; Zhu, G.; Wang, Z. L. *Adv. Mater.* **2012,** *24*, 3356; (c) Pan, C. F.; Wu, H.; Wang, C.; Wang, B.; Zhang, L.; Cheng, Z. D.; Hu, P.; Pan, W.; Zhou, Z. Y.; Yang, X.; Zhu, J. *Adv. Mater.* **2008,** *20*, 1644; (d) Pan, C. F.; Li, Z. T.; Guo, W. X.; Zhu, J.; Wang,

Z. L. *Angew. Chem., Int. Ed.* **2011,** *50*, 11192; (e) Xu, C.; Pan, C. F.; Liu, Y.; Wang, Z. L. *Nano Energy* **2012,** *1*, 259; (f) Pan, C. F.; Luo, J.; Zhu, J. *Nano Res.* **2011,** *4*, 1099.

4. Reinisch, C. M.; Tschachler, E. The Touchdome in Human Skin Is Supplied by Different Types of Nerve Fibers. *Ann. Neurol.* **2005,** *58* (1), 88–95.
5. Pang, C.; Lee, G.-Y.; Kim, T.-I.; Kim, S. M.; Kim, H. N.; Ahn, S.-H.; Suh, K.-Y. *Nat. Mater.* **2012,** *11*, 795.
6. (a) Jiang, F. K.; Tai, Y. C.; Walsh, K.; Tsao, T., Lee, G. B.; Ho, C. M. In *Proc. IEEE Tenth Annu. Int. Workshop on Micro. Electro. Mech. Syst., Nagoya, Japan* 1997; p 465; (b) Lacour, S. P.; Agner, S.; Huang, Z. Y.; Suo, Z. *Appl. Phys. Lett.* **2003,** *82*, 2404.
7. Fan, F. R.; Lin, L.; Zhu, G.; Wu, W.; Zhang, R.; Wang, Z. L. *Nano Lett.* **2012,** *12*, 3109.
8. (a) Clippinger, F. W.; Avery, R.; Titus, B. R. *Bull. Prosthetics Res.* **1974,** *247*; (b) Vedel, J. P.; Roll, J. P. *Neurosci. Lett.* **1982,** *34*, 289.
9. (a) Zhu, G.; Chen, J.; Liu, Y.; Bai, P.; Zhou, Y. S.; Jing, Q. S.; Pan, C. F.; Wang, Z. L. *Nano Lett.* **2013,** *13*, 2282; (b) Zhu, G.; Lin, Z. H.; Jing, Q. S.; Bai, P.; Pan, C. F.; Yang, Y.; Zhou, Y. S.; Wang, Z. L. *Nano Lett.* **2013,** *13*, 847; (c) Nguyen, V.; Yang, R. *Nano Energy* **2013,** *2*, 604; (d) Lin, L.; Hu, Y.; Xu, C.; Zhang, Y.; Zhang, R.; Wen, X.; Lin Wang, Z. *Nano Energy* **2013,** *2*, 75.
10. (a) Zhu, G.; Bai, P.; Chen, J.; Lin Wang, Z. *Nano Energy* **2013,** *2*, 688; (b) Hou, T.-C.; Yang, Y.; Zhang, H.; Chen, J.; Chen, L.-J.; Lin Wang, Z. *Nano Energy* **2013,** *2*, 856.
11. Culbertson, H.; Schorr, S. B.; Okamura, A. M. Haptics: The Present and Future of Artificial Touch Sensation. *Annu. Rev. Control, Robot., Auton. Syst.* **2018,** *1*, 385–409.
12. (a) Han, I. Y.; Kim, S. J. *Sens. Actuators A* **2008,** *141*, 52; (b) Gao, L.; Zhang, Y.; Malyarchuk, V.; Jia, L.; Jang, K.-I.; Webb, R. C.; Fu, H.; Shi, Y.; Zhou, G.; Shi, L.; Shah, D.; Huang, X.; Xu, B.; Yu, C.; Huang, Y.; Rogers, J. A. *Nat. Commun.* **2014,** *5*, 4938.
13. Kim, J.; Banks, A.; Cheng, H.; Xie, Z.; Xu, S.; Jang, K.-I.; Lee, J. W.; Liu, Z.; Gutruf, P.; Huang, X.; Wei, P.; Liu, F.; Li, K.; Dalal, M.; Ghaffari, R.; Feng, X.; Huang, Y.; Gupta, S.; Paik, U.; Rogers, J. A. *Small* **2015,** *11*, 906.
14. (a) Zhong, J.; Zhong, Q.; Fan, F.; Zhang, Y.; Wang, S.; Hu, B.; Wang, Z. L.; Zhou, J. *Nano Energy* **2013,** *2*, 491; (b) Bai, P.; Zhu, G.; Jing, Q.; Yang, J.; Chen, J.; Su, Y.; Ma, J.; Zhang, G.; Wang, Z. L. *Adv. Funct. Mater.* **2014,** *24*, 5807; (c) Yang, W.; Chen, J.; Wen, X.; Jing, Q.; Yang, J.; Su, Y.; Zhu, G.; Wu, W.; Wang, Z. L. *ACS Appl. Mater. Interfaces* **2014,** *6*, 7479; (d) Jung, S.; Lee, J.; Hyeon, T.; Lee, M.; Kim, D.-H. *Adv. Mater.* **2014,** *26*, 6329; (e) Yang, J.; Chen, J.; Su, Y.; Jing, Q.; Li, Z.; Yi, F.; Wen, X.; Wang, Z.; Wang, Z. L. *Adv. Mater.* **2015,** *27*, 1316.
15. Takei, K.; Yu, Z. B.; Zheng, M.; Ota, H.; Takahashi, T.; Javey, A. *Proc. Natl. Acad. Sci. USA* **2014,** *111*, 1703.
16. (a) Chen, M.; Li, X.; Lin, L.; Du, W.; Han, X.; Zhu, J.; Pan, C.; Wang, Z. L. *Adv. Funct. Mater.* **2014,** *24*, 5059; (b) Du, W.; Han, X.; Lin, L.; Chen, M.; Li, X.; Pan, C.; Wang, Z. L. *Adv. Energy Mater.* **2014,** *4*, 1301592; (c) Zhang, H.; Yang, Y.; Su, Y.; Chen, J.; Hu, C.; Wu, Z.; Liu, Y.; Ping, C. W., Bando, Y.; Wang, Z. L. *Nano Energy* **2013,** *2*, 693; (d) Zhang, X.-S.; Han, M.-D.; Wang, R.-X.; Meng, B.; Zhu, F.-Y.; Sun, X.-M.; Hu, W.; Wang, W.; Li, Z.-H.; Zhang, H.-X. *Nano Energy* **2014,** *4*, 123.
17. Kaltenbrunner, M.; Sekitani, T.; Reeder, J.; Yokota, T.; Kuribara, K.; Tokuhara, T.; Drack, M.; Schwoediauer, R.; Graz, I.; Bauer-Gogonea, S.; Bauer, S.; Someya, T. *Nature* **2013,** *499*, 458.

18. Webb, R. C.; Bonifas, A. P.; Behnaz, A.; Zhang, Y.; Yu, K. J.; Cheng, H.; Shi, M.; Bian, Z.; Liu, Z.; Kim, Y.-S.; Yeo, W.-H.; Park, J. S.; Song, J.; Li, Y.; Huang, Y.; Gorbach, A. M.; Rogers, J. A. *Nat. Mater.* **2013,** *12*, 938.
19. Johansson, R. S.; Vallbo, A. B. Tactile Sensibility in the Human Hand: Relative and Absolute Densities of Four Types of Mechanoreceptive Units in Glabrous Skin. *J. Physiol.* **1979,** *286* (1), 283–300.
20. Benight, S. J.; Wang, C.; Tok, J. B. H.; Bao, Z. *Prog. Polym. Sci.* **2013,** *38*, 1961.
21. Lederman, S. J.; Klatzky, R. L. Haptic Perception: A tutorial. *Attention, Percept., Psychophys.* **2009,** *71* (7), 1439–1459.
22. (a) White, S. R.; Sottos, N. R.; Geubelle, P. H.; Moore, J. S.; Kessler, M. R.; Sriram, S. R.; Brown, E. N.; Viswanathan, S. *Nature* **2001,** *409*, 794; (b) Chen, X. X.; Dam, M. A.; Ono, K.; Mal, A.; Shen, H. B.; Nutt, S. R.; Sheran, K.; Wudl, F. *Science* **2002,** *295*, 1698; (c) Ghosh, B.; Urban, M. W. *Science* **2009,** *323*, 1458; (d) Wojtecki, R. J.; Meador, M. A.; Rowan, S. J. *Nat. Mater.* **2011,** *10*, 14; (e) Blaiszik, B. J.; Kramer, S. L. B.; Grady, M. E.; McIlroy, D. A.; Moore, J. S.; Sottos, N. R.; White, S. R. *Adv. Mater.* **2012,** *24*, 398; (f) Odom, S. A.; Chayanupatkul, S.; Blaiszik, B. J.; Zhao, O.; Jackson, A. C.; Braun, P. V.; Sottos, N. R.; White, S. R.; Moore, J. S. *Adv. Mater.* **2012,** *24*, 2578.
23. (a) Xu, S.; Qin, Y.; Xu, C.; Wei, Y.; Yang, R.; Wang, Z. L. *Nat. Nanotechnol.* **2010,** *5*, 366; (b) Wu, W.; Bai, S.; Yuan, M.; Qin, Y.; Wang, Z. L.; Jing, T. *ACS Nano* **2012,** *6*, 6231; (c) Pradel, K. C.; Wu, W.; Ding, Y.; Wang, Z. L. *Nano Lett.* **2014,** *14*, 6897; (d) Kim, D. Y.; Lee, S.; Lin, Z.-H.; Choi, K. H.; Doo, S. G.; Chang, H.; Leem, J.-Y.; Wang, Z. L.; Kim, S.-O. *Nano Energy* **2014,** *9*, 101; (e) Wang, Z. L.; Zhu, G.; Yang, Y.; Wang, S. H.; Pan, C. F. *Mater. Today* **2012,** *15*, 532.
24. Lumelsky, V. J.; Shur, M. S.; Wagner, S. Sensitive Skin. *IEEE Sensors J.* **2001,** *1* (1), 41–51.
25. Dang, W.; Manjakkal, L.; Navaraj, W. T.; Lorenzelli, L.; Vinciguerra, V.; Dahiya, R. Stretchable Wireless System for Sweat pH Monitoring. *Biosensors Bioelectron.* **2018,** *107*, 192–202.
26. Hattori, Y.; Falgout, L.; Lee, W.; Jung, S. Y.; Poon, E.; Lee, J. W.; Na, I.; Geisler, A.; Sadhwani, D.; Zhang, Y. H.; Su, Y. W.; Wang, X. Q.; Liu, Z. J.; Xia, J.; Cheng, H. Y.; Webb, R. C.; Bonifas, A. P.; Won, P.; Jeong, J. W.; Jang, K. I.; Song, Y. M.; Nardone, B.; Nodzenski, M.; Fan, J. A.; Huang, Y. G.; West, D. P.; Paller, A. S.; Alam, M.; Yeo, W. H.; Rogers, J. A. *Adv. Healthcare Mater.* **2014,** *3*, 1597.
27. Stojiljkovic, Z.; Clot, J. Integrated Behaviour of Artificial Skin. *IEEE Trans. Biomed. Eng.* **1977,** *BME-24* (4), 396–399.
28. Halata, Z.; Munger, B. L. Identification of the Ruffini Corpuscle in Human Hairy Skin. *Cell Tissue Res.* **1981,** *219* (2), 437–440.
29. Wang, X., Dong, L., Zhang, H., Yu, R., Pan, C., Wang, Z. L. (2015). Recent Progress in Electronic Skin. Adv. Sci., 2: 1500169. doi: 10.1002/advs.201500169

CHAPTER 13

A Low-Cost Advanced Device for the Detection of Pesticides with NDVI Method

S. SOVAN KUMAR[1], D. PANDA[1], R. R. PADHI[1], V. R. K. PATRO[2], S. R. DASH[1], KANISHK KASHYAP[3], T. K. GIRI[1], D. MISHRA[2], and SANDIPAN MALLIK[1*]

[1]*Department of Electronics Communication Engineering, NIST (Autonomous) Berhampur, Berhampur 760010, Odisha, India*

[2]*Department of Computer science and Engineering, NIST (Autonomous) Berhampur, Berhampur 760010, Odisha, India*

[3]*Department of Electrical Engineering, NIST (Autonomous) Berhampur, Berhampur, Berhampur 760010, Odisha, India*

[*]*Corresponding author. E-mail: sandi.iitkgp@gmail.com*

ABSTRACT

With the growing population in the world, demand for various types of vegetables and crops is increasing. For this reason, farmers are constantly using different fertilizers, pesticides, and other chemicals to increase crop yield. Although this may increase productivity, the quality of the crop is getting decreased day by day due to excessive use of pesticides. While harvesting, the pesticides that were used by the farmers get stored in the vegetables. A large number of fertilizers and pesticides set foot in the food chain and finally cause biological magnifications. People consume the same food and this affects their health in a noxious way. So, there is a need for one device that could detect these chemicals and pesticides. By using this device consumers can get to know the amount of pesticide that a product contains. This chapter consists of an electronic device that can detect the

Intelligent Technologies for Sensors: Applications, Design, and Optimization for a Smart World.
S. Kannadhasan, R. Nagarajan, & Alagar Karthick (Eds.)

pesticides present in vegetables with the help of infrared light. This device would show the amount of pesticide present in the food. The consumer can use the pesticide's data as a benchmark to buy the eatable things that are fit to eat. In this chapter NDVI (Normalized Difference Vegetable Index) method is used to detect the presence of pesticides, and also an application is designed to store and show the data having the amount of pesticide present in a particular edible thing. The data can be easily sharable by using the application to agricultural organizations, food inspection departments and can maintain the pesticide detection records.

13.1 INTRODUCTION

The world population had shown significant growth from 1800 to today. It has grown from one billion in 1800 to 7.7 billion today. According to the UN, from 2019 to the end of the century the world population will grow from 7.7 billion to 11.2 billion.[1] Due to this growing population, the numbers of farmlands are being used for housing and other purposes. Meanwhile, sharp growing demand for foods and vegetables has been seen among the people.[2,3] To satisfy this growing demand for food production, farmers are adding pesticides and other chemicals to increase their productivity. Pesticides cost roughly USD 40 billion to agricultural growers worldwide each year.[3] This proves the fact that how farmers are highly using pesticides to increase food production. These pesticides are harmful to human health as well as to the environment. When people consume the pesticide-mixed foods produced by farmers, then pesticides are also being injected into the human body. Pesticide residues are pesticides that may stay on or in food after they have been given to food crops.[4] The number of pesticides inside vegetables was found to be over the allowable level in the majority of research that are being conducted.[5] Most of these pesticide residues, particularly chlorinated pesticide derivatives, display bioaccumulation and can accumulate to dangerous amounts in the human body and the environment.[6] A pesticide is a mixture of chemicals used to kill pests, which are creatures that are harmful to farmed plants or animals. The phrase refers to a variety of pesticides, including insecticides, fungicides, rodenticides, herbicides, and nematicides.[7,8] So to avoid the harmful effects of pesticide residues on human health there has been a requirement of a new method by which we can reduce pesticide consumption to a large extent.[9] One technique should be devised to verify and test if a certain product is safe to eat or not.[10] Several articles have been published that provide various strategies for implementing various techniques. Pesticide detection techniques include gas chromatography (GC) with ECD, high-performance liquid chromatography (HPLC) in conjunction with mass spectrometry (MS), SVM, electro analytical methods, spectroscopic approaches, biosensors, chemical methods, and flow injection analysis (FIA).[11] Attenuated total internal reflection (ATR), optical screening, some analytical methods, and so on are some other currently available methods to detect pesticides present in the food.[12–14] All of these approaches need a significant amount of time for the process of investigation and yield results after a while. As a result, one solution is necessary that is both cost efficient and yields results in a short period. Bhandari et al. provide a method for analyzing satellite pictures that are based on the NDVI.[15] The NDVI method is used to identify the signatures of various objects; the approach employs multispectral remote sensing. It is also a widely used approach in application areas for measuring crop health. During the past several years a lot of developments have been seen in sensor technologies. So one system is developed in which the analyte could be detected by using a sensor (e.g., IR sensor) and the received analog signal is converted into digital by using an analog-to-digital converter. After passing this signal to the microcontroller the NDVI value is calculated. The NDVI value is calculated by using the following formula:

$$\mathrm{NDVI} = \frac{\mathrm{B} - \mathrm{A}}{\mathrm{A} + \mathrm{B}} \tag{13.1}$$

Here, "A" denotes the red light band and "B" denotes the infrared light band. The values for several objects (vegetables/fruits) were computed and compared to their genuine values, and the error was computed. The microcontroller further displays the NDVI value on an LCD screen and the information regarding the purity of the vegetable.[16] One android application is developed which would collect the data about the NDVI value and purity of one vegetable from the microcontroller. After gathering data from various fruits and vegetables, the data is examined and a judgment is made as to whether the fruit is pure or not.

13.2 HARMFUL EFFECTS OF PESTICIDES

Pesticides are the mixture of the materials used for preventing, destroying, and repelling pests. Pesticides are having a hugely adverse effect on the ecosystem. The risks of the use of pesticides are being surpassed by their favorable effects. Pesticides have harsh effects on far afield species and affect plants, animals ecosystems, and biodiversity. Around 80–90% of applied pesticides can dehumidify within a short gap of operation and it is commonly observed in using sprayers.[17] The dissolved pesticides dissolve in the air and thereafter they damage off target organisms, plants, etc.[18] The unrestricted use of pesticides is causing depletion of the animals and there is also a risk to the endurance of some extinct breeds such as the peregrine falcon, bald eagle, and osprey.[19] With these chemicals to the toxic level of water, air and soil bodies are likewise being polluted. There are three kinds of pesticides in which toxicants are expressed to be most harmful in comparison to fungicides and herbicides. Pesticides insert the natural ecosystems by two distinct means depending upon the solubility of the pesticide. First, the water-soluble pesticides get easily dematerialized in water and arrive at groundwater, rivers, streams, and lakes, where it is causing trauma to nontargeted species and affecting the ecosystem. And the second one is the fat-soluble pesticide that enters the animal body through an amplification process. The pesticides get consumed in the fatty tissues of animals resulting from the existence of pesticides in food chains for further periods.[20] The presence of pesticides in the ecosystem will be resulting in higher toxicity in bodies of higher tropic level species which will finally increase the population of secondary consumers, whereas it decreases the population of primary consumers.

13.2.1 EFFECTS ON ANIMAL

Animals can be affected by pesticides over their direct or indirect implementation including pesticide drift, lower-level toxicity, escape into regional territorial waters. It is attainable that some animals instantaneously could be sprinkled. Pesticides are like synthetic representatives accustomed to destroy both plants and animals. If pesticides kill animals, it can spread to other places through the dead bodies of the animals once they decompose. Contrastingly, pesticides become scattered outside the corresponding area and may turn out to be intensive in the bodies of animals. Last, we found that the effects of pesticides on animals have been executed by the effects of organochlorine germicide which are complex and extremely preserving. The toxicological analysis of the effects of sublethal is a part of organochlorine germicides. This shows that pesticides can change the genetic rate of animals and insects when they are released at very low levels.

13.2.2 EFFECT ON HUMANS

Pesticides incorporate a bunch of chemical families, with thousands of different formulations, hundreds of active ingredients, and many known or suspected unfavorable health outcomes. Furthermore, the active ingredients in pesticides also carry chemicals known as "inerts" such as surfactants, solvents, preservatives, which may have chemical actions different from the active ingredients.[21] The pesticides have adverse effects on the nervous, immune, endocrine, respiratory, and reproductive systems. It has been seen that many health problems are associated with the subjection of pesticides, namely, Endocrine disruption, Hodgkin's disease (HD), Parkinson's disease, non-Hodgkin lymphoma (NHL), reproductive and respiratory disorders. It is also believed that cancer is also one of the causes due to exposure to pesticides, for example, glyphosate is associated with breast cancer. Pesticides can probably lead to acute toxicity, which means that they can cause harmful effects after a short term of inhalation, ingestion, or skin contact. Moreover, pesticides can cause neuropsychiatric sequel because many pesticides underline the change in the functions, namely, (cholinergic critic) of the central peripheral system (CPS), making the cause of suicidal attempts.[22] Table 13.1 presents the different types of health issues due to the large use of pesticides.

13.2.3 EFFECT ON ENVIRONMENT

Pesticides grasp a distinctive position between environmental contaminants because of their high-level biological movement and toxicity. A pesticide is a poisonous chemical substance that is deliberately delivered into the environment in pursuance of preventing, halt, control, and/or kills and demolishes populations of rodents, weeds, insects, fungi, or other harmful pests. The use of pesticides has enlarged many folds above the previous decades. Annually the environment delivered around 5.2 billion pounds of pesticides. Pesticide phrases consist of active ingredients accompanying passive substances, adulterants, and periodically impurities. Once delivered into the environment, pesticides fall through into substances named metabolites that are more poisonous to active ingredients in many situations. Substantial application and disposal of pesticides by institutions, farmers, and people offer various sources of pesticides in the environment. It is practically impossible to restrict the zone of the effect of pesticides. Uniformly,

TABLE 13.1 Different Human Health Diseases Using Different Pesticides.

Sl. no.	Name of pesticides	Harmful health diseases	References
1	2,4-Dimethyl sodium salt and ethyl ester	Cancer	[23,24]
2	Acetamiprid	Hypothermia, tachycardia, hypotension, hypoxia, Anorexia, atrial fibrillation, chromosome breakage, consciousness disorders, hyperglycemia, convlusions	[25]
3	Carbendazim	Anthracnose	[26]
4	Cartap	Hypoxia, prenatal injuries	[27]
5	Chlorpyrifos	Diabetes, lacrimation, excessive salivation, tremors, nausea, diarrhea, Breast cancer, neurological dysfunctions, endocrine disruption, cardiovascular diseases	[28]
6	Clodinafop propagyl	Prenatal injuries	
7	Emamectin benzoate	DNA damage, mitochondrial dysfunction, diabetes mellitus, ischemic heart disease, any chronic respiratory illness, leukemia, non-Hodgkin lymphoma (NHL), brain tumors, and cancers	[29,30]
8	Fipronil	Sweating, vomiting, cough, abdominal pain, agitation, sweating, tonic-clonic convulsions, seizures, paresthesia, pneumonia pharyngeal pain, sensory impairment, nausea, headache, dizziness	[31]
9	Glyphosate	Celiac disease, cancer, kidney disease, Alzheimer's disease, autism, neurodegenerative diseases, endocrine disruption, seizures, Huntington's disease, epilepsy, autism, asthma, Parkinson's disease	[32–42]
10	Imidacloprid	Abdominal pain, headache and diarrhea, cardiovascular effects, mental disorders, dyspnea, and diaphoresis	[43]
11	Lambda-Cyhalothrin	Nerve diseases, skin and eye irritation, convlusions, muscle fasciculation, cell damage, tumors, liver diseases	[44,45]
12	Monocrotophos	Bone diseases, liver injury, diabetes mellitus, dyslipidemias	
13	Paraquat Dichloride	Liver failure, heart failure, kidney failure, muscle weakness, respiratory (breathing) failure, possibly leading to death, seizures, cancer, lung scarring, pulmonary edema	
14	Thiamethoxam	Alzheimer's disease, Parkinson's disease, schizophrenia, breast cancer, hypothermia and respiratory arrest, organ failure, neuropsychiatric disorders, rhabdomyolysis, vomiting	[46–52]

when it is put in a small zone, it scatters in the air, is soak up in the soil, or/ and finally reaches a bigger area. In the agricultural sector, when pesticides are sprinkled, they may get their way by air and end up in another division of the environment like in water and soil. Pesticides that are put straight to the soil may be rinsed out and grasp near the surface water bodies, and may filter through the soil to lower surfaces and pore water.[53] The effects of pesticides on the environment may lie from small deviation in the normal operating process of the ecosystem to the loss of species variegation.

13.2.4 EFFECT ON SOIL

Pesticides are being used to ensure sufficient supplies of crops and conservation of human and domestic health. But the pesticides that are used for crop protection are not free from problems. A large amount of pesticide reaches into the soil regardless of any method of application used. Methods of application may include direct application or arial spray. As a result of using pesticides in the soil a large variety of pesticides are being accumulated. There are pesticides such as methyl bromide that disappears in 2–4 days, whereas there are pesticides such as chlordane and benzene hex chloride that may take a year or even more to disappear. The persistence of pesticides in soil from one season to the next can affect sensitive plants. Leaching and volatilization from the soil and also degradation by soil microorganisms may result in the reduction of major pesticides that disappear quickly. The longer the persistence of pesticides the greater will be the risk. Pesticides are used for controlling the fungi and insects that are harmful to crop plants and their products. However, these chemicals also destroy many beneficial nontarget microorganisms that are known to contribute to developing crop productivity.

13.3 DEVICE PROTOTYPE AND ANALYSIS

For pesticide detection, a 3D design of the device as illustrated in Figure 13.3.1(a–g) shows the top view, front view, and side view, respectively. Figure 13.2 shows the labeled view of the device with details. As shown in Figure 13.2, in this device lid handle will be used to lift the device cover for object placement on the object stand. The insulated lid is made up of an insulating and opaque material to prevent the interference of light and atmosphere with the sensor module to increase the accuracy of the test,

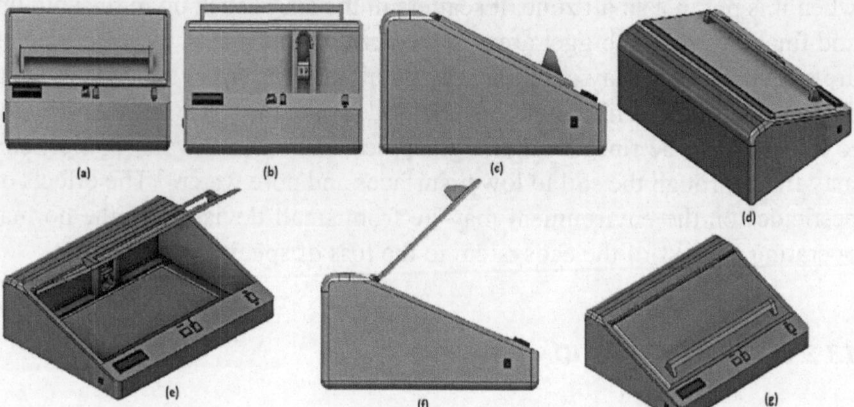

FIGURE 13.1 (a) This figure represents a front view of the device, (b) represents the top view of the device, and (c)–(g) represent all sides of the device.

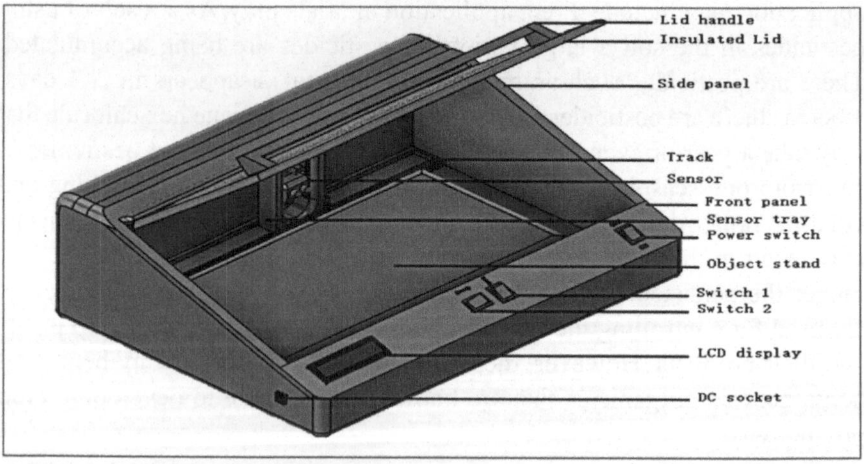

FIGURE 13.2 This figure represents the different parts of the pesticide detection device.

which uses a light-sensitive method to determine the presence of pesticides in a fruit or a vegetable placed on an object stand. A side panel will be useful in providing structural rigidity and support to the body and the front panel of the device as well as a 45° angle increases the accessibility of the object stand for easy placement of objects. The track is provided for the horizontal movement of the sensor try along the stretch of object stand to increase the reach of the sensor module for different types of fruits and vegetables over the stand. A sensor module is always supposed to be in line and proximity with the placed object, to determine the presence of pesticides in a fruit or a vegetable by delicately examining the surface properties of the object without any invasion of the object and thus keeping it usable even after the test. The front panel is housed with many switches for control and manipulation of the device and the sensor block; also an LCD screen is there to interact with the device, sensor and to check test results. The sensor tray is used to carry the sensor block to any arbitrary point that is along the vertical and horizontal stretches of the object stand to get to the closest point of contact possible with the object, where horizontal movement is carried out by the sensor tray using a motor and the track on a device and the vertical movement is carried out by the sensor block using track and motor mounted on the sensor tray. A power switch will be used for controlling the complete power supply of the device off/on. An object stand is used for the placement of fruit and vegetable samples kept for testing. And it also defines the workable limit of the sensor block for a test of pesticides on the object placed on the object stand. Switch 1 controls the movement of the motor controlling the motion of the sensor block along the vertical direction inside the sensor tray. Switch 2 is used for controlling the movement of the motor that moves the sensor tray in the horizontal direction. As a result, the sensor block that is inside of it also moves horizontally along with the sensor tray. The purpose of the LCD is to help the user to interact with the device in controlling the positions of the sensor and its main purpose is to get the test results about the presence of pesticides as output. DC socket is a +12-V DC socket for the whole power supply of the device including sensors, microcontrollers, motors, and display.

13.4 WORKING PRINCIPLE

The flowchart in Figure 13.3(a) and (b) illustrates the method that is pursued detecting the pesticides in fruits and vegetables using the infrared (IR) sensor. The principle encompasses switching on the device using the on/off switch that will start the device, making it ready to use. Now the process is followed by the infrared light that is transmitted to a given subject and the reflected light is received by an IR receiver. Now with the analysis of reflected and transmitted light, the NDVI is calculated with the help of the microcontroller and the obtained result about the quality of the object and the amount of pesticide present in the object is displayed on the screen. Then the user can measure other objects either by clicking the reset switch shown in the device or by switching off the device using the on/off switch.

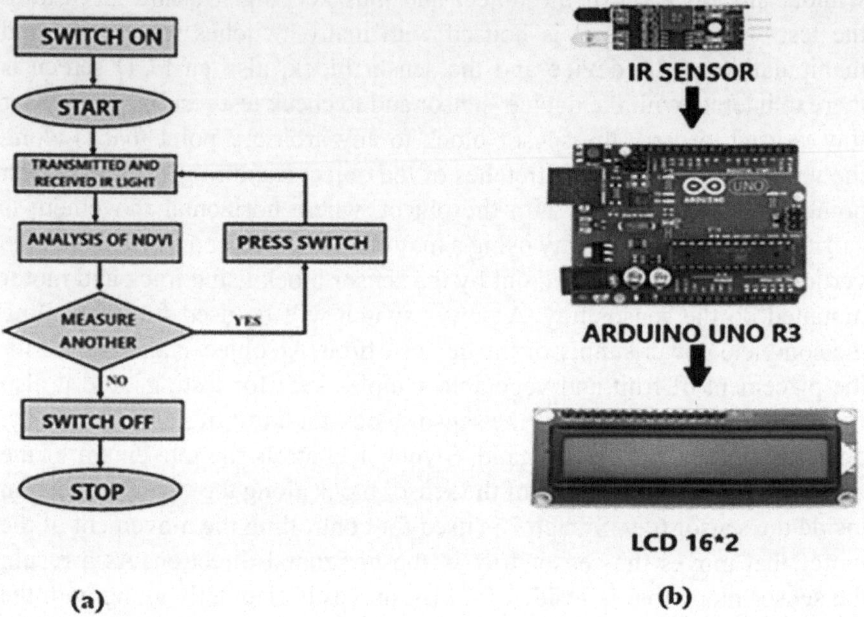

FIGURE 13.3 (a) Flow diagram of working prototype and (b) block diagram of the device.

13.5 METHODOLOGY

This section consists of the implementation of the above-shown flowchart with the android app including all elements attached to it. The development of the app including its design, impression, and user interaction is considered in it. In further part, the device flow and circuit design are described which gives us an efficient idea of the solution of our problem statement.

13.5.1 APPLICATION

The mobile application "LASER" contains 4 pages that consist of an account creation page, login page, daily measurement page, and sample testing page. This application is designed and implemented by using software such as HTML5, CSS3, JS, Node.JS, MongoDB, and Bootstrap.[4]

13.5.1.1 FRONTEND DEVELOPMENT

The frontend of the web application is developed using HTML version 5, CSS version 3, Bootstrap, and JavaScript (JS). The pages are working properly. All the pages are responsive and can adopt according to screen size.

13.5.1.2 BACKEND DEVELOPMENT

For backend development Node.js is used for client–server interaction, collecting the data from a user such as an email, password and storing it into a database. Node.js is also used to manipulate the data in the server. MongoDB is used in the application for creating the database. The data that will be generated from the application will be stored in the created database.

13.5.2 DEVICE DESIGN

In this chapter, the device is implemented on Arduino UNO R3 for the input from the specified IR sensor and displays the result in the LCD (16 × 2 cm^2). The transmitter used here to incident the light on the sample and the receiver is used for determining the value of the reflected ray. The IR sensor is used in many cases to find the obstacle that will detect it with the help of the reflected light from the object in the receiver. Instead of using this module, the transmitter and the receiver are used separately for a wide range of detection and efficient results. The IR transmitter has a wavelength of 700 nm to 1 mm which is much higher than the visible light range for which it is invisible. The block diagram of the device is shown in Figure 13.3(b) which represents the overall flow of the device.

To calculate the transmitted and reflected intensity IR sensor is used. The NDVI is calculated when the IR transmitter transmits the incident ray on the food, the incident ray reflects and the reflected ray is received through the IR receiver. For any food material, the NDVI value lies in the range 0–1, which is a unit less value. By the NDVI formula, the food that has an NDVI value less than 0.33 is indicated as "UNHEALTHY" food and the NDVI value having greater than 0.33 and less than 1 is indicated as "FIT TO EAT" that is also displayed on the LCD screen. As shown in Figure 13.4, IR transmitter and receiver are used there, the analog output is produced there, and it converts to digital by the Arduino that is further used to calculate the NDVI value by using the formula.

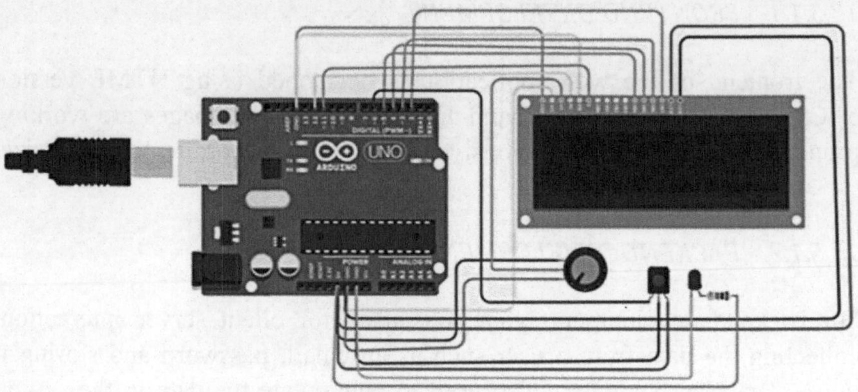

FIGURE 13.4 Circuit diagram for the pesticide detection.

13.6 LASER APPLICATION

Laser application is developed after the analysis of the data, implementation of various algorithms for functions, and the layout for the designing of the application. The device will be connected through the internet using thin peak IoT and then the data from the device will be sent to a web application. By simply downloading the application through any store the user can see all the details on their smartphone or laptop. Users have to first create their account using their name, email, and password as shown in Figure 13.5(a). If the user has already an account, they can log in by giving their password and email id as in Figure 13.5(b). After login, there will be a daily measurement page where the user can check the samples that are given in a graphical form for different vegetables that are tested by the device as in Figure 13.5(c). On the next page that is the sample testing user can see the NDVI value for each of the items that is tested using the device as shown in Figure 13.5(d).

13.7 RESULT AND DISCUSSION

Figure 13.6 represents the prototype under the test condition. Figures 13.7(a) and 13.6(b) represent the real-time LCD that shows the result of two samples taken with imitating which one is fit to eat and unhealthy when connected to the device. After the completion of the device prototype, data are taken from 26 different vegetables with the NDVI values. Figure 13.8 shows the calculated NDVI values (y axis) for the different sample vegetables (x axis).

The blue dotted bar of Figure 13.8 represents the cutoff edge NDVI value (0.33) before which the samples are unhealthy to eat. The outcome of all data is calculated to restore the capabilities of the device.

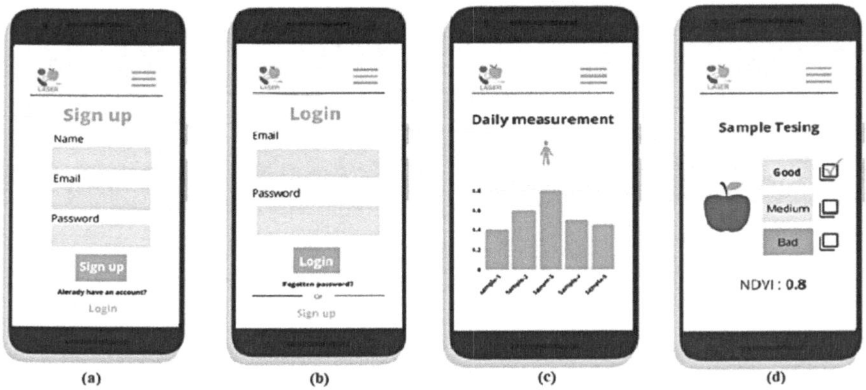

FIGURE 13.5 (a) The signup page, (b) login page, (c) daily measurement page, and (d) sample result.

FIGURE 13.6 Calculated NDVI value of two different vegetables.

13.8 CONCLUSION

In today's scenario considering that the amounts of pesticides in fruits and vegetables are growing, there is a need for a device that can measure/detect the amount/percentage of impurities in the fruits and vegetables. The pesticide-detecting device that is prototyped goes with full affordability; therefore, a variety of consumers can purchase it and use this device to check food impurity levels to make their health condition safe and sound.

FIGURE 13.7 Calculated NDVI value of two different vegetables.

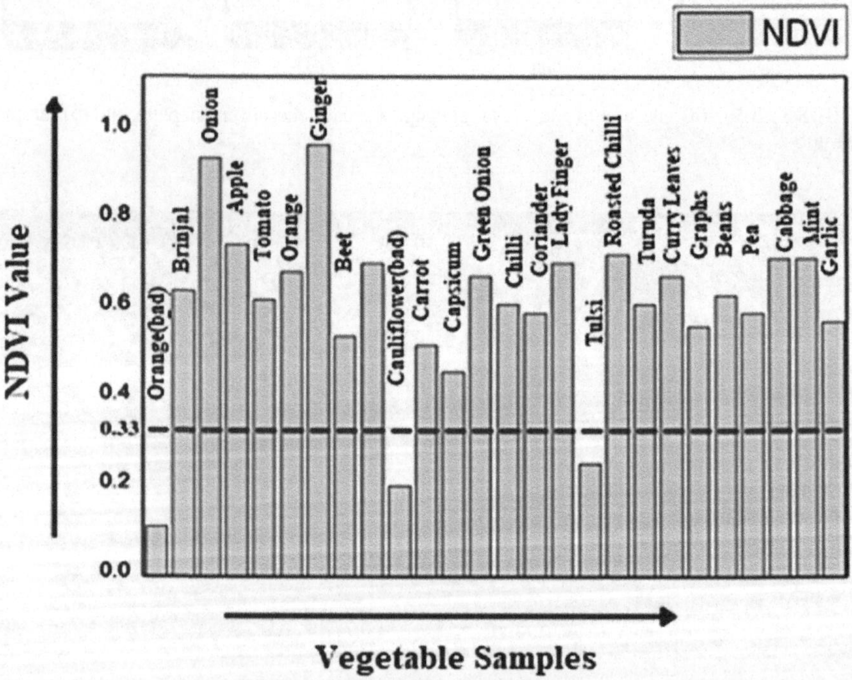

FIGURE 13.8 This figure represents the testing result of 26 sample vegetables with NDVI values.

In the future, this device can increase the use of IR sensors and take their average for NDVI calculation with more calibration, minimal error, and efficient results. Fruit/vegetable samples can be taken and some specifications can be added to check the number of pesticides. Additionally, advanced devices can be integrated for the detection of other harmful chemicals such as fungicides, toxicants, and weedicides.

KEYWORDS

- **pesticides**
- **biological magnifications**
- **electronic device**
- **infrared light**
- **NDVI**
- **innovation**

REFERENCES

1. Roser, M.; Ritchie, H.; Ortiz-Ospina, E. World Population Growth. *Our World Data* **2013**. Published online at OurWorldInData.org. Retrieved from: https://ourworldindata.org/world-population-growth.
2. Manuel, T. The Challenge of Feeding the World. **2019,** *2017*, 3–4.
3. Popp, J.; Pető, K.; Nagy, J. Pesticide Productivity and Food Security. A Review. *Agron. Sustain. Dev.* **2013,** *33* (1), 243–255.
4. McNaught, A. D.; Wilkinson, A. *Compendium of Chemical Terminology*; Blackwell Science: Oxford, 1997; Vol. 1669.
5. Khursheed, K. W.; Manzoor, J.; Anjum, J.; Bashir, M.; Mamta. Pesticides in Vegetables: Their Impact on Nutritional. 2017, No. February 2021. doi: 10.4018/978-1-5225-2970-5.ch006.
6. Crinnion, W. J. Chlorinated Pesticides: Threats to Health and Importance of Detection. *Altern. Med. Rev.* **2009,** *14* (4), 347–359.
7. U. S. Environmental. What Is a Pesticide? Epa.Gov. Retrieved Sept. 2007; p 15.
8. Aktar, W.; Sengupta, D.; Chowdhury, A. Impact of Pesticides Use in Agriculture: Their Benefits and Hazards. *Interdiscip. Toxicol.* **2009,** *2* (1), 1–12.
9. Nicolopoulou-Stamati, P.; Maipas, S.; Kotampasi, C.; Stamatis, P.; Hens, L. Chemical Pesticides and Human Health: The Urgent Need for a New Concept in Agriculture. *Front. Public Heal.* **2016,** *4* (July), 1–8.
10. Verger, P. J. P.; Boobis, A. R. Reevaluate Pesticides for Food Security and Safety. *Science (80–).* **2013,** *341* (6147), 717–718.
11. Bhadekar, R.; Pote, S.; Tale, V.; Nirichan, B. Developments in Analytical Methods for Detection of Pesticides in Environmental Samples. *Am. J. Anal. Chem.* **2011,** *02* (08), 1–15.

12. Ishizawa, H.; Koyama, S.; Yoshinari, H.; Tokutake, S.; Yamada, H. Drifting Pesticide Detection Based on Attenuated Total Reflection of Infrared Spectroscopy. In *SICE Annual Conference 2011*.
13. Tsagkaris, A. S.; Pulkrabova, J.; Hajslova, J. Optical Screening Methods for Pesticide Residue Detection in Food Matrices: Advances and Emerging Analytical Trends. *Foods* **2021,** *10* (1), 88.
14. Samsidar, A.; Siddiquee, S.; Shaarani, S. M. A Review of Extraction, Analytical and Advanced Methods for Determination of Pesticides in Environment and Foodstuffs. *Trends Food Sci. Technol.* **2018,** *71* (July 2017), 188–201.
15. Bhandari, A. K.; Kumar, A.; Singh, G. K. Feature Extraction Using Normalized Difference Vegetation Index (NDVI): A Case Study of Jabalpur City. *Proc. Technol.* **2012,** *6*, 612–621.
16. Singh, T.; Singh, M. Hand Held Device for Detection of Pesticides Using NDVI. *Int. J. Comput. Appl.* **2016,** *154* (9), 29–35.
17. Majewski, M. S.; Capel, P. D. *Pesticides in the Atmosphere: Distribution, Trends, and Governing Factors*; CRC Press: Boca Raton, FL, 2019.
18. STRAATHOF, H. J. M. Investigations on the Phytotoxic Relevance of Volatilization of Herbicides. *Meded. Fac. Landbouwwet. Rijksuniv. Gent* **1986,** *51* (2a), 433–438.
19. Helfrich, L.; Weigmann, D.; Hipkins, P.; Stinson, E. *Pesticides and Aquatic Animals: A Guide to Reducing Impacts on Aquatic Systems*. VCE Publications, Virginia Tech. VCE Publ.: Blacksburg, VA, 2009.
20. Warsi, F. How Do Pesticides Affect Ecosystems. *Pestic* **2015,** 16. Available from http//farhanwarsi.tripod.com/id9.html (accessed Jan 2015).
21. Cox, C.; Surgan, M. Unidentified Inert Ingredients in Pesticides: Implications for Human and Environmental Health. *Environ. Health Perspect.* **2006,** *114* (12), 1803–1806.
22. Abdollahi, M.; Ranjbar, A.; Shadnia, S.; Nikfar, S.; Rezaiee, A. Pesticides and Oxidative Stress: A Review. *Med. Sci. Monit.* **2004,** *10* (6), RA141–RA147.
23. Tomlin, C. D. S. *The Pesticide Manual: A World Compendium*; British Crop Production Council: Farnham, 2009.
24. WHO. *Environmental Health Criteria*, 1998; p 204.
25. Imamura, T.; Yanagawa, Y.; Nishikawa, K.; Matsumoto, N.; Sakamoto, T. Two Cases of Acute Poisoning with Acetamiprid in Humans. *Clin. Toxicol.* **2010,** *48* (8), 851–853.
26. Wang, Y.; Hou, Y. P.; Chen, C. J.; Zhou, M. G. Detection of Resistance in Sclerotinia Sclerotiorum to Carbendazim and Dimethachlon in Jiangsu Province of China. *Australas. Plant Pathol.* **2014,** *43* (3), 307–312.
27. Gupta, V. K.; Siddiqi, N. J.; Ojha, A. K.; Sharma, B. Hepatoprotective Effect of *Aloe vera* against Cartap- and Malathion-Induced Toxicity in Wistar Rats. *J. Cell. Physiol.* **2019,** *234* (10), 18329–18343.
28. Mansour, S. A.; Mossa, A. T. H. Lipid Peroxidation and Oxidative Stress in Rat Erythrocytes Induced by Chlorpyrifos and the Protective Effect of Zinc. *Pestic. Biochem. Physiol.* **2009,** *93* (1), 34–39.
29. Guanggang, X.; Diqiu, L.; Jianzhong, Y.; Jingmin, G.; Huifeng, Z.; Mingan, S.; Liming, T. Carbamate Insecticide Methomyl Confers Cytotoxicity through DNA Damage Induction. *Food Chem. Toxicol.* **2013,** *53*, 352–358.
30. Ly, J. D.; Grubb, D. R.; Lawen, A. The Mitochondrial Membrane Potential ($\Delta\psi m$) in Apoptosis; an Update. *Apoptosis* **2003,** *8* (2), 115–128.

31. Van Der Sluijs, J. P.; Amaral-Rogers, V.; Belzunces, L. P.; Bijleveld Van Lexmond, M. F.; Bonmatin, J. M.; Chagnon, M.; Downs, C. A.; Furlan, L.; Gibbons, D. W.; Giorio, C.; Girolami, V.; Goulson, D.; Kreutzweiser, D. P.; Krupke, C.; Liess, M.; Long, E.; Mcfield, M.; Mineau, P.; Mitchell, E. A.; Morrissey, C. A.; Noome, D. A.; Pisa, L.; Settele, J.; Simon-Delso, N.; Stark, J. D.; Tapparo, A.; Van Dyck, H.; Van Praagh, J.; Whitehorn, P. R.; Wiemers, M. Conclusions of the Worldwide Integrated Assessment on the Risks of Neonicotinoids and Fipronil to Biodiversity and Ecosystem Functioning. *Environ. Sci. Pollut. Res.* **2015,** *22* (1), 148–154.
32. De Roos, A. J.; Zahm, S. H.; Cantor, K. P.; Weisenburger, D. D.; Holmes, F. F.; Burmeister, L. F.; Blair, A. Integrative Assessment of Multiple Pesticides as Risk Factors for Non-Hodgkin's Lymphoma among Men. *Occup. Environ. Med.* **2003,** *60* (9), 1–10.
33. Schinasi, L.; Leon, M. E. Non-Hodgkin Lymphoma and Occupational Exposure to Agricultural Pesticide Chemical Groups and Active Ingredients: A Systematic Review and Meta-analysis. *Int. J. Environ. Res. Public Health* **2014,** *11* (4), 4449–4527.
34. Cattani, D.; de Liz Oliveira Cavalli, V. L.; Heinz Rieg, C. E.; Domingues, J. T.; Dal-Cim, T.; Tasca, C. I.; Mena Barreto Silva, F. R.; Zamoner, A. Mechanisms Underlying the Neurotoxicity Induced by Glyphosate-Based Herbicide in Immature Rat Hippocampus: Involvement of Glutamate Excitotoxicity. *Toxicology* **2014,** *320* (1), 34–45.
35. Castellani, R.; Smith, M. A.; Richey, P. L.; Perry, G. Glycoxidation and Oxidative Stress in Parkinson Disease and Diffuse Lewy Body Disease. *Brain Res.* **1996,** *737* (1–2), 195–200.
36. Wennberg, R. P. The Blood-Brain Barrier and Bilirubin Encephalopathy. *Cell. Mol. Neurobiol.* **2000,** *20* (1), 97–109.
37. Tabrizi, S. J.; Workman, J.; Hart, P. E.; Mangiarini, L.; Mahal, A.; Bates, G.; Cooper, J. M.; Schapira, A. H. V. Mitochondrial Dysfunction and Free Radical Damage in the Huntington R6/2 Transgenic Mouse. *Ann. Neurol.* **2000,** *47* (1), 80–86.
38. Tuchman, R.; Rapin, I. Epilepsy in Autism. *Lancet Neurol.* **2002,** *1* (6), 352–358.
39. Levisohn, P. M. The Autism-Epilepsy Connection. *Epilepsia* **2007,** *48* (Suppl. 9), 33–35.
40. DeLong, G. R. Autism, Amnesia, Hippocampus, and Learning. *Neurosci. Biobehav. Rev.* **1992,** *16* (1), 63–70.
41. Endo, T.; Shioiri, T.; Kitamura, H.; Kimura, T.; Endo, S.; Masuzawa, N.; Someya, T. Altered Chemical Metabolites in the Amygdala-Hippocampus Region Contribute to Autistic Symptoms of Autism Spectrum Disorders. *Biol. Psychiatry* **2007,** *62* (9), 1030–1037.
42. Strange, R. C.; Spiteri, M. A.; Ramachandran, S.; Fryer, A. A. Glutathione-S-Transferase Family of Enzymes. *Mutat. Res.—Fundam. Mol. Mech. Mutagen.* **2001,** *482* (1–2), 21–26.
43. Balani, T.; Agrawal, S.; Thaker, A. M. Hematological and Biochemical Changes Due to Short-term Oral Administration of Imidacloprid. *Toxicol. Int.* **2011,** *18* (1), 2–4.
44. Banz, W. J.; Iqbal, M. J.; Bollaert, M.; Chickris, N.; James, B.; Higginbotham, D. A.; Peterson, R.; Murphy, L. Ginseng Modifies the Diabetic Phenotype and Genes Associated with Diabetes in the Male ZDF Rat. *Phytomedicine* **2007,** *14* (10), 681–689.
45. Kothari, V.; Stevens, R. J.; Adler, A. I.; Stratton, I. M.; Manley, S. E.; Neil, H. A.; Holman, R. R. UKPDS 60: Risk of Stroke in Type 2 Diabetes Estimated by the UK Prospective Diabetes Study Risk Engine. *Stroke* **2002,** *33* (7), 1776–1781.
46. Sheets, L. P. Imidacloprid: A Neonicotinoid Insecticide. *Hayes' Handbook of Pesticide Toxicology*, 3rd ed.; Academic Press: Cambridge, MA, 2010; pp 2055–2064.

47. Gibbons, D.; Morrissey, C.; Mineau, P. A Review of the Direct and Indirect Effects of Neonicotinoids and Fipronil on Vertebrate Wildlife. *Environ. Sci. Pollut. Res.* **2015,** *22* (1), 103–118.
48. Forrester, M. B. Neonicotinoid Insecticide Exposures Reported to Six Poison Centers in Texas. *Hum. Exp. Toxicol.* **2014,** *33* (6), 568–573.
49. Murie, J. Case Reports. *Scott. Med. J.* **2014,** *59* (1), 80.
50. Agarwal, R.; Srinivas, R. Severe Neuropsychiatric Manifestations and Rhabdomyolysis in a Patient with Imidacloprid Poisoning. *Am. J. Emerg. Med.* **2007,** *25* (7), 844–845.
51. Todani, M.; Kaneko, T.; Hayashida, H.; Kaneda, K.; Tsuruta, R.; Kasaoka, S.; Maekawa, T. Acute Poisoning with Neonicotinoid Insecticide Acetamiprid. *Chudoku Kenkyu Chudoku Kenkyukai Jun Kikanshi = Jpn. J. Toxicol.* **2008,** *21* (4), 387–390.
52. Iyyadurai, R.; George, I. A.; Peter, J. V. Imidacloprid Poisoning-Newer Insecticide and Fatal Toxicity. *J. Med. Toxicol.* **2010,** *6* (1), 77–78.
53. Harrison, S. A. The Fate of Pesticides in the Environment. *Agrochemical Fact Sheet# 8*; Penn, USA, 1990; pp 2–8.

CHAPTER 14

IoT-Based Traffic and Router Management System for Drivers

MANOJ KUMAR SAHOO[1*], ASHISH KUMAR DASH[1], SWADHIN KUMAR SENAPATI[2], BORA PAVANI[2], P. DEEPAK[2], SWASTID DASH[2], and G. AKSHAYKUMAR[2]

[1]*Department of School of Computer Science and Engineering, NIST (Autonomous), Berhampur 760010, Odisha, India*

[2]*Department of Electronics and Communication Engineering, NIST (Autonomous), Berhampur 760010, Odisha, India*

[*]*Corresponding author. E-mail: mksahoo@nist.edu, mksahoo@hotmail.com*

ABSTRACT

Surface transport is the backbone of transportation and plays an important role in the development of the society. The industrialization and urbanization is the prime cause of the increase in the number of the vehicles. In recent decays, the numbers are rapid. The utilization of surface transport is done for transport of goods and materials by trains, trucks, and heavy vehicles, whereas the public transport is carried out with train, bus, car, and light vehicles like bikes or two-wheelers. Road accidents and delays due to congested traffic are two negative factors associated with road or surface transports apart from the pollution issue. According to the statics, there is a loss of 15 lakhs of human life and 50 lakhs seriously injured personal become handicapped, which is a huge loss in terms of human value. So traffic management is the biggest issue in urban and industrial areas. The transport delay can be reduced with good infrastructure and good road conditions for transport. With the advantage of technology, this can be reduced to the minimum.

Intelligent Technologies for Sensors: Applications, Design, and Optimization for a Smart World.
S. Kannadhasan, R. Nagarajan, & Alagar Karthick (Eds.)

The IoT platforms allow developing a driver assistance system, which can help the driver to find the free and shortest route. The traffic density can be accessed from the local traffic station via cloud technology. High-speed continuous communication can be achieved by a 5G network. The routes can be identified from the source to the destination place with the help of Google map or a similar platform and estimate time to reach so that the driver should not into the rush or speedy driving mode to avoid the accident.

14.1 INTRODUCTION

The purpose of the Internet of Things (IoT)-based traffic and router management system plays a major role in urban areas. World's urban and industrial areas are facing many challenges because of the rapidly growing population. Because of the growing population numbers of vehicles keep growing on the road and infrastructure is not upgraded so that people are facing many problems during a road trip as well as transportation by road.[1] People who lived in urban and industrial cities are facing traffic congestion, time delay, exasperation, accident, cost, and death.[2,3] Nowadays in urban areas numbers of vehicles are going on increasing due to various industrial sectors and our population, but our road system is not flexible. According to WHO (World Health Organization), approximately 1.3 million (0.13 crore) people die per year due to poor traffic systems. From 2019 onward thousands of people were died due to road accidents in India. According to the Urban mobility Report in 2011 people in India spend more than 30 hours per year in traffic[4] due to traffic congestion. To prevent these kinds of problems we can use modern computational technologies. The IoT is nothing but a wireless network platform that aims to connect a physical device with the internet and provide new opportunities in the IT and communication field. Nowadays, surface transport plays a significant role in the transportation system. A major problem in surface transport is road accidents due to carelessness of the driver and consumption of alcohol is another cause of road accident; congested traffic is another problem in this type of transportation. Also vehicle routing problem is a very big provocation in the regulation of transportation network. The FOG system is a method of sorting out vehicle-routing problem (VRP). The objective of FOG is to sketch easy free route and reduce the social and environmental effects. FOG is to initiate the study of vehicle routing issues. Generally, FOG looks to find a shortest route to fulfil transportation demands. Basically, FOG has comprehensively been used in various industries and firms for better transportation. This study offers to design the structure of IoT-based traffic monitoring system. We are also focusing on reducing the problems in surface transport by the use of IoT network with the driver assistant system (DAS), where we can find the shortest route between the source and the destination. The DAS is directly connected to the local traffic monitoring system which will provide the information about traffic-free route on that urban area. And the route can be selected through the Google map which gives the estimated time with distance between sources to destination which will help the high-speed driving mode of a driver.

14.2 LITERATURE REVIEW

From past few decades, there is rigorous growth of human population with increase of vehicular demand. Many strategies were planned to tackle traffic congestion. In the present scenario, time became a very important factor in everyone's life. In metropolitan cities everyday a person spends hours in traffic, which in turn delays his/her plans or speeds up his/her vehicle to complete plans in stipulated time. Earlier, there was use of magnetic sensors at every convergence to update about traffic flow through optic fibers, at backend software in computer systems an allotted person used to analyze the data automatically to proceed for further adjustments in every second. But these plans take time for proceedings as there was use of optic fibers but nowadays by the use of a sensor without any interconnections and human interference we can tackle with traffic problems. IoT is featured to lead an effortless life. A driver can easily tackle with traffic issues by knowing minutes before about clogging in any route and also blockage free route to reach destination as early as possible.

14.3 WORKING PRINCIPLE

Nowadays, the traffic congest problems are playing a very major role in urban areas, due to industries and public sectors the road transport is busy. In the local urban areas we are facing traffic congestion issue, thus daily we are losing much time in traffic, which will affect our daily time survival. So here we introduce a IoT-based router management system. The whole block diagram represents an overall structure of IoT-based router management system for drivers. Here mainly two advance technologies are used which are cloud technology and fog technology. First, application (IoT device) turns on. Then after the driver will set the source and destination of his/her journey through the Google map, now complete route map data for the journey get selected by the driver. This complete data is stored in the cloud; on the other hand, local traffic monitoring system collects all the data regarding routes and traffic conditions in the respective urban area. This local traffic monitoring system is operated as well as updated by local traffic police or local traffic stations. This local traffic-monitoring system data is also stored in cloud. For this router management idea the fog computing plays a vital role. From the cloud the data (route between source and destination) transfer to fog computing. If any problem occurs between our source and destination route, like traffic-congested problem or route-damaged problem in that scenario the fog computing will take the data and it gives traffic free alternate route until the problem is not solved. Then the cloud is also updated. If there is no problem between source and destination route, then fog computing follows the same rout as selected before by the driver. Figure 14.1 shows the flowchart of driver assistant system.

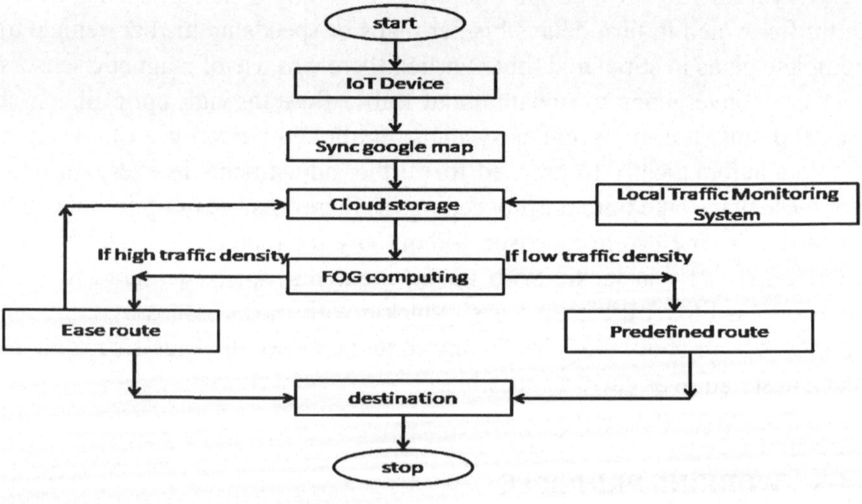

FIGURE 14.1 The overall block diagram of driver assistant system.

14.4 PROPOSED SYSTEM

14.4.1 DRIVER ASSISTANT SYSTEM

In this modern era of rapid development and industrialization an increased number of casualties in road accidents has become a major issue in almost all parts of the world. So, the need for proper road safety and vehicle security is very much essential in the developing countries. Continuous efforts are being put by all governments to provide solutions to accidental threats and road safety. The researches are set up for vehicle infrastructure and security advancement like communication among the vehicles about any threat during crossing; examples of such research programmers are Prometheus European program[5] and the IVHS American program.[6] They focus on improving abilities of the vehicle to the next level by implementing few advanced systems, that is, automatic blocking system (ABS), automatic obstacle warning system. It is possible to show warning to the user and also control the vehicle on receiving the warning. The status of the traffic in all possible routes to the destination can be collected using RFIID technology and stored in the high-speed servers setup in different areas. The data collected in cloud storage is used to guide the drivers. The driving assistance systemizes the data collected over the cloud storage and processes the data to help the driver to reach his/her destination in a faster and safer way by providing the information about the density of vehicular flow in all the possible routes to predict the traffic congestion and provide the best route. This system also ensures the safety of the passengers by keeping the driver updated about the road crossings, road structure, turnings before they are very close. The sensors set up in vehicles determine the speed and distance of the passing by vehicles and warn the pilot if there is any threat of accidents. This system can also handle the speed of vehicle automatically.

14.4.2 LOCAL TRAFFIC-MONITORING SYSTEM

The traffic-monitoring system cites on technologies that are mapped to gather data of the traffic for roadway usage, for example, the number of vehicles. It takes a part as prime role to analyze the transportation strategy and to set schemes for the requirement of future transportation.[7] It is an essential instrument for state Departments of Transportation (DOTs) to carry out many processes like finding out road improvement ideas, for the development of road network, evaluation of economic benefits, to carry out new development plans etc. The systems are classified into: over-road, in-road, and aerial-sensor-based approaches. In-road-based system requires to sawcut the footway and embeds sensors in the pavement of the roadway. Those sensors comprise inductive loop detectors[8] and magnetometers.[9] Installation of these sensors is time taking and requires much human involvement, and the working of these sensors is to get involved in roadway surfacing or pavement drop. Overroad-based sensors are installed (e.g., video cameras,[10] ultrasonic sensors[11]) on the top of the road or besides the road. These systems can be simply affected by weather conditions and are costly. Aerial-based approaches[12] use satellites or aircraft to catch road pictures, and to uproot traffic details via image processing.

IoT changed the world by bridging billions of physical and virtual objects with use in particular plans to Internet. The unexpected traffic decline in quick navigating routes and highways, classified on scarce visibility is one of the major issues for road accidents. It can also cause if work is in progress on roads, more traffic etc. Incomplete visibility of drivers may be for different causes like tight corners, tunnels, and fog. The IoT cloud system focused on traffic monitoring and alerting drivers. Thus, there is increase of interest in vehicular cloud computing, representing best of our knowledge; no scientific works are planned using IoT-based cloud system for traffic monitoring and alert notification work on big traffic data processing using mobile sensors precisely installed on vehicles. In these circumstances, development in networks makes intelligent transportation system (ITS) services more feasible. In this write-up, we focus on the replacement of ITS solution to address an issue by considering GPS-based mobile traffic sensors that are installed in private/public transport or volunteer vehicles. This part focuses on the structure of the IoT-based Cloud system for traffic monitoring and alert notification. An IoT cloud is defined for storage, processing, and networking potentiality on different bases to fixed and mobile sensing issues to postulate Platform as a Service (PaaS), infrastructure as a Service (IaaS), and also Software as a Service levels a new intersecting type of service level called IoT as a Service (IoTaaS).[13,14] In IoT-based cloud system, mobile sensors are focused on a tracking device that is put in vehicles that assemble position data. Therefore, it is feasible to calculate the velocity, acceleration, and so on. Local traffic monitoring system can be accessed easily for solution to any problem.

14.5 CLOUD TECHNOLOGY

Nowadays, data in IoT are stored and processed highly in cloud servers. Mostly cloud provides flexible computing platforms that are on demand mode, consequently deducting the investment required to build the applications for storage. Cloud computing is becoming one of the next powerful tools for users who evacuate respective data as well as applications to the remote "Cloud" and users can access in a easy and convenient manner. Data in many fields and from different sources are retained via services by using network. For example, Google map is a simple data selection service[15] that provides a simple route map interface that can be used to select the route between our source and destination. In our paper the total map data is stored in the cloud. The traffic management system deals with many users for services like storage of traffic data. Figure 14.2 shows how the traffic data and map data is stored in the cloud.

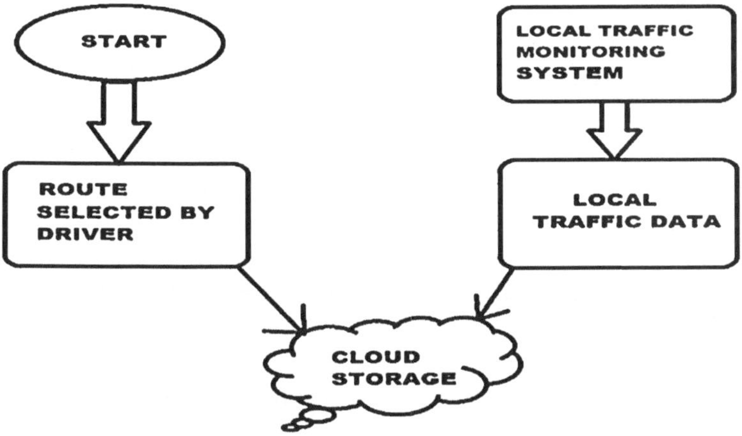

FIGURE 14.2 This figure shows how we store the total traffic data as well as map data in cloud.

The traffic management systems focused on cloud computing and it depends on the service contributor and the user. All the service providers like traffic strategy database, ATS, traffic strategy agent database, and the trial of average traffic scenes are covered core in the systems. The data of local traffic management systems is stored in the cloud. In terms of The smart traffic management system, clouds supply agent-distribution maps and traffic strategy data. It effects the manual traffic decisions taken by the local traffic police because of the traffic strategy performance and growth of smart traffic clouds, many traffic management systems will bridge to share the cloud data to save resources. Further, new traffic strategies can be modified into mobile agents to continuously upgrade with the transportational development science.

Data of traffic blocking monitoring system will be stored in the cloud from the local traffic monitoring system and share the data to the server of fog for detection, processing, and visualization. In this design, regarding high traffic density from the cloud, the user gets notified.

This system works on step by step. First, road traffic information data will collect from local traffic monitoring system in the respective areas; it will give information about traffic on different routes. Then the raw data will be transferred to the fog for the calculation of number of vehicles in a certain route. Afterward it verifies the data to find where the vehicle congestion occurs and it will suggest a traffic-free route. For the detection of HIGH congestion or any route damage, it will send an alternative path to the cloud in view of a map.

14.6 FOG TECHNOLOGY

Fog networking or fogging is a modern technology that uses IoT or edge device to carry out a significant amount of computation, data, and communication locally and routed over the Internet. Fog computing is an intermediate computing structure situated in between cloud and edge devices that enlarge the services supplied by the cloud to the edge device.[16–18] Fog computing has the dominance of low latency and location awareness, the rising requirements of overall connectivity, and ultra-low latency challenge traffic management for urban areas. A fog network is a collection of several fog nodes. Any device with data storage, network connectivity, and computing can be a fog node. These nodes can be routers, switches, or embedded servers, etc. The fog nodes can be placed anywhere either on road, along with a railway track, or an oil rig. Fog computing provides low latency that means it is geographically closer to users and provides instant responses. These nodes send data to the cloud for analysis and longer-term storage. The fundamental concept of fog technology is constant and a real-time monitoring and preparing of a system and an instant response in case of any system failure.[19,20] Therefore, this application can be helpful for a diversity of critical monitoring applications. Figure 14.3 shows the design of fog computing.

14.7 HARDWARE DESIGN

In this hardware project here we are using a GPRS module and Raspberry pi with an inbuilt GSM module. This schematic diagram in figure 14.4 explains that at the beginning the user's mobile phone is connected with the GSM by Bluetooth or internet. After that through GPRS (General packet radio service) the IoT application can detect the current location of a user. Then the user sets the destination. With the help of Google map, someone can find route between source and destination. The data will be stored in cloud, through GSM and the raspberry pi can communicate with cloud.

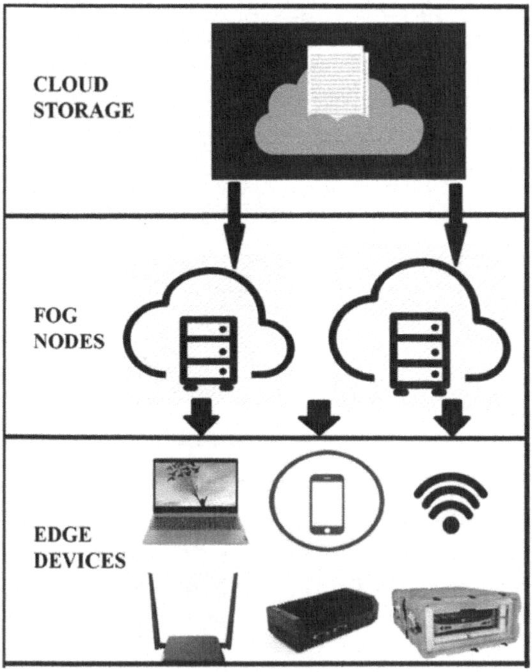

FIGURE 14.3 Three-layer architecture of fog computing.

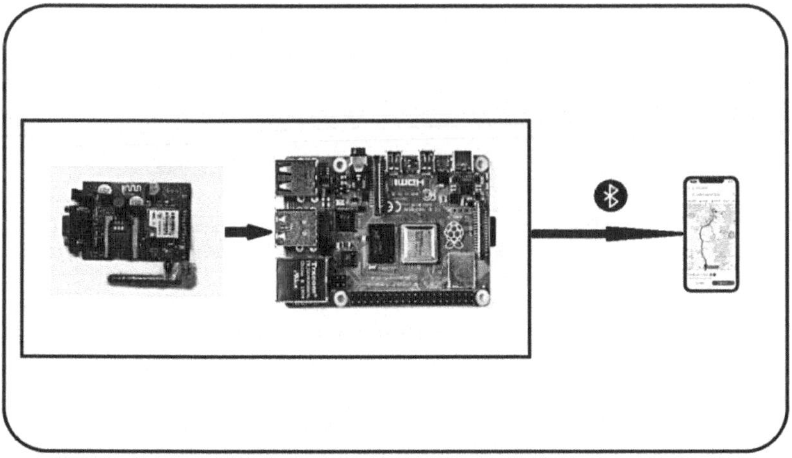

FIGURE 14.4 A schematic flow diagram of the IoT device.

14.8 APPLICATION

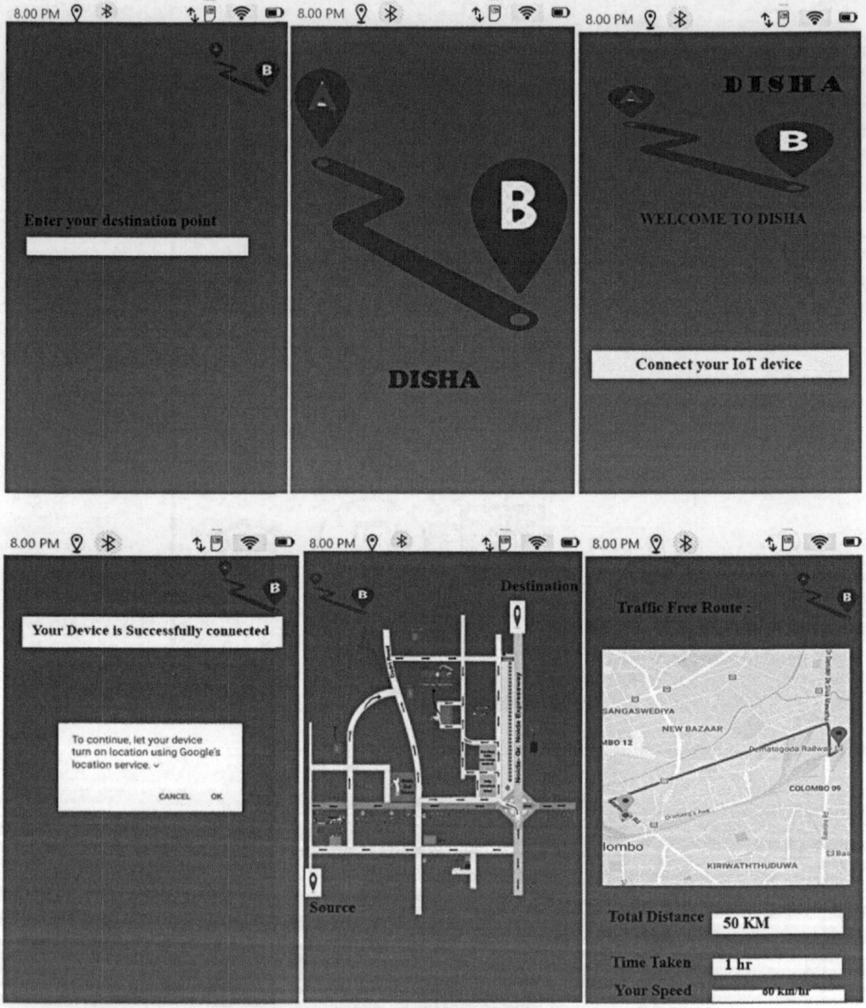

FIGURE 14.5 The overall architecture of smart traffic and routing system application.

Here, we introduced an IoT-based router management system application where the drivers can easily find the traffic-free route; Figure 14.5 shows that, in the first stage the application will turn on and after that it will search for the Bluetooth device whether it is on or off. After successfully being connected with a Bluetooth device, the application chooses the current location of the user. After that the user has to enter the destination point, where the user wants to reach. If there is any route problem or traffic congestion problem occurring in between our current location and our selected destination, then this IoT device suggests a traffic-free alternate route for the user, so that we can reduce the traffic congestion in urban areas. This application as shown in flowchart (Figure 14.6) will highly oblige to the emergency vehicles like ambulance, police service.

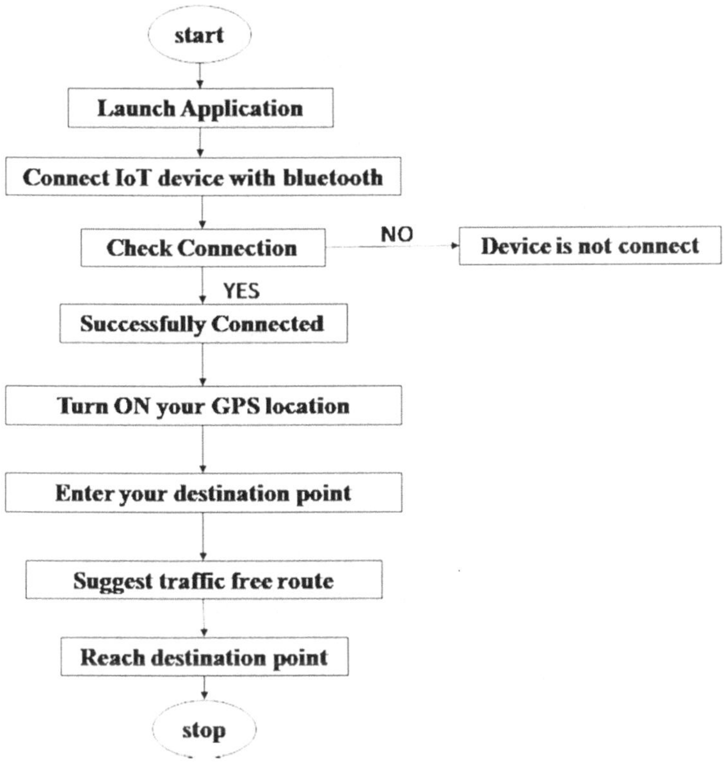

FIGURE 14.6 The block diagram for working principle of smart traffic and routing system application.

14.9 RESULT

The proposed IoT-based traffic and router management system for drivers is lined up and examined over traffic system of our locality and the proceedings were carried on using disha application. The outcomes of the idea were recorded and screenshots are also taken as shown in Figure 14.7; we are notified about the connection of the device, tracing of current location, questioned on destination place, and finally shown detailed road map with time, speed, distance parameters. Last, we stepped on roads to seek solution on practically implemented issues and it was fully prosperous.

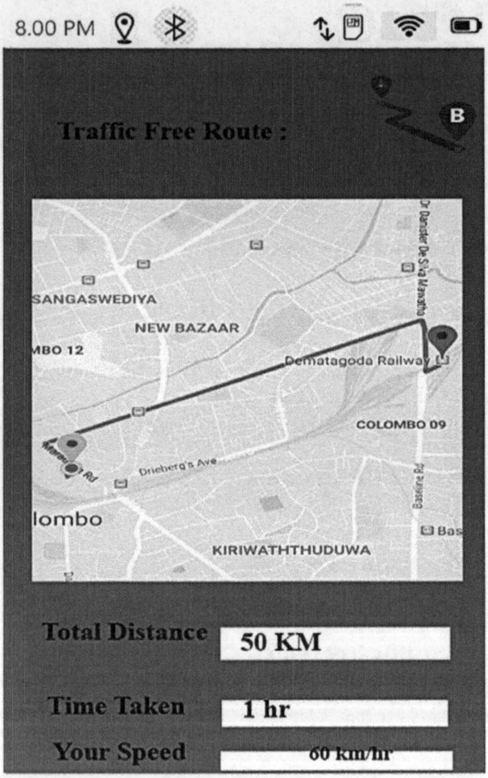

FIGURE 14.7 The overall view of traffic free route.

14.10 CONCLUSION

The write-up focused for relevant solution on traffic congestion by implementing IoT features using the GPS module and raspberry pipe. Traffic congestion is a major issue in a very developing city. In this era of inventions, IoT is the best invention that decreases human consultancy. An application can inform or alert about a traffic issue to reach a destination without any ease and it will suggest a noncongested route to reach the destination as early as possible. The detailed working of the application will be based on the FOG system to detect nonarduous route, which in turn will collect information cloud storage and local monitoring system.

KEYWORDS

- **IoT**
- **surface transport**
- **driver assistant system**
- **could technology**
- **innovation**

REFERENCES

1. Uddin, A. Traffic Congestion in Indian Cities: Challenges of a Rising Power. In *Kyoto of the Cities, Naples*, 2009.
2. Knorr, F.; Baselt, D.; Schreckenberg, M.; Mauve, M. Reducing Traffic Jams via VANETs. *IEEE Trans. Veh. Technol.* **2012,** *61* (8), 3490–3498.
3. Hasan, M. M.; Saha, G.; Hoque, A.; Majumder, M. B. Smart Traffic Control System with Application of Image Processing Techniques. In *2014 International Conference on Informatics, Electronics & Vision (ICIEV)*, 2014; pp 1–4.
4. Schrank, D.; Lomax, T.; Eisele, B. *TTI's 2010 Urban Mobility Report*; Texas Transportation Institute, The Texas A&M University System, 2011.
5. Malaterre, G.; Fontaine, H.; Van Elslande, P. Analysis of Driver Needs Using Accident Reports: An a Priori Evaluation of PROMETHEUS Functions. *Arcueil INRETS Res. Rep.*, no. 139, 1992.
6. Ramaswamy, D.; Medanic, J. V.; Perkins, W. R.; Benekohal, R. "Partitioned Strategies for Lane Assignment in Multi-Lane AHS. In *Proc. 1994 33rd IEEE Conf. Decis. Control* **1994,** *3*, 2938–2943.
7. Mimbela, L.-E. Y.; Klein, L. A. *Summary of Vehicle Detection and Surveillance Technologies Used in Intelligent Transportation Systems*, 2007.
8. Liu, H. X.; Sun, J. Length-Based Vehicle Classification Using Event-Based Loop Detector Data. *Transp. Res. C Emerg. Technol.* **2014,** *38*, 156–166.
9. Yang, B.; Lei, Y. Vehicle Detection and Classification for Low-Speed Congested Traffic with Anisotropic Magnetoresistive Sensor. *IEEE Sens. J.* **2014,** *15* (2), 1132–1138.
10. Tian, B.; *et al.* "Hierarchical and Networked Vehicle Surveillance in ITS: A Survey. *IEEE Trans. Intell. Transp. Syst.* **2014,** 16 (2), 557–580.
11. Oudat, E.; Mousa, M.; Claudel, C. "Vehicle Detection and Classification Using Passive Infrared Sensing. In *2015 IEEE 12th International Conference on Mobile Ad Hoc and Sensor Systems*, 2015; pp 443–444.

12. Valavanis, K. P.; Vachtsevanos, G. J. *Handbook of Unmanned Aerial Vehicles*; Springer: Berlin, 2015; Vol. 2077.
13. Celesti, A.; Mulfari, D.; Fazio, M.; Villari, M.; Puliafito, A. "Exploring Container Virtualization in IoT Clouds. In *2016 IEEE International Conference on Smart Computing (SMARTCOMP)*, 2016; pp 1–6.
14. Celesti, A.; Fazio, M.; Giacobbe, M.; Puliafito, A.; Villari, M. Characterizing Cloud Federation in IoT. In *2016 30th International Conference on Advanced Information Networking and Applications Workshops (WAINA)*, 2016; pp 93–98.
15. Guide, D. *Amazon Simple Storage Service*, 2008.
16. Dastjerdi, A. V.; Gupta, H.; Calheiros, R. N.; Ghosh, S. K.; Buyya, R. Fog Computing: Principles, Architectures, and Applications. *Internet of Things*; Elsevier: Amsterdam, 2016; pp 61–75.
17. Yi, S.; Li, C.; Li, Q. A Survey of fog Computing: Concepts, Applications and Issues. In *Proceedings of the 2015 Workshop on Mobile Big Data*, 2015; pp 37–42.
18. Paranjothi, A.; Tanik, U.; Wang, Y.; Khan, M. S. Hybrid-Vehfog: A Robust Approach for Reliable Dissemination of Critical Messages in Connected Vehicles. *Trans. Emerg. Telecommun. Technol.* **2019,** *30* (6), e3595.
19. Tarantilis, C. D.; Kiranoudis, C. T. Using a Spatial Decision Support System for Solving the Vehicle Routing Problem. *Inf. Manage.* **2002,** *39* (5), 359–375.
20. Ng, K. K. H.; Lee, C. K. M.; Zhang, S. Z.; Wu, K.; Ho, W. A Multiple Colonies Artificial Bee Colony Algorithm for a Capacitated Vehicle Routing Problem and Re-routing Strategies under Time-Dependent Traffic Congestion. *Comput. Ind. Eng.* **2017,** *109*, 151–168.

CHAPTER 15

Intelligent Big Data Analytics: A Perspective for Online Education System

MANOJ KUMAR SAHOO[1*], BHABANI SANKAR GOUDA[1], PRIYANKA PRATIHARI[1], ANAND KUMAR SATAPATHY[1], SHYAM SUNDAR PRADHAN[2], and SONALIKA NAYAK[3]

[1]*Department of Computer Science and Engineering, NIST (Autonomous), Berhampur 760010, Odisha, India*

[2]*Department of Information Technology Engineering, NIST (Autonomous), Berhampur 760010, Odisha, India*

[3]*Department of Electronics Communication Engineering, NIST (Autonomous), Berhampur 760010, Odisha, India*

[*]*Corresponding author. E-mail: mksahoo@nist.edu, mksahoo@hotmail.com*

ABSTRACT

Education is the foundation of modern-day society. India being the land of knowledge, our education system is one of the oldest systems. With the change in time, the education system modifies itself for the betterment of society. The current educational system is introduced by the English, during their colonial rules, where the education is imparted to the students in an institute via classroom teaching. Previously to this, the students go to the GURUKUL where the learning was carried out traditionally. The education system introduced by the English is a more organized and modernized one, which spread education, in a better way to the bottom of the society. The education system is consisting of institutes, the teachers, the students, and

Intelligent Technologies for Sensors: Applications, Design, and Optimization for a Smart World.
S. Kannadhasan, R. Nagarajan, & Alagar Karthick (Eds.)

the policymaker, which includes Government and Educationists. A good education policy is a help to build the society and the nation, as the students are the future of the nation. Science and technology play a major role in the betterment of the quality of education to the student. The computer and the internet are an integral part of our educational process. The use of the computer and the internet make it easier to access learning materials across the globe, which helps to increase in efficiency of the teaching profession and the students. The visualization and presentation make it more effective. The internet is also making it possible to access expert teaching professionals for a remote place. Currently, in the COVID-19 pandemic situation, the internet provides an alternative platform for teaching or educating the student. The online education process, which initially limited to the seminar, webinar. In the COVID-19 situation, all the government or private offices, business organizations, and educational institutes are closed, the online process is adapted as an alternative. The online teaching process of education is well-adapted, where all activities are carried out using the internet. To make it more effective, we need to define and use a database. The database will contain all the information so that the persons involving can access and use it for their purpose like a teacher looking for attendance of the student, internal assessment, etc., the student can access the classroom materials and other information similar to the management level personal look information to make smooth running of the process in their institutes. The database is needed to be intelligent by using data analytics, which analyses the data or information from the database, it can provide proper information to the faculty member, the students, and management personal to make better decision and carry out action, which makes more efficient and can make the online education to be robust.

15.1 INTRODUCTION

Education is the foundation of modern day society. By the passing of time education system is changing day by day. The main aim of this new education system is to provide the best and quality education. In every single aspect of technology, digitalization plays an important role and it affects us to the millions. But now the education system has been changed and introduced in the English language which is a more organized and modernized one, which helps in spreading of education in a better manner to the bottom of society. The education system consists of institutes, teachers, students, and policymakers, it may be led by the Government or a particular person. A good education system helps the society and the nation builders, as the students are the future and backbone of the country. To walk hand in hand with the developed technology, we have to achieve the conceptual growth in a deeper manner, which must consider that discerned approach in learning should be essential. The smart classroom is a fully digitalized system that holds the same mission and vision as a traditional classroom. Due to the drastic change in technology, it became easier to lead a life smoothly and comfortably. In the internet world, the online education is possible nowadays. With the help of technology, different learning system has been created, for which several methods are used and the computerized activities are developed to facilitate the learners to access it without any physical presence, which is mainly the computer application. Nowadays, the demand of IT world is increasing and it requires a strong developed technology for manipulation, management of data, and resource acquisition.[1] If we develop a system by utilizing the distributed properties infrastructure design in the Hadoop, integrating with the in-memory processing and the analysis is done by the Spark tool. The Spark implementation is completely different, if we look forward for the distributed the processing with multiple nodes, with each single node having a well-defined for a specific task to perform, the FP-growth algorithm is characterized in parallel for the different cluster nodes or the machines.[4] The development of the distributed system is limited to the resources like computing and storage, and needs to solve with enhancements in performance and processing, needs to be cost-effective one. The recent development in cloud computing and data science, is more suitable for cost-effective robust solution as SaaS model (System and as Service) to this kind of problem. The SaaS model (System and as Service) for the e-learning or the e-education by cloud computing in the educational institution and training organization can be developed, executed, and implemented on the highly performed application at minimum cost. It can be implemented in multidevices, such as PCs, smartphones, laptops, and tablets by developing the framework, programming language, and operating system which enhances the role of a teacher. Then after, the big data technologies will help the operating system to act automatically and adapt the condition according to the user.[5] With the advance methods and techniques, tools, architectures, and technology make the big data to more effective, robust than before, and is a cost-effective and secure system, as could platform data security be quite high and access is easier and faster. This helps to handle huge amount of data with faster speed, processing, and storage. In the big data platform, the data can be used as data science or the technology. The data science domain will consider the data evaluation, data exploitation, and technique acquisition, whereas the data technologies have the responsibility of the packages, software application, tool and the utility of the algorithms and the framework. A recent trend shows its advantages as in many applications, it is based on these platforms and technologies. Selecting useful data for learning is a great challenge, mainly to reach the demands of the amount of data production in online learning platforms. To solve this problem, advanced technology, methods, and techniques are provided through the big data ecosystem as a software library in the machine learning algorithm. The main motive of the project is to develop a smart classroom that can be a perspective for online education, in which the learning quality of the students can be improved. The adaptive learning and collaborative learning are the main aim of the smart classrooms.

15.2 BIG DATA

Big data is described as the fundamental tenet of information systems in the field of research, which provides accurate information to the accurate receiver with exact volume and quality at the right time. The term "big data" works on the area of technological developments based on data storage and data processing, which handles the exponential growth of data volume present in any type of format in gradually decreasing order of periods.[1] Big data forms a platform to provide the opportunity not only for handling the operations but also to use and add the large amount of data coming from various resources, such as social sites, images, videos, and other communication resources.[2] It is a methodology of data analysis which is enabled by advanced technologies in new generation and architecture which have the capacity of capturing the data, storage, and analysis in high velocity.[3] The data source includes the database of e-mail, sensor-generating data, smart devices output.[4] In big data, data are not stored in a structured database but the database adds unstructured data having no standard format.[5] Figure 15.1 shows the flowchart of big data intelligent data analysis.

With the need for the storage and high-speed processing in the big data platform, the cloud is a smarter and cost-effective solution. In the case of smaller and medium type business or organization, it may be bit expensive in the initial phases, but the data and operation security make it to be flexible.

The distributed nature makes it to be more secure, robust, and effective. The SaaS model is well-adapted one in recent times in small and medium-scale organizations.

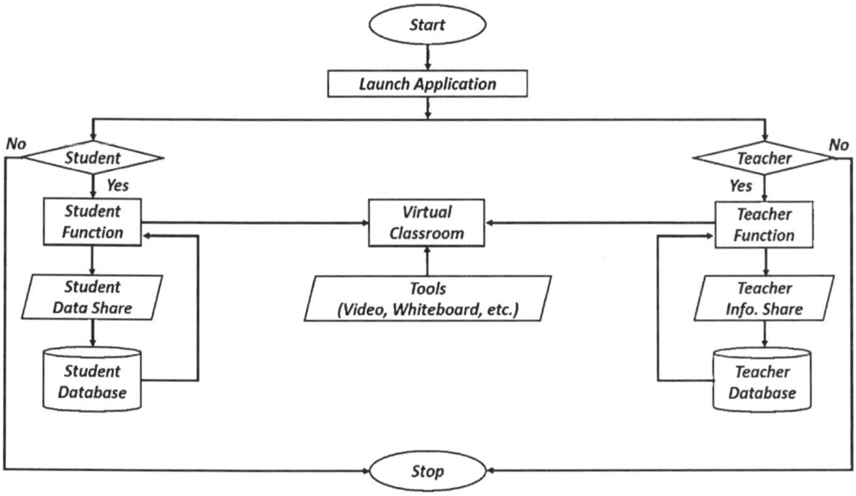

FIGURE 15.1 Flowchart of big data intelligent data analysis.

15.3 TYPES OF BIG DATA

In big data, the information can be used as data, these data can be classified as below:

15.3.1 STRUCTURED DATA FORMAT

Structured data is the basic form, where the normal data are accessed, processed, and stored in a fixed format. The development of such kinds of techniques in computer science helps over a period to achieve success in an advanced format and also it delivered value out of it.[9] Nowadays, we are facing an issue when the size of data growth increases to a huge extent, mainly the size is in the range of multiple zettabytes. From the figure given below, one can easily get the meaning of big data and can also understand the challenges involved in its process and storage.

15.3.1.1 EXAMPLE OF STRUCTURED DATA

An example of structured data using student table:

Student _ID	Student_Name	Gender	Department
18452	Manoj Roy	Male	Computer Science
18232	Pratibha Sahu	Female	Electronics and Communication
18121	Raj Kumar Rao	Male	Electrical
18004	Prakash Sen	Female	Computer Science

15.3.2 UNSTRUCTURED DATA FORMAT

Any raw data having an unknown form or structure is defined as unstructured data. As it requires the huge amount of size, the unstructured data challenges in the process of deriving value from it. A heterogeneous data source contains a combination of images, videos, and text files, etc. Although there is availability of wealth from the organization, we will not be able to derive value as it contains the raw facts or unstructured formats.

15.3.3 SEMI-STRUCTURED DATA FORMAT

The data stored in the semi-structured are of both forms. We can find that semi-structured data is structured in form but it is not in a defined manner, for example, the table of relational DBMS. The example of a semi-structured can be represented in an XML file.

15.3.3.1 EXAMPLE OF SEMI-STRUCTURED DATA FORMAT

<rec><name>ram</name><sex>Male</sex><age>22</rec>
<rec><name>gopal><sex>Male</sex><age>24</rec>
<rec><name>Sita</name><sex>Female</sex><age>25</rec>

15.4 BIG DATA IN CLOUD COMPUTING PLATFORM

The cloud technology initially offers the storage in distributed place across the globe, later, the addition of the computing opens up for software application runs in IaaS, PaaS, and SaaS structure, which are more friendly for the small to medium-sized businesses to use the techniques of big data analytics. Nowadays cloud computing is on high demand as it computes all the resources through network which are often provided by a management or an outside entity.[11] There are huge number of architectures, and developed models are present in cloud computing, these models and architectures are able to be used in designing approaches of other technology. Due the cloud computing, the small to middle scale businesses can afford to meet their big data needs. The term big data has been derived as the data system will not be able to store and analyze the larger data sets.[12] The data are not structured traditionally, so the data sets are larger in size. The data can be accessed from many new sources which include internet accessing sensors, social media, and e-mails.

To store the data in a house as a storage is called the storage attached network. By using storage attached network devices, the several systems are attached with a computer, and the configuration of which is made by storage attached network pods. Several storages attached network pods are attached to each other through the computer used as storage attached network devices. The storage attached network stores all the small to large expensive prospect data. In very low cost, we can get the furnished cloud services with necessary required storage space. MapReduce is a method to analyzing the data using paradigm of programming.[13] In the MapReduce paradigm, a query is made and the data are wrapped according to the key values related to the query, as a result, the data set gets reduced and is answering the query.[12] The huge amount of data is required to get analyzed using the MapReduce paradigm. Then the mapping is done concurrently by each device of storage attached networks, it is the process of parallel mapping. The deployment of MapReduce as the parallel processor is expensive. So, this factor deployment depends upon the functionality, the client, and the cloud service provider, which are described below in a schematic diagram. So, this factor deployment depends upon the functionality, the client, and the cloud service provider, which are described below in a schematic diagram, shown in Figure 15.2.

15.5 BIG DATA ANALYTICS

Big data analytics is all about analyzing large sets of data. It is used in guiding technologies on very diverse and large data sets which includes structured, unstructured, and semi-structured data from different sources, and in different volumes which range from terabytes to zettabytes. It is used on those data sets that cannot be processed and managed by traditional databases with low latency.[12] Big data have the characteristics which include "3Vs," that is, volume, velocity, and variety.[10]

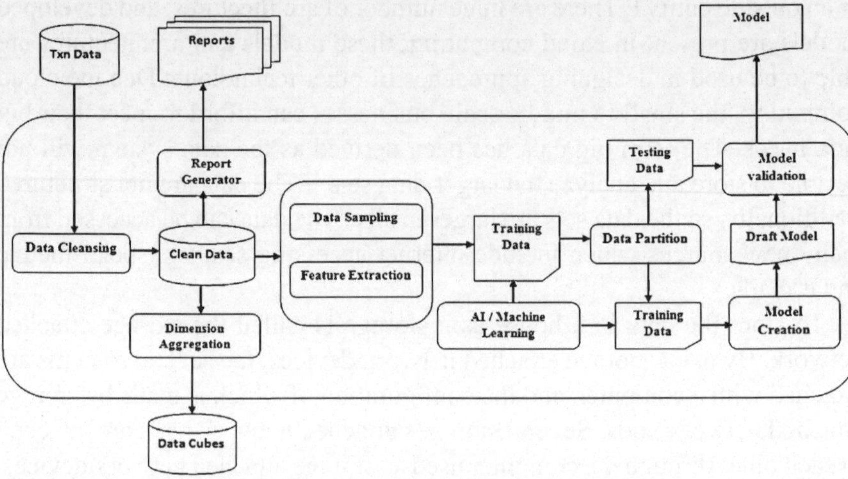

FIGURE 15.2 Intelligent big data education—schematic diagram.

The analysis has given a helping hand to the researchers, analysts, and business personnel to make better and faster decisions from the data which were previously unusable or inaccessible as shown in Figure 15.3. Businesses have the option of using advanced analytical techniques, such as machine learning, mining of data, natural language processing, which allow them to discover previously untapped sources of data. When we talk about the implementation of this analysis on e-learning,[7] then we come across different technical processes.

Several new technologies for storage and processing data have been introduced in the development of big data. Apache Hadoop, Apache Spark, SQL databases, MapReduce, etc. are some of the technologies which have proved to be useful in the development of big data ecosystems.[14]

15.6 TECHNOLOGIES AND COMPONENTS OF BIG DATA

We activated a cluster of nodes to host the technologies that connect them with each other through specific network protocols that help in the transfer of the data during analysis. These are divided into three parts which are shown in Figure 15.4.

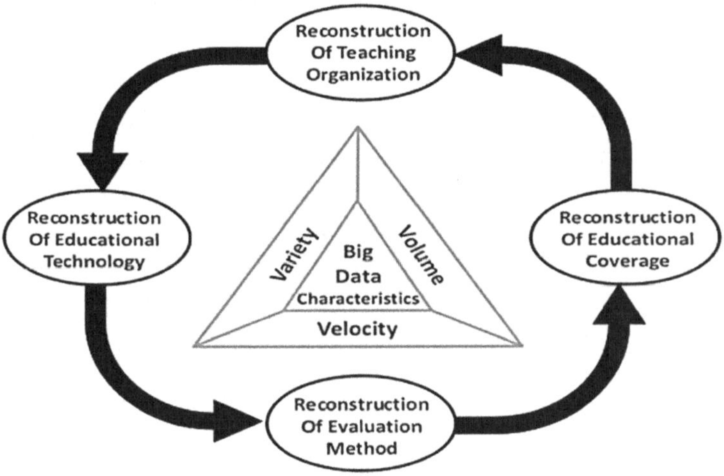

FIGURE 15.3 Big data education—schematic diagram.

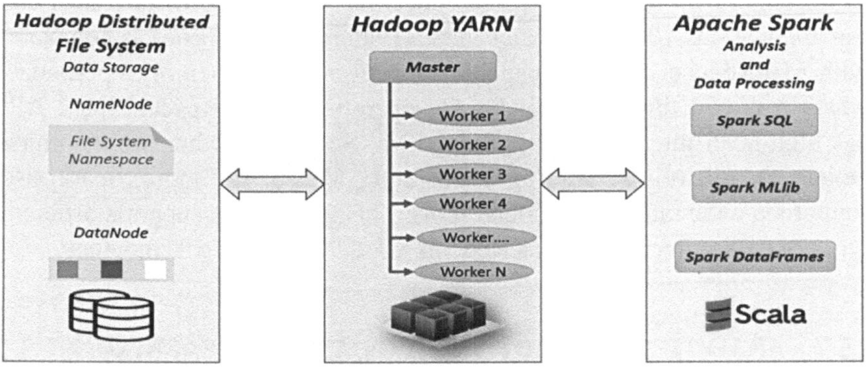

FIGURE 15.4 Big data technologies.

In addition to handling big data processing, Apache Hadoop also manages its storage. Using this framework, large amounts of data can be stored in a distributed manner and is also helpful for parallel processing. The Hadoop framework is used over various e-learning platforms for their projects related to big data.[15] HDFS and the framework are included.[11] Those who need to process large quantities of data can use MapReduce[10] which can process a large amount of data in parallel.

The HDFS works on the storage of data and the file management system. Commodity hardware is specifically designed for deployment with it. HDFS provides continuous access as well as fault tolerance for high-throughput data storage. It has two main components including Name Node, where the Name System caches the names of files or directories and the Data Node caches information about the data residing in distributed nodes and stores, reads, replicates, and deletes a part of the data.[9] When we talk about the processing of data with MapReduce, it processes the data which are in several types (structured, semi-structured, and unstructured) and differ in their structure and size which are huge and consumes a lot of time to be analyzed. It is a programming model that assists in the generation and processing of huge data sets as well as an associated implementation. The affiliations of MapReduce are compiled of a Map () as a function which performs sorting and filtering, Counting the number of students is performed by the Reduce () function. MapReduce divides the data and spreads those data in the Data Nodes.

Additionally, Apache Spark is a powerful framework used for processing large amounts of data in a fast and general way. This framework which made for the network which offers application to have high performance both when it comes to batching as well as to interactive processing.[1] The Apache Spark works similar to the Hadoop MapReduce but the Apache Spark cluster can run applications 100 times faster than a Hadoop cluster faster as compared with MapReduce. Spark is capable of performing in-memory operations, which result in fewer read and write operations. The Apache Spark API has high-level interfaces for Python, Java, Scala, and other programming languages. As well as offering Spark SQL, MLlib, and more, Spark also supports a wide range of high-end tools. This application supports different types of environments, such as Hadoop, Cloud Computing, standalone.

15.7 ANALYTICS USING IOT AND BIG DATA IN E-LEARNING

Big data analytics and the Internet of things are changing the way we learn in every sector, including e-learning. Using this asset intelligence, the professionals in the field of e-learning will be beneficial to enhance their skills. Our goal is to create e-learning courses that are informative, engaging, and engaging for learners.

The platform which is designed to be used for learners and teacher, administrator, and IoT-centered online learning provides the specific devices, management of data analytics part and the tools and software part to support learning. The network infrastructure includes an application programming interfaces, distributed resources, and a high-speed network, in addition to a virtual database. Platforms integrated with the analysis of the big data support software development and Internet of things network environment. This approach to e-learning involves the management of web courses, contexts, information, and knowledgeable processes. There are software and tools to perform data analytics in order to process those information as per the requirement of the teacher or advisor. Due to this methodology, we are able to use that platform in multiple devices. IoT on smart devices is centered on data. Data from the sensors are gathered and communicated by the IoT sensors and the data are integrated for analysis which helps in the process of eE-learning.

15.8 AN ARCHITECTURE FOR DISTRIBUTED SYSTEMS

Learning resources are generally tailored for scholar based on examining their learning activities over time. By analyzing these recommendations, knowledge can be gained (Figure 15.5). This document describes the architecture as it contains of historical information imported from the e-learning platform database of the program. Data collected from the learners through interaction with the online platform must first be loaded into the database.[16] To load, load, query, and store data, the Spark Framework uses the Spark SQL library, which is part of the Hadoop cluster and runs on the Yarn Resource Manager. The MLlib library in Apache Spark has been a key development in supporting parallel FP-growth in Scala due to the use of this library.[13] In addition to the parallel algorithm, the results of FP-Growth have also been used to examine the data. Finally, the results can be presented to the user by a recommendation engine, thus guiding them and suggesting school materials they would find useful. Our system recommends courses that may suit the learner's cognitive level and interests. Therefore, the learner may search through the courses that may suit his or her needs.

The system also uses Zabbix, an open-source monitoring tool. Using it, you can monitor massive environments like clusters and grids, and take advantage of efficiency analysis, such as usages of CPU, memory consumption, and data storage in each node.[13] Further, the software allows the visualization of information about using bandwidth, for example, is a way to measure network traffic and traffic volume.

15.9 BIG DATA COMBINED WITH ONLINE LEARNING TO CREATE A HYBRID LEARNING SOLUTION

A recent study found that cloud computing combined with big data technology has become fundamental component of any successful technological system, and that includes e-learning platforms. In order to take advantage of the significant amount of data that are generated by a system of this type, the system must be integrated, because installing the software and hardware can hinder processing. The challenge of finding useful information from learning data has grown significantly throughout the years, especially those data which are being generated from the e-learning paths. In response to this problem, the big data ecosystem facilitates the application of different methodologies and techniques via computer codes which are specially designed in the form of frameworks and application programming interfaces which are easily accessible and extremely powerful. Technologies of this kind allow e-learning professionals to continue enriching and improving their implementation of plans through the preparation and analysis of large amounts of data in a distributed manner.[17]

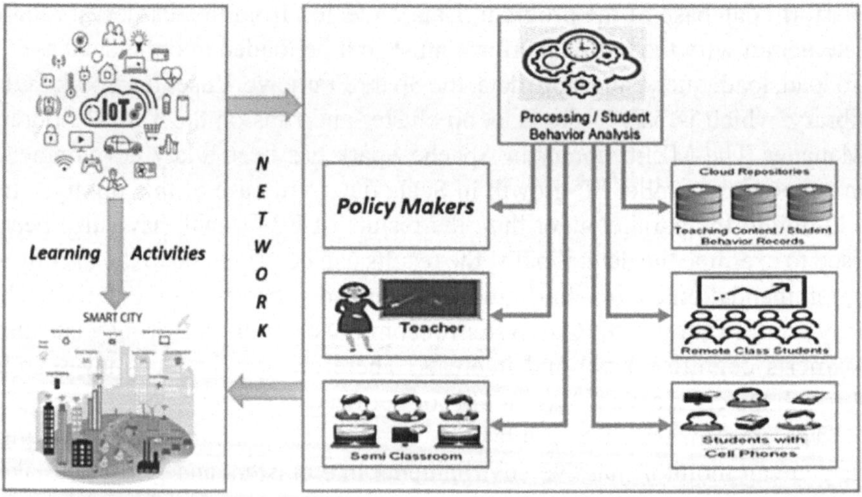

FIGURE 15.5 Block diagram of e-learning using big data.

Infrastructure is the top-most layer of the presented solution, which is the lowest level. A cloud infrastructure is made by different technical resources. Cloud services deliver these resources. By providing these technologies, e-learning systems can be operated in an environment with low technology costs and a favorable learning environment. This layer is scalable in terms of its resources. With this mechanism, users are much more flexible than with traditional hosting technology that is based on limited server resources. This application provides a variety of benefits, including efficient, adaptable, and error-tolerant infrastructure.

A distributed computing platform consists of storage technologies distributed across multiple servers, large data processing on machines with huge parallelism, superior analytics, and visualization. Various big data technologies fall under this category, which can be divided into the following:

File system that is distributed among the nodes within a cluster that provides redundancy and high availability by storing data in a copied or in the same manner. As a distributed data management solution, HDFS continues to be the most popular open-source solution (Figure 15.6).

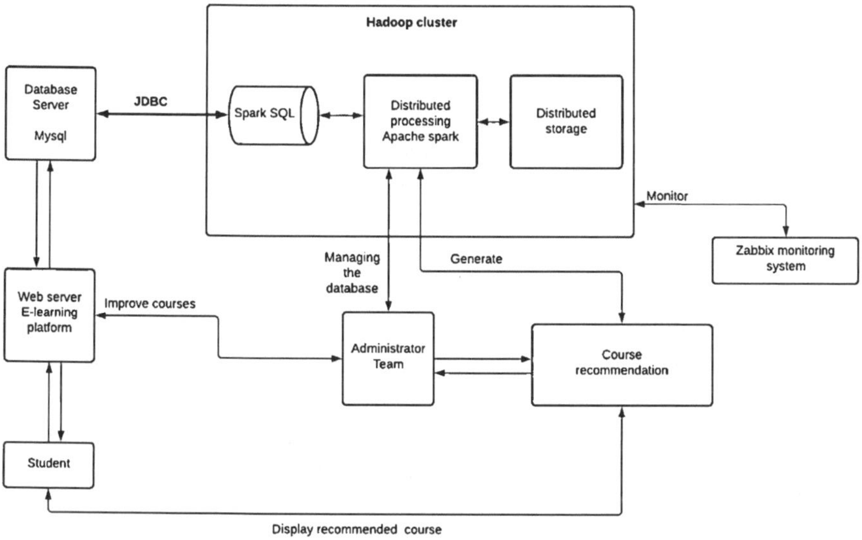

FIGURE 15.6 Distributive system.

A NoSQL database management system represents the next generation of databases.[19] The flexibility, types, and quantities of data they provide allow them to provide capabilities beyond relational databases, which are limited by RDBMS. CouchDB Cassandra including HBase (column-oriented) are examples of distributed databases that do not conform to rigid schema rules, unlike relational databases.

A distributed processing concept with the combination of predictive analysis infrastructure is used to handle large data sets across a cluster of machines. The Apache founded a concept Hadoop which is arguably the most famous example of distributed system which is an open-source application.[20] They also provide mathematical methods for analyzing giant databases via machine learning predictive models by applying mathematical methods to computers. As well as providing high-performance tools for designing and implementing large-scale predictive investigation applications, Apache Spark also provides memory-based powerful designs and solutions for large-scale predictive analysis.

E-learning systems are the third level.[21] This layer represents the e-learning system as a whole, encompassing tools, such as learning management systems, system to manage the content, distance learning system. These data will prove to be extremely helpful in creating resources with personalized learning by adapting educational content to meet the individuals' needs, as well as creating a learning environment that meets the standards of each learner. For the e-learning system to implement this modified mechanism, it must utilize the resources that are useful and present in the lower layer which needs to apply machine learning algorithms parallel to the learning data in order to exploit advanced predictive models.

A general definition of online learning refers to a platform that consists of hardware, software, and users. In addition to the memory, bandwidth of the internet, and operating system provided by the services of the cloud computing, there is also cloud hardware. As far as software is concerned, it is an e-learning system.[20] Information is communicated, stored, and processed by users through the system. There are three main users which include the system administrator or the policy maker, the advisor or teacher, the students or learners.[15] The cluster may also be expanded to include one or more individuals who are responsible for the installation and maintenance of the cluster's architecture which is distributed broadly, and who also develop and deploy models which are analytical in nature and implement techniques related to data mining.

15.10 CONCLUSION

Education is the foundation of modern society. Due to India's heritage of knowledge, its educational system is one of the oldest in the world and continues to evolve to better serve society as time passes. In colonial times, the English instituted the current education system, where students are instructed in classrooms in a school system. Students are a nation's future, so a sound education policy is essential. Science and technology are vital components of an education designed to benefit students. Online learning is a devoted, computing-based environment dedicated with new communication and information technologies, such as big data, are being introduced. Massive parallel processing, distributed storage, and predictive analysis are just a few of the many tools and systems available when it comes to big data. Students generate huge amounts of data that can be processed and analyzed using the technologies described here. Distance learning platforms can be greatly improved by using frameworks that offer greater quality and accessibility. Consequently, Models we use enable each student to get the most out of online learning, as a result, we believe our model will significantly enhance the online learning field. As a result, we can make better decisions, improve learning methods, and offer a wide range of programs. Using technologies based on big data in a system of online learning, in this chapter, we want to develop an adaptive learning solution that can match individualized instruction to the preferences and needs of each student. A machine learning algorithm is used to implement the proposed course recommendation engine, specifically the association rules technique, which identifies interesting associations between historical enrolment data and course recommendations. Machine learning techniques are used by the class recommendation engine, specifically the association rules method to reveal the relationships between student enrollment data and class recommendation results. Besides process decentralization and analyzing of data, it demonstrates that our system is effective as it relates to recommendation of courses that are relevant and of high quality and performance within predetermined during frames. Implementing our system utilizes the distributed infrastructure of Hadoop as well as Apache Spark, which performs a rapid process that operates in-memory and analyzes data in depth. It has been decided to adapt a completely different method of implementing the Spark application, namely, to distribute processing and data processing among multiple nodes, with each instance performing different tasks, to run the running FP-growth algorithms on multiple clusters of computers simultaneously. As a result of distributing the storage and computing industries system, the insufficient resource problem is solved, the performance is improved and processing costs are reduced. In order to make online education to function efficiently, it has to provide correct information for faculty, students, and managers to make proper decisions and actions accordingly.

KEYWORDS

- **big data**
- **data analytics**
- **IoT**
- **intelligent education system**
- **online education**
- **remote education system**
- **innovation**

REFERENCES

1. Bouslama, F.; Kalota, F. Creating Smart Classrooms to Benefit from Innovative Technologies and Learning Space Design. In *2013 International Conference on Current Trends in Information Technology (CTIT)*, 2013; pp 102–106.
2. Guri-Rosenblit, S.; Gros, B. E-learning: Confusing Terminology, Research Gaps and Inherent Challenges. *Int. J. E-Learn. Dist. Educ.* **2011,** 25, (1).
3. Udupi, P. K.; Malali, P.; Noronha, H. Big Data Integration for Transition from e-Learning to Smart Learning Framework. In *2016 3rd MEC International Conference on Big Data and Smart City (ICBDSC)*, 2016; pp 1–4.
4. Melton, J. ISO/ANSI Working Draft: Database Language SQL (SQL3). *ISO/IEC SQL Revis*; New York Am. Natl. Stand. Inst., 1992.
5. Schermann, M.; et al. Big data. *Bus. Inf. Syst. Eng.* **2014,** *6* (5), 261–266.
6. Armbrust, M.; et al. I. Stoica in Drugi. A View of Cloud Computing. *Commun. ACM* **2010,** *53*, 50–58.
7. Carraro, G.; Chong, F. *Software as a Service: An Enterprise Perspective*, 2006.
8. Kim, S.-T. *Dr. Seang-Tae Kim, Leading Korea's e-Government Advances*. *J. E-Govern.* **2011,** *34* (165), 165.
9. Pinnell, C.; Paulmani, G.; Kumar, V. Curricular and Learning Analytics: A Big Data Perspective. In *Big Data and Learning Analytics in Higher Education*; Springer: Berlin, 2017; pp 125–145.
10. Kusuma, S.; Viswanath, D. K. IOT and Big Data Analytics in e-Learning: A Technological Perspective and Review. *Int. J. Eng. Technol.* **2018,** *7* (18), 164–167.
11. Aggarwal, A. Identification of Quality Parameters Associated with 3V's of Big Data. In *2016 3rd International Conference on Computing for Sustainable Global Development (INDIACom)*, 2016; pp 1135–1140.
12. Dahdouh, K.; Dakkak, A.; Oughdir, L.; Messaoudi, F. Big Data for Online Learning Systems. *Educ. Inf. Technol.* **2018,** *23* (6), 2783–2800.
13. Dean, J.; Ghemawat, S. MapReduce: Simplified Data Processing on Large Clusters. *Commun. ACM* **2008,** *51* (1), 107–113.

14. Dorsey, A. D.; Clements, K.; Garrie, R. L.; Houser, S. H.; Berner, E. S. Bridging the Gap. *Appl. Clin. Inform.* **2015,** *6* (02), 211–223.
15. Zikopoulos, P.; Eaton, C. *Understanding Big Data: Analytics for Enterprise Class Hadoop and Streaming Data*; McGraw-Hill Osborne Media: New York, NY, 2011.
16. Mohanur Jagadeesan, P. R. *A Framework Design to Improve and Evaluate the Performance of Security Operation Center (SOC)*; National College of Ireland: Dublin, 2020.
17. Ibriz, A.; Benslimane, M.; Ouazzani, K. Didactics in Online Learning Technical Courses: A Case Study Based on Activity Theory. *Int. J. Comput. Sci. Inf. Technol.* **2016,** *7*, 849–854.
18. Dahdouh, K.; Dakkak, A.; Dakkak, A.; Oughdir, L.; Abdelali, I. Improving Online Education Using Big Data Technologies. *The Role of Technology in Education*; 2020. doi:10.5772/intechopen.88463.
19. Sikal, R.; Sbai, H.; Kjiri, L. Configurable Process Mining: Variability Discovery Approach. In *2018 IEEE 5th International Congress on Information Science and Technology (CiSt)*, 2018; pp 137–142.
20. Sarker, D. K.; Hossain, N. I.; Jamil, I. A. Design and Implementation of Smart Attendance Management System Using Multiple Step Authentication. In *2016 International Workshop on Computational Intelligence (IWCI)*, 2016; pp 91–95.
21. Benslimane, M.; Ouazzani, K.; Tmimi, M.; Berrada, M. Proposal of an Approach of Online Course Design and Implementation: A Case Study of an Algorithmic Course. *Int. J. Comput. Technol. Appl.* **2016,** *7*, 7.
22. Manyika, J.; et al. *Big Data: The Next Frontier for Innovation, Competition, and Productivity*; McKinsey Global Institute: Brussels, Belgium, 2011.

CHAPTER 16

Automatic Fault Detection, Locating, and Monitoring in Distribution Lines Using LabVIEW

KUNJABIHARI SWAIN[1], SANDIPAN MALLIK[2*], KANISHK KASHYAP[1], SUMANJIT PATTANAYAK[1], and ARPITA BEBARTA[2]

[1]*Department of Electrical and Electronics Engineering, National Institute of Science and Technology, NIST (Autonomous), Berhampur 761008, Odisha, India*

[2]*Department of Electronics and Communication Engineering, National Institute of Science and Technology, NIST (Autonomous), Berhampur 761008, Odisha, India*

**Corresponding author. E-mail: sandi.iitkgp@gmail.com*

ABSTRACT

With the increasing demand for electricity for various applications and purposes in any country, it is required to coherently develop and equip the already existing infrastructure with advanced power protection, management, and handling systems. Distribution lines are constantly exposed to lots of disturbance by nature due to natural disasters, such as cyclones, storms, or heavy rainfall. Any minor fault in the line or short circuit may lead to severe damages in distribution line, ultimately leading to expensive equipment failures in feeder stations. By using an automatic system, it will be possible to computerize the process entirely and precisely locate the nature along with the intensity of the fault and its exact location in a particular area and provide that information to the serviceman or the person responsible for the

Intelligent Technologies for Sensors: Applications, Design, and Optimization for a Smart World.
S. Kannadhasan, R. Nagarajan, & Alagar Karthick (Eds.)

management of distribution lines of that area by sending them an automatic system-generated SMS. This system consists of LabVIEW, a virtual software that uses programming in a local power distribution unit. It is interfaced along with an array of fault detectors in the distribution line, which detects different types of interruptions in line, such as short circuit, drop in voltage, excessive voltage, heating due to higher load are sent to local power distribution units using a wireless communication device like a GSM module, which is connected with the software to every local monitoring stations, which keeps monitoring the parameters and scans for any abnormalities or fluctuations. If any such failure occurs due to various reason within a transmission network or any specific substation, the autonomous fault detection system quickly generates and transmits an SMS to the repairing team and prompts the team to cut off the power of a particular area. This will terminate any possibilities of electrocution, and many other fatal accidents can be prevented. Also, previously knowing the nature of the fault will warn servicemen about the severity of the situation, for which they can be prepared beforehand. This system also helps the consumers be comfortable. There will be no requirement of approaching an electrician to complain to him about the failure, and the delay in these repairing works will be ideally minimized.

16.1 INTRODUCTION

Today, economic and technological growth is increasing rapidly and requires a higher demand for efficiency and energy in the field of electricity. This rapid growth has generated a number of opportunities for developers and researchers to design a set of new products and facilities that can meet the new reality and upcoming needs of customers.[1] Distribution systems represent the final connection between the customers and utilities. Practically, the distribution network is operated by a simple design having less cost along with very well fortified schemes and the least short current. To provide continuous supply, the normal working of all the component power stations and connected loads are required. Thus, distribution networks possess lower reliability that if a breakdown of any component may result in disruption for the loads continuously downstream in a faulty zone.[2] Electrical disturbances may result in numerous economic impacts, such as loss in production, apparatus damage, cost to restart, and raw material wastage. Also, disturbance in electricity can also influence uncomfortable temperature at home and workplace and is the risk to health.[3] Not only that, fault in distribution systems can affect the quality of power in terms of propagation of disturbance and service continuity.[4]

Hence, they are also responsible for industrial production disturbance along with economic and information loss, among others. Statistics shows that most of the failures in distribution lines are due to accidents, component degrading, and weather conditions. It has been recorded that on the medium voltage of 11-kV network, the faults resulted in 74% of customer minutes lost, on an average of 20 min over per user, annually in the United States. Therefore, distribution arrangement results in the maximum unattainability of power supply by the users.[5] Developing a self-healing capability smart grid after keeping the past painful lessons in mind caused due to earlier large-scale power blackouts is of greater urgency. Self-healing is nothing but the ability of distribution arrangement to tackle when it faces any fault, and it can automatically and intelligently perform the suitable actions for securing itself in the best steady-state so that it can perform the basic operation without disrupting and constraints.[6] As a result, organizations are looking forward to a solution that automatically turns the switch to protect a network by enhancing customer average interruption duration index (CAIDI), system average interruption duration index (SAIDI), and system average interruption frequency index (SAIFI) they all collectively reduce restoration cost, decrease restoration times and boost organization reputations.[7] Reliable and continuous supply of energy with minimal outages is the main objective of the working of an electrical power system. An efficient fault location (FL) immediately and accurately helps to reduce the needed time to catch the problem and repair the arrangement and also speed up the power supply reestablishment over the network restoration.[8] When there is a cut in a power line, that time, the sensor will sense out the power cut and notify the signal to the Wi-Fi module about cut off in that power line. Then the Wi-Fi will send those data to the Amazon Web Services (AWS) cloud for further notification to the nearest electric board regarding that power line cut. This system can be useful to both rural and urban areas when the power line is damaged. Many researchers, after implementing the concept of power distribution by smart grid and such arrangement also target a control function, such as protection, self-healing, and DG control. So, this framework would be able to coordinate the main subject and constraints at multiple stages and can respond in a flexible way to the challenging conditions.

16.1.1 ORGANIZATION OF THE CHAPTER

The conceptual framework of this system is completely derived and designed by analyzing the problems and solutions dealt with by transmission lines due to lack of smart automation; issues and solutions are described in Sections 16.2 and 16.3, respectively. Section 16.4 covers the device's working principle. Section 16.5 explains the methodology of this technique, which also explains the role of involved hardware in the device's circuitry, along with a scope of installment of the device on the existing power transmission framework. Section 16.6 discusses the result of the prototype, which is a snapshot of the system-generated message. Section 16.7 concludes.

16.2 PROBLEMS DUE TO LACK OF SUCH SYSTEMS

Electrical machines, networks, and types of equipment are always subjected to different types of faults during their operational period. A fault is basically an unstable and unfavorable condition for an electrical system that includes the electrical failure of equipment, such as generators, transformers, and other rotating machines. It changes the characteristic value like impedance of the machines from their existing values to different values. The fault involves conducting path failure and insulating failure in the system. As a result, short circuits and open circuits of conductors happen. There are many reasons for failure in a power system. Some of the possible reasons of such faults and failures are the following:

- *Weather conditions*: The natural calamities, such as cyclone, flood, and other natural effects, such as a thunderstorm, torrential rain, salt depositions over transmission lines, ice accumulation on the transmission line can cause electrical faults, which can highly interrupt the power supply and damage the electric installation.
- *Human errors*: Human errors like a selection of improper rates of devices and equipment and electrical conductors after any maintenance, energizing the circuit while under supervision have also caused electrical faults.
- *Equipment failures*: Due to insulation failures, aging, malfunctioning of winding or cables, different electrical pieces of equipment, such as a transformer, motor, generator, switching devices cause short circuit fault.
- *Smoke or fires*: Due to smoke particles, occasionally interfering with the transmission lines can result in the air being ionized, resulting in a spike between the transmission lines, conductors, and insulators. It will cause the reduction of the insulating capacity of the insulator due to high voltages.

- *Damage of transmission line*: The faults are also caused due to falling of trees on the transmission lines, perching of birds on the lines, accident of vehicle with poles or towers of transmission line, etc.[13]

Using an autonomous fault detecting and warning system, one can automate and precisely observe the nature and intensity of the fault and the exact location in a particular area and can provide that information to the serviceman or the person responsible for that area by sending them an SMS. Also, many other failures happened due to the lack of this kind of intelligent locating and monitoring system. Some of the main failures are described below.

1. **High maintenance time:** Without this kind of intelligent system, the serviceman or the person responsible for that area cannot find the exact position or location where the fault occurred. So, it takes much more time to find the fault and solve it. For this type of problem, the time required for maintenance is very high. Sometimes, people in some undeveloped rural areas cannot get electricity for many days due to this problem.
2. **Danger to operating person:** As the operating person is unaware of the exact position or location where the fault occurred, it can also cause shocks to individuals. Depending upon the current and voltage at fault location, the severity of the surprises is predicted, and the shocks may lead to that person's death.
3. **Overcurrent flow:** Due to the occurrence of a fault, a considerable current finds a significantly less-resistive path. As a result, a huge amount of current is drawn out from the supply lines, causing damage to the insulation and other different components and the equipment.
4. **Loss of equipment:** Short circuit fault may create a heavy current, resulting in severe damage to components. Being brunt ultimately results in the improper working of many devices and equipment. The heating of rotating machines is due to the unbalancing of supply current and voltages caused by a short circuit.
5. **Electric fires:** In the day of today's life, it can be observed in the news about the firing of builds and large shopping complexes. It is that because of short circuit that causes plasma flash, and sparks ionize the air present in between conductors, thus leading to an electric fire. There is a chance of spreading fire to the other part of the system.
6. **Disturbs interconnected active circuits:** Faults that occur in the distribution line disrupt the area where they occur and affect the entire dynamic network consisting of many interlinked circuits connected to the faulty lines.

7. **High maintenance cost:** As the fault locations are not found quickly, the equipment and components which are related or connected to the faulty network can be damaged due to overheating, and the interconnected circuit may be disturbed. To solve the problem, the maintenance cost is very high, and sometimes it requires the complete replacement of the equipment, which leads to high maintenance costs.
8. **Fatal accident:** At the location where the fault occurs, the local people are unaware of the fault. If some person came in contact with the fault, it may cause a severe accident and lead to the death of that person. Sometimes, it is noticed that different birds are found dead on the distribution lines. This is due to such kinds of faults. So nowadays, another intelligent technology is developed for automatic fault detection, location, and monitoring of the distribution lines to detect the fault as soon as possible and solve it.[14]

16.3 ADVANTAGES OF SUCH SYSTEMS

In this era of technological advancement, where everything has standard features like easier to use, faster and more effective communication, and less wastage. This modern-day innovative technology has many scopes of improvement and can evolve into new products using some creative approaches. Due to the rapid global increase in energy requirement, the transmission system needs to be improved and updated to meet the electricity demand for growing populations, thus demanding increased industrial output. And to ensure an uninterrupted and efficient power supply, continuous maintenance and flawless operations of transmission lines are very crucial. But there is no significant advancement in our transmission lines which are one of the most necessary needs of the hour; we keep on using the traditional method of transmission and distribution lines with a least or no modification. As a result, we come across a lot of fatal accidents reported in the newspaper and over social sites about the death due to electrocution and faults or errors in these transmissions lines. Using automatic fault detection, by a detecting and warning system, it is possible to completely automate and precisely reveal the nature and intensity of the fault and its exact location in a particular area to the technical crew and hence help save transformers from damage and disasters and will also replace the traditional method of finding the faulty network, equipment and the time elapsed in precisely locating a fault will be significantly reduced.[15] There are numerous advantages of using such kinds of intelligent locating and monitoring systems. Some of the main benefits are listed below.

1. GPS module present in the transmission poles provides information about the exact fault point using communication devices like GSM modules, which are connected by software to every local monitoring unit. This keeps monitoring the fault by cross-checking with its already saved (voltages and currents) values every instant, and if any area happens to be under a faulty condition, the system automatically generates the SMS for the electrician or any distribution station and also cuts off the power supply of a particular area. This will result in a significantly less response time for the technical team to identify the faults and thus can prevent transformers from any severe damage, and in the case of disasters, chances of electrocution and many other fatal accidents are also less.
2. Automatic fault detection, locating, and monitoring system is capable of locating symmetric and asymmetric faults—as minimum sensors are mounted in the transmission line, these are very easy to install, and these can be used over the existing transmission poles with a bit of modification—as a result, there will be a minimal installation cost.
3. In the power distribution line, most of the current and voltage distortion is caused by fault and unethical human interventions. Faults that occur in a transmission line can cause disturbance in the electric supply and will disturb the entire energized, interconnected circuits to the faulty network. As a result, we suffer a power cut very frequently, and once it is encountered, it usually takes a lot of time to get repaired, whereas by using an automatic fault detection, warning, and monitoring system, this time can be minimized ideally.
4. As this system gives the exact location of the fault to the technical crew or the service station immediately by providing the map coordinates through SMS, which is automatically generated by the system, the types of equipment or the components related to the fault can be repaired as soon as possible. The features can be prevented from getting damaged due to overheating and burning, and the interconnected circuit of the transmission lines will be saved.
5. The maintenance cost is significantly less because the servicemen can shut down the power supply as soon as they see a fluctuation in the voltages and current values and can arrive at the wrong location as more quickly as they get the message of fault or error in the transmission line from the intelligent fault detection, locating, and monitoring system.

6. This system also helps the consumers be comfortable. There will be no requirement of contacting the service team to inform about the faults. The delay in these repairing works will be ideally minimized.
7. Using this system, we can detect and reduce the chances of electrocution and many other fatal accidents causing problems, such as falling of trees on the transmission lines, perching of birds on the lines, accident of vehicles with poles or towers of the transmission line.
8. The automatic fault detection, locating, and monitoring system has a simple architecture. It uses a GSM module, GPS module, and Arduino UNO that are very fast and easy to run using the LabVIEW software installed in the substations. This device provides high accuracy while testing the data.[16]

16.4 WORKING PRINCIPLE

LabVIEW software is used to distinguish and display the different values: voltage, current. It also gives us information and location about the fault, which is transferred to the GSM module and sends the message to the repairing team responsible for that area. The fault detection device consists of two central units that are (1) transmitting unit and (2) receiving unit.

16.4.1 TRANSMITTING UNIT

Transmitting unit is installed over various locations where it is connected via GSM module to the receiving unit located at the substation using GSM module which facilitates wireless communication. This unit is linked with a driver circuit which is further linked with circuit breakers for isolating faulty networks from the healthy area. It is mainly divided into two sections: potential transformer and current transformer for measuring voltage and current values, both connected and powered by 230 V AC power supply.

16.4.1.1 VOLTAGE MEASURING UNIT

A 230/(0–6)-V stepped down potential transformer is used for measuring the supply voltage. By scaling down it in a range of 0–6 V, a microcontroller can easily filter it for faults by observing the net fluctuations in the given field.

16.4.1.2 CURRENT MEASURING UNIT

Here, a current transformer (CT) 5 A is used to step down the current, which is then converted into voltage with the help of a shunt resistor. Later, the converted volt is then rectified by a precision rectifier. This circuit is designed to keep a watch on the supply current.[9]

16.4.1.3 ARDUINO

The primary function of the Arduino microcontroller in the transmitting unit is to keep checking the voltage from the step-down transformer continuously and current from the current measuring unit, then to calculate the actual line voltage by using a predefined formula, then to send the data, which include voltage present and the location coordinates, to the GSM module, which will then transmit the data to the nearest/defined/local substation where receiving unit will be analyzing the data. If the fault is detected, using a GSM module, an alert message will be sent to defined/set phone number which will be containing information about fault type and the location coordinates of the fault.

16.4.1.4 GPS MODULE

The GPS module used here is the NEO-6M having an external antenna to enhance the performance. This module will communicate to nearest power station via a GSM module integrated with it, GPS module will be mainly used in pin pointing the exact location of the faulty network in terms of latitude and longitude coordinates.

16.4.2 SUBSTATION OR RECEIVING UNIT

The substation unit will consist of three main GSM modules: They are essentially identical to an ordinary mobile phone, including the need for a SIM to identify themselves to the network. A sim900A Quad-Band GSM module, which works on commands written in LabVIEW blocks.[10]

16.4.2.1 UART-TTL CONVERTER

This tool is excellent for embedded systems, and it requires a serial connection with the computer. This board can be attached simply to any

USB bus, and it will appear as a standard COM port. In any event of a short circuit, this can actively guard the computer as it transmits single bit at a time and also with a specific data rate, so, in this serial communication method, the (TTL) level always remains between 0 V and Vcc which is mostly 5 or 3.3 V.[11]

16.4.2.2 LABVIEW

In receiving unit, LabVIEW software is the base of all the operations which are happening in the unit when the transmitted data are being continuously received and analyzed for any transmission fault. So, this analyzing part of data is done by this software where it checks for different voltage and current values and compares it with the previously fed data (which were earlier recorded under normal parameters) which is already saved, and if in any case, the LabVIEW code detects any abnormalities indicating some kind of fault, then an alert message is transmitted to the specific phone number depending upon the location coordinates of the fault, the number is filtered from a set of saved numbers which are previously fed in the system along with their respective coordinates. And then, if recently received data show any fault, LabVIEW program proceeds to interrupt the supply of that specific phase under fault. Thus, the nature and intensity of faults along with their location can be easily located and tackled. This proposed prototype of a fault system is developed only for single-phase use to detect all possible faults relating to single-phase like Line to Ground Faults. Still, such systems can be further installed in a single device, which can be used for a line-to-line type fault analysis for all three phases.

16.5 METHODOLOGY

The diagram of a local distribution system is represented in Figure 16.1 along with the device part, which, when implemented with the android app, including all the assets attached, gives an end-to-end user interface, making it easier for working with the control system. Also, the development of the app, including its design, look and feel, and user interaction, is marked upon with it. The device flow and implemented prototype are described later, giving us the leading solution for our problem statement.

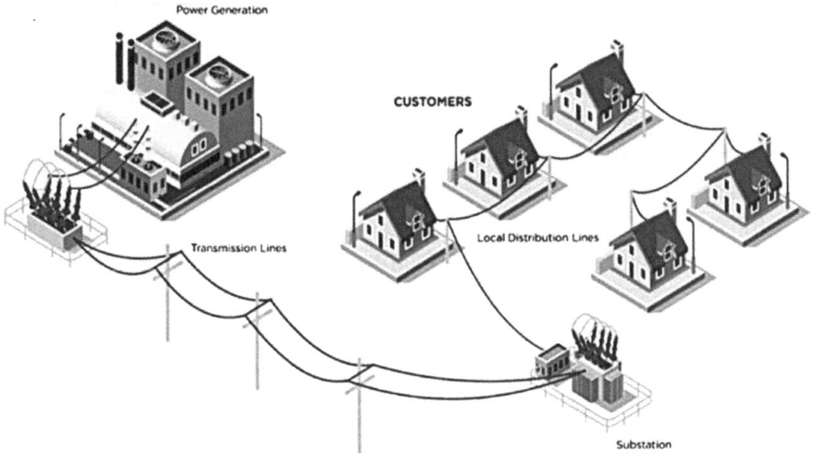

FIGURE 16.1 Local distribution systems.

16.5.1 LAYOUT MECHANISM/PRINCIPLE

The significant steps in this layout include

- We are taking continuous voltage and current readings from the distribution lines.
- Comparing these values with the preset values in the Arduino to check for faults.
- When a fault is identified, the transmitting unit will send the location and values of voltage and current to the substation or receiving unit indicating the nature of fault then.
- The fault type will be determined using LabVIEW, and the fault parameters will be directly sent to the station/assigned person. At standard conditions, the transmitting unit measures the voltage and current values at every instant. The microcontroller circuit compares the predefined value from the present value. If there is any abnormality among current and voltage values, then the measured V–I value and the map coordinates will be sent to the local transmission station. Simultaneously, the faulty network is quickly isolated from the unaffected section by circuit breakers and relays controlled by the microcontroller circuit when any abnormality happens in any network or any specific station. The transmitting unit transmits the data to the

nearest station, the data will have voltage, current, and the exact location code/coordinates of the fault. These data consist of the voltage and current values with the latitude and longitude location of that place. The data are then fed to LabVIEW software, where the data get distinguished into different values, and then received voltage, current values are displayed. Also, it gives geographical information about the location of faults and the type of the faults. It shows the popup menu, it notifies the type of faults and where they occurred. After informing the faults, it automatically generates the text message, which includes location of the fault, fault type, voltage value, and current values in LabVIEW, and it is transferred to the GSM module which reverts the data to the service person of that area. Comparing with other protective devices, this smart system segregates the faulted network from the unaffected section. Also, it sends data to assigned person, thus there will be no manual involvement to notify about the fault. This can save lives from many accidents like electrocution and other fire hazards.

16.5.2 HARDWARE/DEVICE

The made device for agreeing to solve our problem statements mainly consist of the two main units that are transmitting unit (for sending alert messages and V–I parameters to the substation) and receiving unit (for receiving the transmitted data from faulty area/network). One unit will be used to receive the transmission signal and other unit which is receiving signal will flash an alert in the app and if needed will activate some specific circuit breakers.

16.5.2.1 DESIGNED PROTOTYPE

Transmitting units will be attached to preexisting power distribution installments, and will be connected with local substation via a GSM module. The transmitting unit of the device is connected with the driver circuit, which is then combined with the circuit breakers for separating/isolating the failed network from a healthy network. In transmitting unit, potential and current transformer are used for calculating current and voltage magnitudes and a GPS module is used for determining the coordinates. The transmitting unit is considered further and associated with many devices, such as voltage measuring unit, current measuring unit, Arduino (UNO), GPS module. Now, with the help of these, the whole transmitting unit is incorporated to perform the actions, each having its specialized functions to perform upon.

Figure 16.2 represents the block diagram and the schematic flow for the transmitting unit where the potential transformer and precision rectifier combine to form the voltage measuring unit. Then, the current transformer, shunt resistor, and precision rectifier combine to form the current measuring Unit. Now, coming to the substation or receiving unit, it mainly consists of GSM module, UART-TTL converter, computer with LabVIEW enabled.

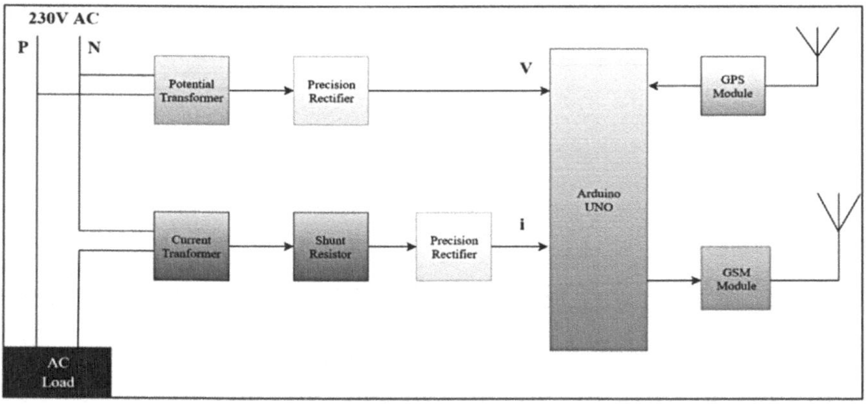

FIGURE 16.2 Block diagram of transmitting unit.

Figure 16.3 represents the block diagram and the schematic flow for the transmitting unit where the processing and receiving part until the phone's alert message will be performed. This module can be easily interfaced with devices that do not have standard serial ports. The primary identification of types of faults and location of faults is made by LabVIEW software. It consists of geographical coordinates, current, voltage, and it also stores the complete database of concerned mobile numbers of the electricians or repairing team for respective areas. As of now, only a single-phase system is used, so it deals only with the line to ground faults, which can be further enhanced and developed for a three-phase-based transmission system, then also considering line-to-line faults.

16.5.2.2 INSTALLMENT ON THE EXISTING FRAMEWORK

The device, when installed with a new framework, will initially be facing no interventions. Still, when an existing set of transformers, supply network, and lines already exist, the device will be incorporated/fixed on the transformers to understand its in-built structure design, power distribution network, pole connections, and specific location intervals to minimize the cost. Now, for most households, apartments will be a very cheap/cost-effective method to install with a supervised support team. Also, the device's design is straightforward so that it can be easily fixed and maintained in the case of any damage without any complexities. This device can be easily installed on those structures and centrally linked with the application. After installation, it will be convenient to handle the device on its existing framework effectively in an exact manner.

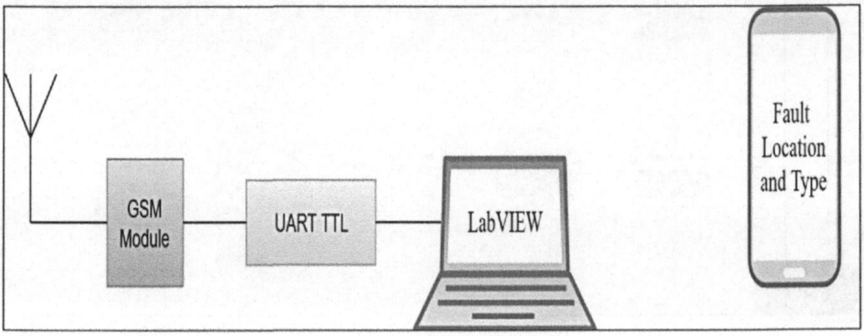

FIGURE 16.3 Block diagram of substation and receiving unit.

The main advantage of mounting the transmitting part of the fault detecting system over preexisting infrastructures is that it is helpful in many ways, like minimizing the initial investment. It would be the least as the already existing poles, and transformers for the transmission system will be utilized. Next is the easy maintenance of the devices as they are mounted on very accessible and already known locations, it will be easier to do any kind for repairing works in the transmitting unit. Also depending upon the probability of occurrence of the fault, it will be used to determine the number of devices required in any specific region and also at a definite interval as shown in Figure 16.4.

16.6 RESULT AND DISCUSSION

The distribution lines have always been bared to natural disasters, which tends to make them faulty. These faults lines must be removed. Otherwise, this will result in significant damage to distribution lines. Presently, there is not any real-time system for fault monitoring. It just notifies about the fault to the control panel of the substation unit. Thus, this insufficient information is not of that much help to the technicians.[17]

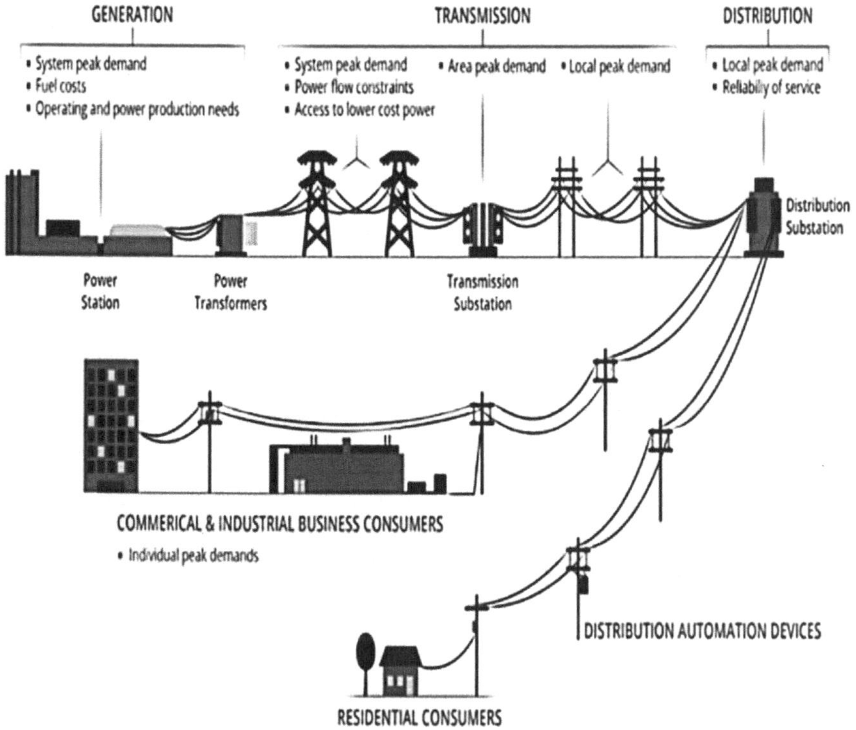

FIGURE 16.4 Distribution systems mounted with fault detecting devices.

A model has been designed using LabVIEW, which notifies the operators about the location of the fault, identifies the type of fault, and rectifies the output. A fault detector and indicator for monitoring the current in the power line have been mounted adjacent to the power lines. When a fault occurs, the steady(stable) grid/station is separated from the network or area, causing the fault by using crucial components like circuit breakers and relays controlled by a microcontroller when an unfavorable situation appears. Here, the GSM modem sends a warning message instantly as there is any fault detection to the controlling station stating the recorded values of current and voltage along with the same pole to pole location. The GPS and GSM modules are interfaced with Arduino UNO so that we can accurately locate the coordinates from the GPS unit via GSM to the registered mobile number it sends the message to other networks.[18] The fault detector is always in on state and thus can identify faults whenever there is any such occurrence. Information is recorded and analyzed in an electronic format, which can easily be interpreted and understood by a transmission line. That data could by anything like total stations or map coordinates or photogrammetric, electronic topographical maps, etc.

The true accomplishment of this fault monitoring and detecting system will result in achieving the desired objective to analyze the nature of fault caused due to various reasons of the harsh environment of nature like lightning, thunderstorm, hailstorm, fog, smog, wet insulators, and pollution which causes very poor contact between insulators and conductors. And also eliminates its causes with highest possible efficiency in very less time ensuring the comfort and well-being of the customers. The idea put forward here gives us a clarified and exceptionally dependable approach to find the fault in the three-stage transmission lines and furthermore it keeps monitoring the faults every second, which will minimize the chances of death due to electrocution.[19]

Figure 16.5 shows a sample message which is automatically generated by the fault detecting device, after which it is transmitted to substations and concerned persons by accessing their pre-stored data from the cloud to contact them for repair.

16.7 CONCLUSION

The distribution lines are continuously exposed to lots of disturbances from nature. Due to natural disasters, such as storms or heavy rainfall, any minor fault or shortfall can cause severe damage to distribution lines, which may ultimately lead to expensive feeder equipment failures. There is no messaging system in today's technology for relaying messages to service technicians when a fault is detected. At present, fault locating strategies just notify the control panel of the fault and give no other information to service technicians. As a result of an inadequate information passing system, there is an increased chance of accidents and there is a loss of life and property as a result. Therefore, special equipment is needed that finds the fault location area and notifies the service technician immediately. With the use of an automatic fault detecting and warning system, we can automate the entire process to pinpoint the exact nature and intensity of the fault in a certain area and then send a text message with that data to the serviceman or the person responsible for the area. In this system, we have used LabVIEW, which facilitates and uses programming in the sub-feeder/distribution units, which is connected with many fault detectors mounted on the structures involved in the distribution line. A wireless communication device, such as a GSM, is used to transmit interruptions to the substation, which is then connected by software to a local distribution station. With the GSM and GPS module integrated with Arduino, we received accurate coordinates from the GPS unit via GSM to the registered mobile number. The potential transformer and current transformer have been interfaced, and we are getting accurate AC voltage and current readings. The readings of current and voltage will be continually compared with these set values. When any deviation is observed for any fixed period, the GSM module will send a link to the substation to perform the subsequent task. This system keeps monitoring the faults every second. If any area is malfunctioning, it will automatically generate a message to the servicing company and turn off the power accordingly, which will minimize the chances of death due to electrocution. Using this system, consumers will also find themselves more comfortable. They will not be required to call servicemen for complaints about problems, and the delay in the repairing process will be minimized to the greatest extent possible.

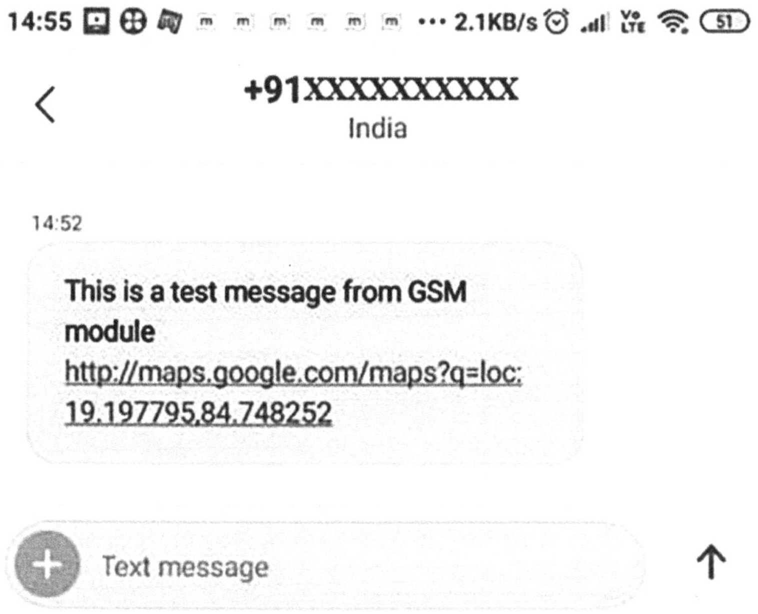

FIGURE 16.5 Sample text transmitted by the device.

KEYWORDS

- **advanced power protection**
- **automation**
- **autonomous fault detection**
- **distribution lines**
- **LabVIEW**
- **innovation**

REFERENCES

1. Kezunovic, M. Smart Fault Location for Smart Grids. *IEEE Trans. Smart Grids*, **2011,** *2* (1), 11–22.
2. Coelho, A.; Rodrigues, A. B.; Da Silva, M. G. Distribution Network Reconfiguration with Reliability Constraints. *Proc. Int. Conf. Power Syst. Technol.* **2004,** *2*, 1600–1606.
3. Linares, P.; Rey, L. The Costs of Electricity Interruptions in Spain. Are We Sending the Right Signals? *Energy Policy* **2013,** *61*, 751–760.
4. Suarez, I. D.; Caicedo, G. C.; Vargas-Torres, H. R. Review of Stable and Transient Fault Location Techniques for Distribution Systems. Part II: Steady-State Knowledge-Based and Transient Analysis Techniques. In *International Conference on Renewable Energies and Power Quality* 2010; pp 1–6.
5. Kavousi-Fard, A.; Niknam, T. Optimal Distribution Feeder Reconfiguration for Reliability Improvement Considering Uncertainty. *IEEE Trans. Power Del.* **2014,** *29* (3), 1344–1353.
6. Zidan, A.; El-Saadany, E. F. A Cooperative Multiagent Framework for Self-Healing Mechanisms in Distribution Systems. *IEEE Trans. Smart Grid* **2012,** *3* (3), 1525–1539.
7. Angelo, C.; Selejan, P. Technologies of the Self-Healing Grid. In *Proc. 22nd Int. Conf. Electr. Distrib. (CIRED)*, Stockholm, Sweden, 2013; pp 1–4.
8. Decanini, J. G. M. S.; Tonelli-Neto, M. S.; Minussi, C. R. Robust Fault Diagnosis in Power Distribution Systems Based on Fuzzy Artmap Neural Network-Aided Evidence Theory. *IET Gener., Trans. Distrib.* **2012,** *6* (11), 1112–1120.
9. Horn, D. *How to Test Almost Everything Electronic*; McGraw-Hill/TAB Electronics: New York, 1993; pp 4–6.
10. Oancea, C. D. GSM Infrastructure Used for Data Transmission. *7th Int. Symp. Adv. Topics Electr. Eng. (ATEE)* **2011,** *12* (14), 1–4.
11. Vimalraj, S.; Gausalya, R. B. GSM Based Controlled Switching Circuit between Supply Mains and Captive Power Plant. *Int. J. Comput. Eng. Res.* **2013,** *3*, 4.
12. Sujatha, M. S.; et al. On-line Monitoring and Analysis of Faults in Transmission and Distribution Lines Using GSM Technique. *J. Theor. Appl. Inf. Technol.* **2011,** *33* (2), 258–265.

13. Adem, S. H. Fault Detection, Protection and Location on Transmission Line: A Review. In *Mas 13th International European Conference on Mathematics, Engineering, Natural and Medical Science*, Tartan University, Afghanistan, 2020.
14. Kumar, B. R.; Mohapatra, A.; Chakrabarti, S.; Kumar, A. Phase Angle-Based Fault Detection and Classification for Protection of Transmission Lines. *Int. J. Electr. Power Energy Syst.* **2021,** *133* (December), 107258.
15. Aboshady, F. M. Modified Distance Protection for Transmission Line with Hexagonal Phase-Shifting Transformer. *Int. J. Electr. Power Energy Syst.* **2021,** *134*, 107379.
16. Fahima, S. R.; Sarker, S. K.; Muyeen, S. M.; Das, S. K.; Kamwa, I. A Deep Learning Based Intelligent Approach in Detection and Classification of Transmission Line Faults. *Int. J. Electr. Power Energy Syst.* **2021,** *133*, 107102.
17. Panth, D. Reasons for Failure of Transmission Lines and Their Prevention Strategies. *Int. J. Electr., Electron. Data Commun.* **2014,** *2*, 1.
18. Shekar, C. Transmission Line Fault Detection & Indication through GSM. *Int. J. Rec. Adv. Eng. Technol.* **2014,** *2*, 5.
19. Pereira R. A. F.; Kezunovic M.; Mantovani J. R. S. Improved Fault Location on Distribution Feeders Based on Matching During-Fault Voltage Sags. *IEEE Trans. Power Deliv.* **2009,** *24* (2), 852–862.
20. Ayon, J. S.; et al. *Remote Monitoring of a Power Station (Voltage Monitoring) Using GSM*; Bangladesh University of Engineering & Technology, 2014.
21. Kleene, S. C. Representation of Events in Nerve Nets and Finite Automata. *Ann. Math. Stud.* **1956,** *34*, 3–41 (retrieved June 17, 2017).
22. Gebhardt M.; Weinmann F.; Dostert K. Physical and Regulatory Constraints for Communication over the Power Supply Grid. *IEEE Commun. Mag.* **2003,** *41*, 5.
23. Ferreira, K. J. *Fault Location for Power Transmission Systems Using Magnetic Field Sensing Coils*; ECE Department of Worcester Polytechnic Institute, April 2007.
24. Gupta, V.; Trivedi, U. C.; Buch, N. J. *Solid State Electronic Fault Current Limiter to Limit the Fault Current in Power System*; Electrical Research & Development Association: Adodara, 2010.
25. Lin, Y. J.; Latchman, H. A.; Lee, M. Katar, S. A Power Line Communication Network Infrastructure for the Smart Home. *IEEE Wireless Commun.* **2002,** 9 (6), 104–111.
26. Jayant, N. Special Issue on Gigabit Wireless. *Proc. IEEE* **2004,** *92*, 2.

PART III

Optimization of Intelligent Technologies for Sensors

CHAPTER 17

Performance Analysis of Clustered Routing Protocol for Wearable Sensor Devices in an IoT-Based WBAN Environment

J. VIJITHA ANANTHI* and P. SUBHA HENCY JOSE

Department of Biomedical Engineering, Karunya Institute of Technology and Sciences, Coimbatore, India

**Corresponding author. E-mail: vijithaananthi@karunya.edu.in, vijithaananthij@gmail.com*

ABSTRACT

Wireless body area network (WBAN) has played an important role in healthcare applications in recent decades. In healthcare applications, the internet of things has a huge impact on getting medical data to the appropriate systems. In a wireless body area network, security and energy consumption are the most critical considerations. While transferring the patient health records, there is a possibility of third-party access and it leads to false communication. Due to the smaller size of medical sensor devices, there is a possibility of high energy consumption in wireless body area networks. To address these concerns, a new Clustered based security and energy-efficient (CSEE) routing protocol was proposed. The use of clustering techniques helps to alleviate security and energy consumption issues. The five wearable medical sensor devices such as EEG, Oximeter, EMG, Vision, and blood pressure sensor and two different controller central nodes are placed on the human

Intelligent Technologies for Sensors: Applications, Design, and Optimization for a Smart World.
S. Kannadhasan, R. Nagarajan, & Alagar Karthick (Eds.)

body. The acquired medical data will be sent over Bluetooth to the personal device assistant. The same information will be sent to other devices via the Internet of Things. Simulation results analyzed the performance of proposed and existing techniques in an IoT-based WBAN environment. The proposed Clustered based security and energy-efficient (CSEE) routing protocol improves the delivery rate, throughput, and minimizes drop rate, energy consumption, and end-to-end delay when compared to the existing technique.

17.1 INTRODUCTION

Wireless body area network (WBAN) has played an important role in healthcare applications in recent decades. In healthcare applications, the Internet of Things (IoT) has a huge impact on getting medical data to the appropriate systems. In a WBAN, security and energy consumption are the most critical considerations. While transferring the patient health records, there is a possibility of third-party access and it leads to false communication. Due to the smaller size of medical sensor devices, there is a possibility of high-energy consumption in WBANs. To address these concerns, a new clustered-based security and energy-efficient (CSEE) routing protocol was proposed. The use of clustering techniques helps to alleviate security and energy consumption issues. The five wearable medical sensor devices such as EEG, oximeter, EMG, vision, and blood pressure sensor and two different controller central nodes are placed on the human body. The acquired medical data are sent via Bluetooth to the personal device assistant. The same data are sent to other devices via the IoT. Simulation results analyzed the performance of both proposed and existing techniques in an IoT-based WBAN environment. The proposed CSEE routing protocol improves the delivery rate and throughput, and minimizes drop rate, energy consumption, and end-to-end delay when compared to the existing techniques.

WBAN is derived from the wireless sensor network and it consists of medical sensor devices deployed on the human body. This supports the real-time healthcare monitoring system and transmits the medical data to the doctors. The medical sensors are very portable and smaller in size and it is placed on the human body. This network is most helpful nowadays due to COVID-19 and lockdown issues who need doctor's attention without being hospitalized. The process of monitoring may be done at home without visiting the hospital and the sensed data will be transmitted to the doctor. This helps to get advice from doctors via the Internet. In case of any emergency conditions, the necessary action can be taken without any delay. There are three different phases based on the data communication in the WBAN. Intra BAN helps to connect the sensor nodes and the controller nodes to personal device assistants via Bluetooth. Inter BAN connects the personal device assistant to the Internet. Collected sensor information will be transmitted to the Internet from a personal device assistant. WBAN sends the required information to doctors or caretakers through the gateway node. IoT is used to connect things via the Internet and it will be more helpful for medical applications. IoT helps to transmit the medically sensed data to more users such as ambulance, cloud storage servers, caretaker, hospital, doctor, and nurses, etc. Security is a major concern, because it contains patient's medical data, which need to be transmitted with high authenticity.

Figure 17.1 shows that the integration of WBANs and the IoT. This integration achieves high network usage with high network efficiency. Sensor nodes are placed in the human body and they may be wearable and implanted based on the requirement. IEEE 802.15.6 is the international standard for WBAN and this helps to maintain the wireless communication between the inside and outside of the human body. The most important challenging issues in the WBAN are as follows:

1. Energy,
2. Mobility,
3. Security,
4. Cooperation,
5. Networking, and
6. Quality of Service (QoS).

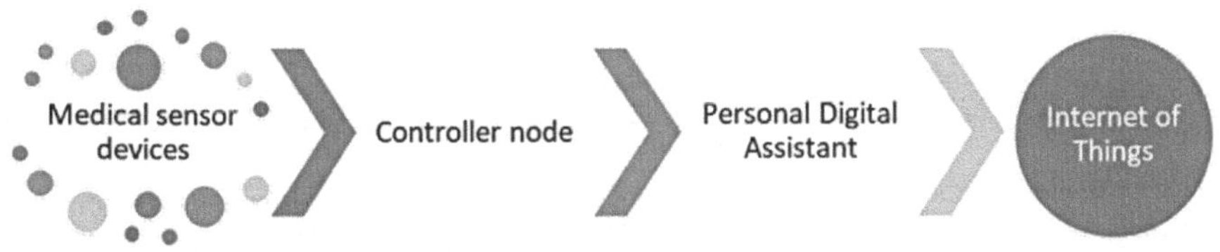

FIGURE 17.1 Integration of WBAN and IoT.

The energy level of the medical sensor node is very low because of its smaller size. Mobility needs to be maintained because the medical sensors are deployed in the human body. The human body is movable to all places and the connection is to be maintained properly for efficient communication.

Security is the most important thing and sensed data contains a patient's health record. There is the possibility of third-party access and editing of the health record. That leads to wrong communication to doctor and is not safe for patient's life. Cooperation is the most important cause for the controller node and it should collect the medical sensed data from all medical sensors. Networking helps to communicate between the patient to doctor through the Internet and gateway node. Quality of Service (QoS) identifies the network performance and analyses the overall performance.

There are five different types of wearable sensors used in this research work, which includes:

1. EEG,
2. Oximeter,
3. EMG,
4. Vision, and
5. Pressure.

An electroencephalography (EEG) sensor is a voltage differential measurement. It is easy-to-use and single-channel sensor. This sensor helps to identify the brain activity of the human body. The electrodes are placed on the scalp of the human body and it measures the electrical potentials of the scalp region concerning the reference electrode which is placed near the back of the ear. The difference between the scalp and reference electrodes will eliminate the unwanted signals. It has a high amplification gain. To remove the noise signal due to high amplification gain, the patient should be away from the lighting things or high-power areas from the household environment. A pulse oximeter sensor is used to measure how well the patient heart is pumping oxygen throughout the entire body. It uses high-quality LEDs to eliminate the inferences that occurred due to second frequencies. A small beam of light will be passed throughout the body and it will measure the amount of oxygen in the blood. This result provides a very fairly accurate result. It varies from ±2%. For example, the result obtained is 92% and the accurate value will range from 90% to 94%. Electromyography (EMG) sensor measures the muscle activity of the human body. This sensor helps to identify neuromuscular abnormalities. The patient should avoid wearing metal objects while during this test. It will produce the results when the muscle is in active mode. If it is in stable mode, the result will appear neutral. A vision sensor is used to capture the image in front of the human body. A pressure sensor is used to identify the blood pressure of the human body. This is an automatic blood pressure that detects the pressure in an artery and gives output. It is a non-invasive method and easier to use.

17.2 RELATED WORK

Dhanvijay et al.[1] proposed the constraint application protocol-based secure aware mobility management protocol and it was designed in an IoT-based environment. This protocol focused on mobility control in WBANs and maintained secure communication using encryption-based key management techniques. It technically analyzes the total transmission delay, hop count, node delay, and wireless link delay. The proposed approach focused on the handover process between the sensor node and wireless client via a gateway. The performance of the proposed and existing techniques was analyzed in the results section. When compared to existing techniques, the proposed COSAMP protocol performs very well and improves network efficiency. Şen et al.[2] suggested the inter-WBAN geographic routing algorithm for corona-affected patients. This WBAN senses the respiration rate, blood saturation level, and temperature of the patient continuously. This technique scans the status of masked patients and will be operated by using the IoT software. Experimental results compared the performance of the AODV routing protocol and the proposed geographical routing algorithm in terms of packet drop rate and delivery rate. Izza et al.[3] explained the issues related to RFID-based IoT routing protocol for WBAN. To overcome the security issues, the authors proposed the RFID and security-based IoT routing protocol with the help of the elliptic curve cryptography mechanism. This provides security features and maintains data security while communicating. Ullah et al.[4] discussed the lightweight and provable secure cross-domain encryption scheme for WBANs. This helps to achieve security with high confidentiality and authenticity. The proposed scheme provided less computational cost and communication overhead. Soni et al.[5] introduced the secure and light-weight health authentication and key agreement protocol for WBANs. This proposed routing protocol focused on the identification of denial-of-service attacks during data communication. This scheme reduces the computational overhead and computational cost.

Olatinwo et al.[6] suggested the integration of IoT with the WBAN for high productivity in the health care field. Different stack layers in networks are supported for QoS enhancement and network connectivity. This approach improved the energy efficiency, channel utilization rate, data transmission, and delay. Majeed et al.[7] explained the process of the remote patient monitoring system with the integration of IoT and cloud computing. This technique improved the delivery rate and reduced the delay while transmitting the patient information to the hospital. Chowdhury et al.[8] proposed the dynamic resource assignment scheme with the help of cluster and aggregate sensor nodes with the aid of WBANs. This method helps to reduce the delay overhead and transmit the data without any backoff delay. Hariharan et al.[9] explained the process of development of sensor networks to body area networks and the importance of WBAN in healthcare applications. Rokonuzzaman et al.[10] proposed the real-time village doctor app by using android software. This application helps the patients and supplies the needs automatically. Poongodi et al.[11] presented the wearable sensor devices in healthcare applications and the performance of localization systems in body area networks. This approach analyzed the performance of packet drop rates in both indoor and outdoor environments. Selem et al.[12] introduced the mobTHE protocol which helps to tracks the communication between the two coordinator nodes such as the serving coordinator node and neighbor coordinator node. A handover mechanism was implemented to maintain the communication between the two coordinator nodes. This mobTHE protocol improved the overall network lifetime of the WBAN, throughput, and residual energy. Khan et al.[13] explained the combination of IoT and WBAN success in healthcare application. This scheme improves the high energy resources utilization and high network efficiency with the help of cooperative-enabled efficient routing protocol. Ullah et al.[14] focused on the resource-constrained WBAN with less memory and less energy utilization. This technique used the certificateless signcryption scheme for secure communication for IoT-based WBAN environments. Olatinwo et al.[15] explained the process of communication in WBAN with two different phases such as content phase and the transition phase. The content phase allocates the schedule with the help of time division multiple access schemes. The transition phase transmits the information based on the scheduled time. This helps to avoid collisions and increase the resource utilization rate.[16]

17.3 COOPERATIVE-BASED ENERGY HARVESTING SCHEME (CEHS)

This cooperative-based energy harvesting scheme (CEHS) is applied in the combined WBAN and IoT for less energy consumption and high network efficiency. Energy harvesting is integrated with the sensor node and that helps to improve the network lifetime of the WBAN. Link statistic is added to the concern which helps to maintain the link while transmitting the data between sensor nodes. This routing protocol performs based on multihop communication. Data forwarder node has greater path cost estimation model and this helps to balance the energy utilization of sensor node and that helps to improve the network performance. The path cost estimation function calculates the distance, energy, hop count, and node congestion level.

Figure 17.2 shows that the existing technique integrates the WBAN to the IoT cloud platform. Here, different types of sensors are fixed in the body. The sensors include sensors for blood pressure, vision, EEG, lactic acid, EMG, ECG, motion, respiration, glucose, and pulse oximeter. All these sensor nodes are connected to the controller central node. This controlled central node are located at the right- and left-side hip of the human body and maintains the line of sight between the two central nodes to avoid collision and delay. Controller central node communicates the collected sensor data to the personal digital assistant via Bluetooth. The collected sensor data will be transmitted from a personal digital assistant to an IoT with a cloud platform through the communication network.

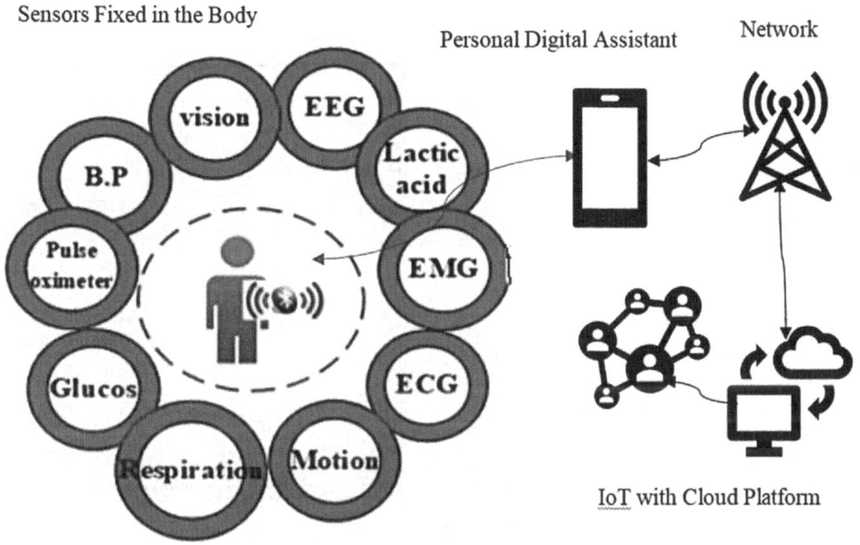

FIGURE 17.2 Cooperative-based energy harvesting scheme (CEHS).

The existing technique performs through the following three different phases:

1. Data initialization and sensing phase,
2. Forwarder node selection phase, and
3. Cooperative effort phase.

This existing technique integrated a WBAN with IoT and it contains 10 heterogeneous sensor nodes with the wearable and implanted sensor nodes. These sensor nodes are deployed in the human body and the central node controller is also connected in the human body. This central controller node transmits the collected sensor details to the personal digital assistant for the further communication process.

This existing technique is focused on the following parameters:

- Network stability,
- Network lifetime,
- Energy consumption,
- Energy reservation,
- Line of sight,
- Duplicate data transmission,
- Node deployment, and
- Network efficiency.

In WBANs, network stability and network lifetime are the most important issues and it should be improved by using the IoT. Energy consumption and energy reservation arise due to the small size of sensor nodes and it has very limited bandwidth. Energy consumption can be reduced with the help of CEHS and each node is deployed with a harvesting scheme. Due to the improper line of sight, there is a possibility of path loss and end-to-end delay in the WBAN and that can be reduced by the proper fixing of the central controller node. Here, all the sensor nodes continuously monitor the patients and transmit the data. There is a possibility of duplicate data transmission and that leads to high energy consumption and this can be reduced by coordinator nodes. Network deployment should be focused on when it is placed in the human body. This helps to improve the network efficiency of the WBAN. Receiver signal strength indicator (RSSI) plays a vital role in the calculation of the location of the sensor node and the distance between the neighbor nodes to the coordinator node.

The cooperative routing protocol helps to analyze the network performance by using the quality-of-service parameters such as packet delivery ratio, end-to-end delay, throughput, and signal-to-noise ratio. The issues obtained from the existing methodology include cooperative scheme, which provides lesser energy consumption and higher network efficiency. The major drawback in this existing approach is that there is no security concern in data communication and that should be improved by using the clustering technique.

17.4 PROPOSED METHODOLOGY

CSEE routing protocol is the proposed model for WBANs with IoT environments. This proposed approach is deployed in the WBAN environment and it ensures a secure mechanism while data transmission. The clustering technique plays a vital role to increase the energy resources and security between the sensor nodes.

Figure 17.3 explains the process of the proposed network architecture and it consists of three different phases. The first phase is the sensor node deployment phase and this approach uses the five different wearable sensors in the human body. The two controller nodes are used to collect the sensed data and transmit it to the personal device assistant. CH denotes the cluster head and CM denotes the cluster member. Clustering formation is the second phase and this helps to increase security and reduce energy consumption. The last phase is the secure data transmission phase via the IoT.

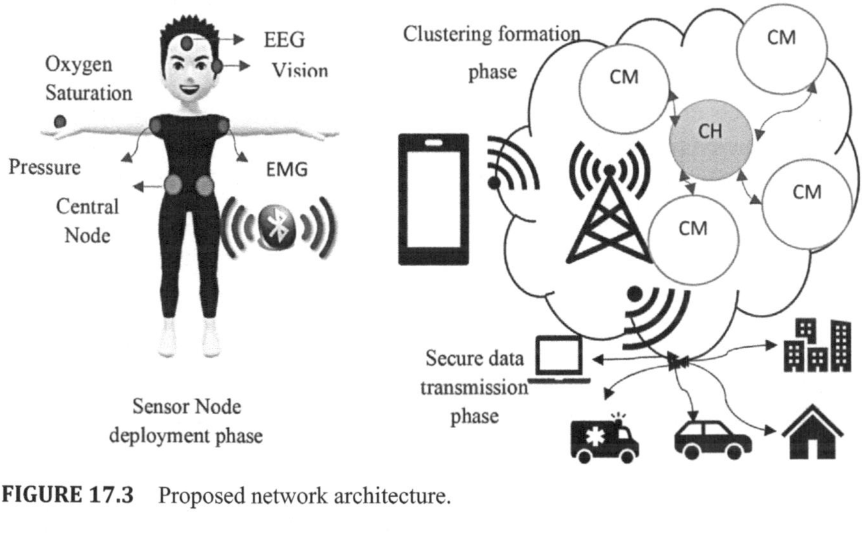

FIGURE 17.3 Proposed network architecture.

The proposed technique will be performed in the following three different phases:

1. Sensor node deployment phase,
2. Clustering formation phase, and
3. Secure data transmission phase.

17.4.1 SENSOR NODE DEPLOYMENT PHASE

In this phase, the proposed technique uses the five-wearable sensor with two central nodes. The five wearable sensors are EEG sensor, oximeter sensor, vision sensor, EMG sensor, and pressure sensor. The EEG sensor is placed on the patient's head, oximeter sensor is placed on the right hand finger, vision sensor is placed on the face, EMG sensor is placed on the left arm, and pressure sensor is placed on the right arm. Two central nodes are placed on the left and right stomach of the human body and it needs to maintain the line-of-sight path to avoid collision and delay.

An electroencephalography (EEG) sensor monitors the electrical activity of the scalp and it is placed on the head. It is a non-invasive method. It measures the voltage fluctuations of brain neurons and it helps to analyze human brain activity. Oximeter sensor is the non-invasive method and it monitors the oxygen saturation level of the human body. It should be placed on the thin part of the human body. It passes light wavelength to the body part via photodetector. By varying each wavelength, it will determine the oxygen level in blood. Electromyography (EMG) sensor monitors the electrical activity of skeletal muscle and it is a non-invasive method. This sensor helps to identify muscular disease. Vision sensor helps to identify the nearby objects and it is placed on the face. A pressure sensor helps to analyze the blood pressure in arteries and it is a non-invasive method. This pressure creates changes in the arteries and that helps to analyze the blood pressure. Table 17.1 shows that the sensor types and their considerations.

TABLE 17.1 Sensor Types and Its Considerations.

Sensor type	Localization	Identification	Type	Power consumption
EEG	Head	Brain activity	Wearable	Low
Oxygen saturation	Right hand finger	Blood oxygen	Wearable	High
EMG	Left arm	Muscular disease	Wearable	High
Vision	Face	Object	Wearable	High
Pressure	Right arm	Blood pressure	Wearable	High

After the deployment of all the sensor nodes, the sensor starts to sense the data from the human body and it will transmit to the central node. The central node is wearable and it is placed on the left and right stomach of the human body. The central node collects all the sensor information and transmits the required information to the personal digital assistant via Bluetooth.

17.4.2 CLUSTERING FORMATION PHASE

The clustering formation phase creates the cluster head and cluster members to reduce the energy efficiency and improve the security technique. The cluster head controls the cluster members. If any of the new members should join the network, it needs to get permission from the cluster head. The cluster head controls and monitors the entire cluster members. Cluster members should be controlled by the cluster head and there is less chance of third-party access and security-related issues. Energy consumption is also controlled by the cluster head. The energy level of cluster members will be reported to the cluster head. If any one of the energy levels of a cluster member goes down, the neighbor node will be replaced at that place. This replacing procedure helps to create efficient communication. If the cluster head energy level goes down, the cluster head position will be shifted to the neighbor cluster member which has a high energy level.

17.4.3 SECURE DATA TRANSMISSION PHASE

After the successful data transmission between cluster members, sensed data will be transferred to end users such as hospitals, doctors, nurses, caretakers, ambulances, cloud storage servers, and specialists based on the patient's emergency case.

Algorithm: Clustered-based security and energy-efficient (CSEE) routing protocol

Input: Create G (Vc, Ec) € Body area network; G(Vi, Ei)
€ Connectivity network

Place Vcs, Vcc→ Patient Body \\ Vcs—Medical sensor vertex;
Vcc—controller central node

Vcc collects D(s)¥Vms \\ D(s)—Sensor data

Vcc connects to PDA \\ PDA—Personal digital assistant device

PDA connects to G(Vi, Ei)

Create Vic as cluster head; Vic+1; Vic+2… Vicm as cluster members \\ m=no. of neighbor nodes

To maintain security

Vicm ¥ Belongs to Vic \\ Vicm—Cluster member of connectivity vertex; Vic—Cluster head of connectivity vertex

Vicm sends Join/Leave request to Vic

Vic Monitors Vicm \\ to avoid attackers

To maintain energy

Vic calculates EL(Vicm) \\ Energy level of Vicm—Cluster member of connectivity vertex

If any EL(Vicm) >EL(Vicm+1), Then Vicm+1 be replaced \\ EL—Energy level

If EL(Vic) <<, Vicm to be replaced as Vic

Vic transmits D(S) to V(IoT) \\ D(S)—Sensed data; V(IoT)—Internet of Things vertex

Output: Vc→Vic→Viot

The aforementioned algorithm states the entire process of the proposed CSEE routing protocol. The input is given as two different networks: one is G (Vc, Ec), which are used to collect the medical sensor devices to the human body, and the another one is G (Vi, Ei), which is used to transmit the collected data to things using the Internet. Initially, Vcs-medical sensor node and Vcc-controller central node are placed on the human body. Vcc sends the collected data to a personal digital assistant via Bluetooth. PDA connects to G (Vi, Ei) which is used to transmit the sensed data through the Internet. Here, cluster formation will be created to avoid security issues and energy consumption issues. Consider Vic as cluster head and Vic + 1; Vic + 2 … Vicm as cluster members. The cluster head (Vic) controls and monitors the cluster members. Cluster members should send join/leave requests to be in the network. It gives less chance of security threats in the WBAN. Cluster head (Vic) calculates and monitors the energy level of cluster members (EL(Vicm)). If any one of the cluster members' energy level is lower than the other cluster members' energy level, the lower energy level cluster member will be replaced by a new one. This helps to avoid transmission delays and rate drops caused by the energy consumption issue. Because the cluster head's energy level is low, the neighboring cluster member will be replaced as the new cluster head and the monitoring process for efficient communication will continue. The final output is sensed data that is transmitted from the controller node to the cluster node and then connected to the things via the Internet.

17.5 RESULTS AND DISCUSSION

The experimental results are done by using a network simulator. The result section analyzes the performance of the existing CEHS and proposed CSEE routing protocol with two different scenarios. The two different scenarios are CEHS in the body area network environment and it is denoted as CEHS-B and the technique in the IoT environment is denoted as CEHS-I. Likewise, the proposed routing protocol performed in the BAN environment is denoted as CSEE-B and the techniques implemented in the IoT environment are denoted as CSEE-I. The 500 nodes are used to analyze the performance and bandwidth 2 Mbps used for data transmission. The omnidirectional antenna is used for effective communication. The network performance has been analyzed with the different quality of parameters such as delivery ratio, throughput, delay, energy consumption, and drop rate. Each node has 20 J and the simulation time is 100 s.

Packet delivery ratio is defined as the ratio of the number of packets transmitted to the difference between the transmitted and received. Throughput is calculated based on the number of successfully received packets concerning the simulation time. Delay is calculated based on the time taken to transmit the medical dataset from one sent to another end. Energy consumption is calculated based on the consumed energy level of each node concerning the simulation time. Drop rate is calculated based on the number of packets dropped concerning the number of packets transmitted.

Figure 17.4 shows that the packet delivery ratio analysis and the delivery rate are very low in the existing CEHS in both IoT and BAN environments compared to the proposed CSEE in both IoT and BAN environment. The proposed CSEE achieves high delivery due to the clustering technology.

Figure 17.5 shows that the drop rate analysis and the drop rate are very low in the proposed CSEE compared to the existing system in both IoT and BAN environment. The proposed scheme avoids security threats and energy consumption issues and that leads to the dropping of rate reduction.

Figure 17.6 shows that the throughput analysis and the proposed system achieve the maximum of 1.9 Mbps throughput from the 2-Mbps bandwidth allocation. The proposed CSEE routing protocol achieves high throughput with the help of clustering technology.

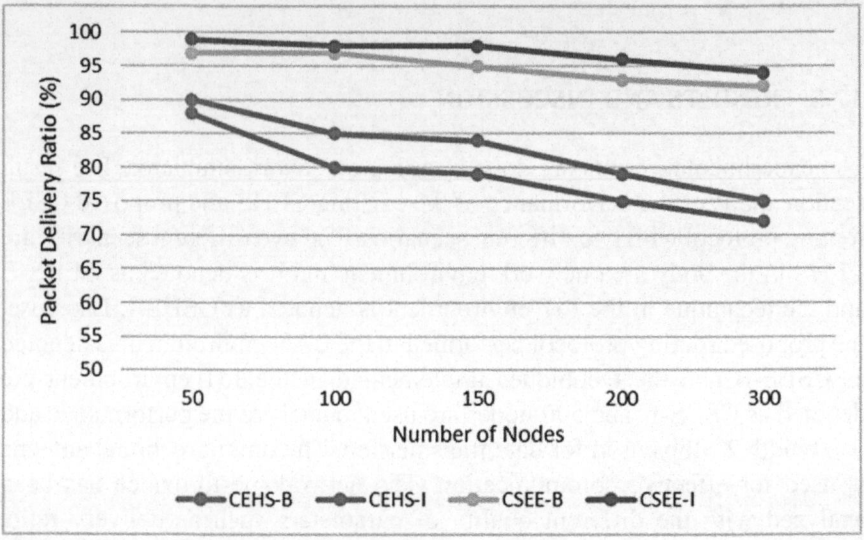

FIGURE 17.4 Packet delivery ratio analysis.

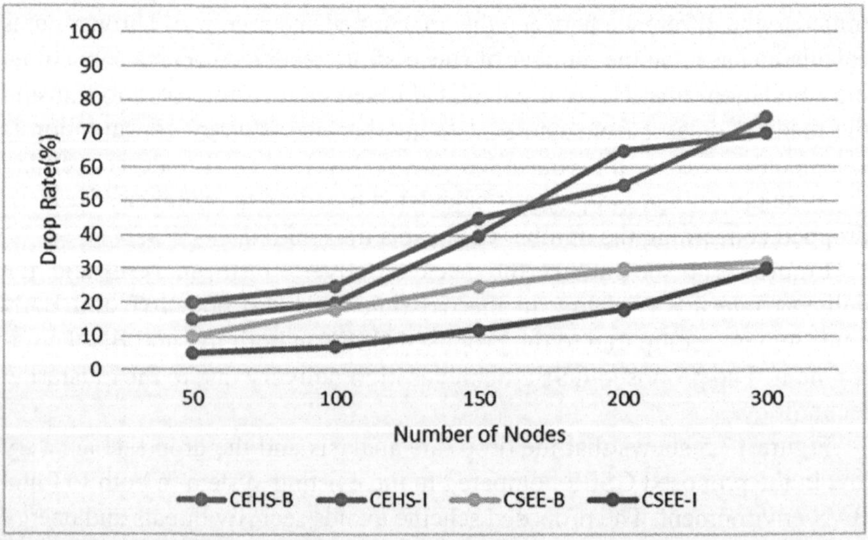

FIGURE 17.5 Drop rate analysis.

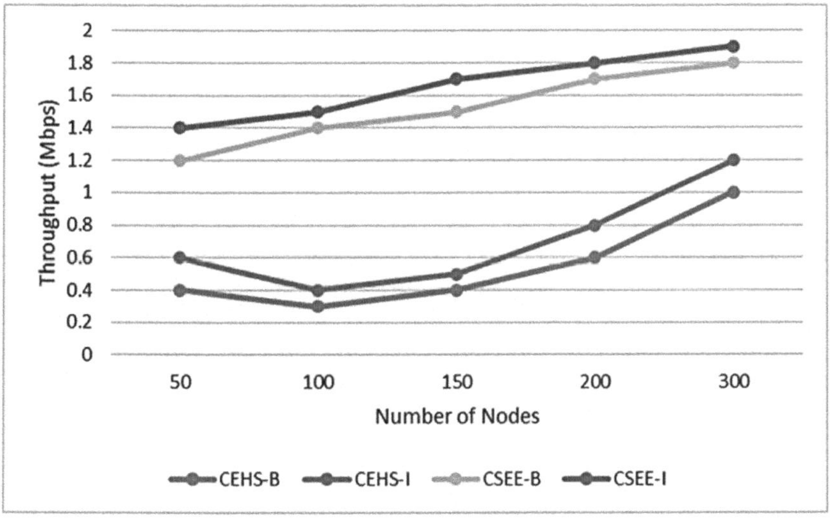

FIGURE 17.6 Throughput analysis.

Figure 17.7 shows that the end–end delay analysis and the proposed system have lesser delay compared to the existing technique. The proposed CSEE techniques achieve lesser delay due to the clustering head and they help to avoid the security and energy consumption issues.

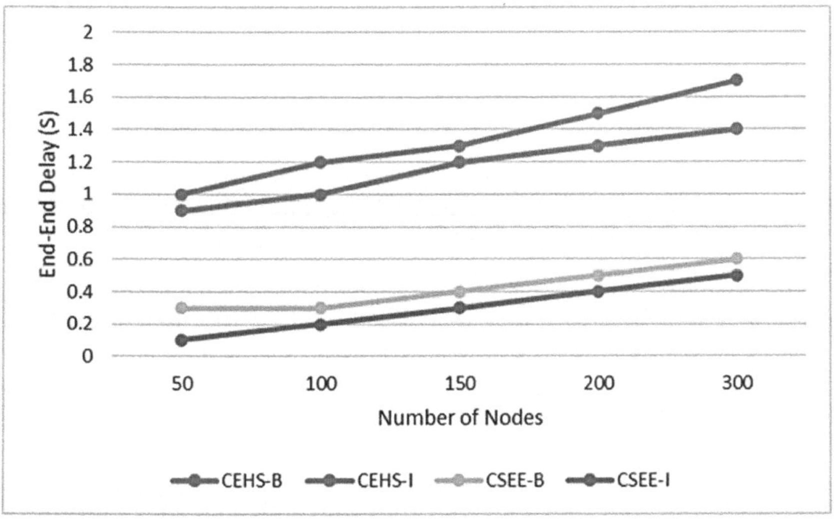

FIGURE 17.7 End–end delay analysis.

Figure 17.8 shows that the energy consumption analysis calculates the consumed energy level of each different set of nodes from 50 to 300. The consumed energy level is calculated based on the difference between the given energy before the simulation starts and residual energy after the simulation ends. Compared to the existing technique, the proposed CSEE routing protocol consumes a lesser energy level because of the clustering technique.

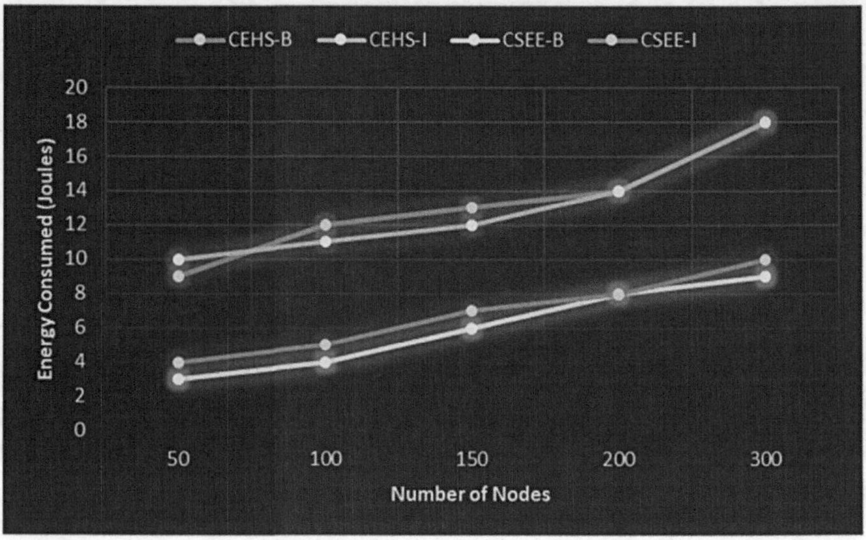

FIGURE 17.8 Energy consumption analysis.

17.6 CONCLUSION

Finally, the proposed CSEE routing protocol improved the overall network performance compared to the existing CEHS routing protocol in IoT-based WBAN environments. The overall network performance has been calculated based on the performance analysis of QoS parameters such as packet delivery ratio, drop rate, delay, throughput, and energy consumption. The proposed CEHS routing protocol reduces safety threats during medical data communication between doctors and patients. Furthermore, the proposed scheme reduces the possibility of network performance degrading as a result of energy consumption and security threats in IoT-based WBAN environment. The proposed CSEE routing protocol improves the delivery rate and throughput while decreasing drop rate, delay, and energy consumption.

KEYWORDS

- **wireless body area networks (WBAN)**
- **Internet of Things (IoT)**
- **cluster technique**
- **cooperative node**
- **sensor devices**

REFERENCES

1. Dhanvijay, M. M.; Patil, S. C. Optimized Mobility Management Protocol for the IoT Based WBAN with an Enhanced Security. *Wireless Netw.* **2021,** *27* (1), 537–555.
2. Şen, S. S.; Cicioğlu, M.; Çalhan, A. IoT-Based GPS Assisted Surveillance System with Inter-WBAN Geographic Routing for Pandemic Situations. *J. Biomed. Inform.* **2021,** *116*, 103731.
3. Izza, S.; Benssalah, M.; Drouiche, K. An Enhanced Scalable and Secure RFID Authentication Protocol for WBAN within an IoT Environment. *J. Inf. Sec. Appl.* **2021,** *58*, 102705.
4. Ullah, I.; Zeadally, S.; Ul Amin, N.; Khan, M. A.; Khattak, H. Lightweight and Provable Secure Cross-Domain Access Control Scheme for Internet of Things (IoT) Based Wireless Body Area Networks (WBAN). *Microprocess. Microsyst.* **2021,** *81*, 103477.
5. Soni, M.; Singh, D. K. LAKA: Lightweight Authentication and Key Agreement Protocol for Internet of Things Based Wireless Body Area Network. *Wireless Pers. Commun.* **2021,** 1–18. doi: 10.1007/s11277-021-08565-2.
6. Olatinwo, D. D.; Abu-Mahfouz, A. M.; Hancke, G. P. Towards Achieving Efficient MAC Protocols for WBAN-Enabled IoT Technology: A Review. *EURASIP J. Wireless Commun. Network.* **2021,** *1* (2021), 1–47.
7. Majeed, J. H.; Aish, Q. A Remote Patient Monitoring Based on WBAN Implementation with Internet of Thing and Cloud Server. *Bull. Electr. Eng. Inform.* **2021,** *10* (3), 1640–1647.
8. Chowdhury, M. A Dynamic Resource Assignment Scheme with Aggregation Node Selection and Power Conservation for WBAN Based Applications. *Int. J. Sensor Networks* **2021,** *35* (4), 207–220.
9. Hariharan, U.; Rajkumar, K.; Ponmalar, A. WBAN for e-Healthcare Application: Systematic Review, Challenges, and Counter Measures. In *2021 International Conference on Computer Communication and Informatics (ICCCI)*, IEEE, 2021; pp 1–7.
10. Rokonuzzaman, M.; Hossain, M. I.; Islam, T.; Sarkar, P. P.; Islam, M. R.; Amin, N. Design and Implementation of Telehealth Device: Linking IoT Sensors to Cloud Networks. In *2020 IEEE-EMBS Conference on Biomedical Engineering and Sciences (IECBES)*, IEEE, 2021; pp 281–285.
11. Poongodi, T.; Rathee, A.; Indrakumari, R.; Suresh, P. IoT Sensing Capabilities: Sensor Deployment and Node Discovery, Wearable Sensors, Wireless Body Area Network

(WBAN), Data Acquisition. In *Principles of Internet of Things (IoT) Ecosystem: Insight Paradigm*; Springer: Cham, 2020; pp 127–151.

12. Selem, E., Fatehy, M.; Abd El-Kader, S. M. mobTHE (Mobile Temperature Heterogeneity Energy) Aware Routing Protocol for WBAN IoT Health Application. *IEEE Access* **2021,** *9*, 18692–18705.
13. Khan, M. D.; Ullah, Z.; Ahmad, A.; Hayat, B.; Almogren, A.; Kim, K. H.; Ilyas, M.; Ali, M. Energy Harvested and Cooperative Enabled Efficient Routing Protocol (EHCRP) for IoT-WBAN. *Sensors* **2020,** *20* (21), 6267.
14. Ullah, I.; Zeadally, S.; Ul Amin, N.; Khan, M. A.; Khattak, H. Lightweight and Provable Secure Cross-Domain Access Control Scheme for Internet of Things (IoT) Based Wireless Body Area Networks (WBAN). *Microprocess. Microsyst.* **2021,** *81*, 103477.
15. Olatinwo, D. D.; Abu-Mahfou, A. M.; Hancke, G. P. A Hybrid Multi-Class MAC Protocol for IoT-Enabled WBAN Systems. *IEEE Sensors J.* **2020,** *21* (5), 6761–6774.
16. Ananthi, J. V.; Subha Hency Jose, P. A Review on Various Routing Protocol Designing Features for Flying Ad Hoc Networks. *Mobile Computing and Sustainable Informatics*; Springer: Singapore, 2022; pp 315–325.

CHAPTER 18

Recent Trends in Wearable Sensor Technology for E-Health Monitoring

RENJITH V. S.* and P. SUBHA HENCY JOSE

Department of Biomedical Engineering, Karunya Institute of Technology and Sciences Coimbatore, India

Corresponding author. E-mail: notify_renjithvs@yahoo.com

ABSTRACT

Technology has advanced significantly in the previous decade, and hence, there is a pressing need to change from conventional health system to preventive medicine and health management to improve individual/patient quality of life while lowering healthcare costs. The presence of individual/patient is needed in the current system for successful evaluation, health management, and preventive medicine. Since some symptoms may not manifest themselves in diagnosis, continuous and periodic monitoring over a long time is required to gain a deeper understanding of individuals/patients. As a result, there is an emerging demand for monitoring healthcare system (wearable) that can be accessed at any time and from any location. Embedded systems, information and communication technologies (ICTs), micro-electro-mechanical technologies, and nanotechnology have advanced to the point where individual/patient can now benefit from wearable-based sensors to screen/monitor parameters of health in real time, thanks to breakthroughs in advanced wireless communication technology. Bluetooth, ANT++, ZigBee, near-field communications, and IEEE 802.11 are some of the recent advances in wireless communication systems that are enhanced and got adapted to portable devices such as healthcare equipment and smart gadgets.

Intelligent Technologies for Sensors: Applications, Design, and Optimization for a Smart World.
S. Kannadhasan, R. Nagarajan, & Alagar Karthick (Eds.)

In addition, major advancements in micro-electromechanical technologies, wireless communication system associated with wearable sensors, and future scope of the wearable sensors and sensor networks in the field of E-Health are discussed, allowing for the development of small, lightweight wearable sensors.

18.1 INTRODUCTION

Wearable sensors have grown increasingly popular in a variety of real-world applications such as security, medicine, commerce, and entertainment since they are extremely beneficial for receiving accurate and reliable information regarding the activities of a person. The advancements in wearable devices with sensors listed above have revolutionized our life activities, daily societal interactions, and so on. Wearable sensors are most commonly used in health monitoring in recent years, where state-of-the-art applications for tracking physical activity have been developed. Blood pressure (BP), brain activity, heart rates, body temperature (BT), muscular mobility, and other vital standard data can be monitored using tiny sensors worn or attached on the body. Wearable sensors are becoming increasingly popular, particularly in sports training. Previously, only laboratory-based investigations could assess levels of glucose, perspiration rate, and energy levels, but today all of these measurements can be made utilizing wireless wearable sensor nodes. Patients recovering from surgery can benefit from the use of wearable sensors, allowing them to remain at home with sensor patches on them for daily instructions. Body-attached sensor nodes are remotely monitored by doctors and the data can be used to observe physical activities during the time of rehabilitation. Healthcare devices can now make multi-purpose, real-time detections with the improvements in the Internet of Things (IoT). Many researchers have provided in-depth analyses of different sensors and devices, including silk fibroin, power source devices, electro-chemical sensors, metal nanowires, and physical sensors, as well as the efficient software programs that permit us to interpret the data, were given. Wearable health-monitoring devices can track real-time data regarding a person's physical state and activities involved. A number of flexible sensors are used for a wearable sensor-based system that can be embedded in textile fibers, clothing, and elastic bands, or fastened straight to the human body for monitoring health state. Body temperature, Electromyogram (EMG), Electrocardiogram (ECG), heart rate (HR), arterial oxygen saturation (SpO_2), electro dermal activity (EDA), respiration rate (RR), and blood pressure (BP) are among the physiological indications that the sensors can measure. Furthermore, micro-electro-mechanical systems (MEMS)-based microscopic motion sensors, including gyroscopes, accelerometers, and magnetic field sensors, are commonly employed for monitoring mobility and activity data. Continuous physiological and functional signal monitoring could help with the early detection and diagnosis of a number of diseases such as cardiovascular, neurological, and pulmonary. Posture analysis and Gait pattern, fall detection, and sleep evaluation could all benefit from real-time observing of an individual's mobility activities. Actuators, electronic and MEMS sensors, signal processing units, and wireless communication modules are commonly found in health-monitoring devices that can be worn. Sensor measurements are relayed to a neighboring processing node through an appropriate protocol for communication, particularly a short-range and low-power wireless medium, in a Wireless Body Sensor Network (BSN), such as Near Field Communications (NFC), ZigBee, Bluetooth, or ANT.[1]

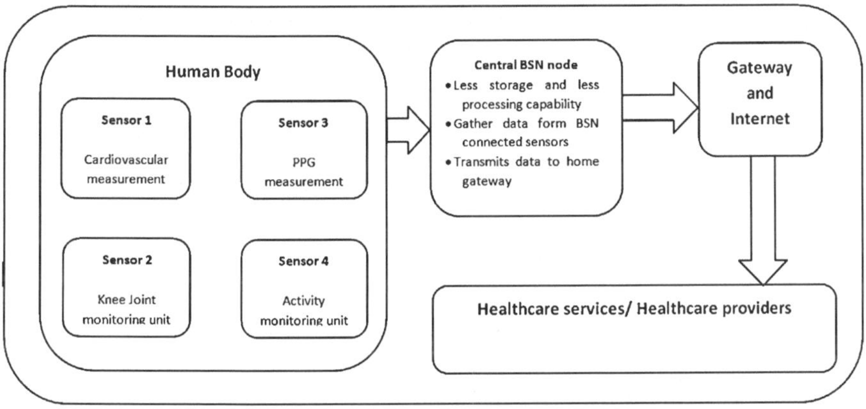

FIGURE 18.1 A general overview of the remote health-monitoring system.

A processing node can be a smartphone, PDA, computer, or Field Programmable Gate Array (FPGA) or a microcontroller, performs complex processing, decision algorithms, and analysis, as well as storing and displaying the findings of the user. It acts as a gateway to send the evaluated data remotely to healthcare facilities or healthcare staff through the internet. Figure 18.1 depicts a generic model that provides an overview of a remote health-monitoring system, while the actual system implemented may differ based on the application needs. Some systems, for example, can be built with only a few sensors, each of which has the ability to send data immediately to a neighboring gateway. With body sensor network (BSN) in certain systems, the sensors are connected with the central BSN node collecting data from the sensors and conducting minimum processing before transmitting it to the advanced processing system unit.

The wearable sensor system must be flexible, chemically inert, nontoxic, hypoallergenic, and compact in size to the human body. Furthermore, in a multi-sensor network system, in which the central node is responsible for managing a large volume of data from multiple sensor nodes, hardware resources are a major challenge. It also has a substantial impact on system power consumption, which must be reduced to lengthen long-term battery life for operation. The recorded and analyzed physiological data is finally sent over the internet to a remote healthcare facility. As a result, to protect the confidentiality of sensitive personal and medical information, it is also required to use a secure communication route. For improved data security, Secure Sockets Layer (SSL), Public Key Infrastructure (PKI), and appropriate authorization and authentication procedures could all be employed. Finally, to achieve general acceptance of the system for omnipresent health monitoring, it must be affordable and user-friendly. As a result, integrating multiple electrical and MEMS constituents while maintaining efficient data processing, measurement accuracy, low power consumption, information security, and user comfort is a significant design problem for wearable health-monitoring systems. This overview provides a summary of the current research and development level in wearable devices for monitoring health in this study by comparing and summarizing the most important achievements in this field. Wearable health-monitoring devices must meet certain medical and ergonomic requirements for long-term monitoring. Moreover, hardware resources are a significant barrier in a multi-sensor body sensor network system, as the central node must manage a large volume of data from multiple sensor nodes. Finally, to achieve general acceptance of the system for omnipresent health monitoring, it must be affordable and user-friendly. As a result, integrating multiple MEMS and electrical components while maintaining effective data processing, measurement accuracy, information security, and low-power consumption. While designing a sensor module, user comfort is a significant design problem for wearable health-monitoring systems.

Transportation standards for low-power wireless communication have advanced dramatically during the last decade. Low power consumption, small size, and low cost of transmitters are the three main specifications/standards.

The aforementioned technologies have become a reality thanks to advancements in ZigBee, IEEE 802.11, and Bluetooth. Low-power, low-value devices can now be made due to the IEEE 802.15.4a ultra-wideband radio impulse standard. Cloud-based healthcare solutions are ideal for extending health care to rural locations. In addition, a wearable sensor added with environmental sensors can be utilized to monitor applications on home-based such patients/individuals who have been discharged from the hospital for rehabilitation. These sensors keep track of the patient's vital signs, movements, and data analysis methods to determine his or her condition. Blood pressure, pulse rate, and muscle mobility are all physiological parameters that must be taken throughout rehabilitation. Parameters gathered during these procedures can be used to provide diagnostic values and health status to better know the progression of health.

18.2 E-HEALTH SENSORS AND ACTIVITY MONITORING

Medical technology will soon assist us in making fitness results, redefining the relationship among patient–doctor, and lowering medical costs. Electronics that can be worn, such as smartwatches, eyewear, and tracking gadgets are quickly gaining popularity, and technology associated with the medical system will soon assist us in making fitness decisions. A number of researchers are interested in human activity monitoring, and it is a hot topic that necessitates the consideration of wireless communication protocols, methodologies for extracting significant relevant data, and thus design for lightweight, small sensor nodes with minimal power consumption for usage with smartphone technologies. Different varieties of health-monitoring systems and sensors may be employed depending on the subject. The primary concept is to take raw sensor data from appropriate sensor nodes, preprocess it, and then display it. Furthermore, for processing data, a device may be able to send data wirelessly to a nearby data collector. Industry breakthroughs have aided the advancement of unique, well-engineered sensors that can use less power, handle computational complexities, and sense data fast. Human body temperature is the most useful physical parameter tracked by wearable sensors. Variations in skin temperature are a sign of what's going on and can lead to major health issues such as cerebral strokes, hypertension, hypotension, and heart attacks, as well as characterize human body functions. Accelerometers based on piezoelectric, piezo resistive, and capacitance technology are utilized to monitor body movement analysis and fall detection. Electrocardiogram (ECG) sensors placed on the bodies discover heart abnormalities that are chronic. Capacitive sensors in the textile class monitor the human body's capacitance, which could reveal human activities such as arm motions, heart rate, breathing, and eating. Falls can be detected using a mix of sensor networks and home-based robots; here, a network architecture for sensors containing body-worn and ambient sensors is dispersed throughout the environment. Large bodily sensor systems consisting of a film of polyvinylidene fluoride make constitute receptors of sensors in wearable sensors which assist Braille readers. The sensors are attached over a fingertip and shifted across Braille, in order to attain the output. The primary goal of a developing wireless body area network (WBAN) is to acquire a power management mechanism.

The sensor architecture for networks is important for continuous monitoring of human activity system fluctuations. When creating a network, we must consider factors such as scalability, cost, performance, of additional sensor nodes, ease of configuration, processing configuration, and low-power consumption. A large range of IEEE protocols are currently accessible, including:

ZigBee network: Data rates d 250–300 kbps, bandwidth d 2.4 GHz, ranged 100 m, standard IEEE 802.15.4, ranged 100 m, standard IEEE 802.15.4, ranged 100 m, standard IEEE 802.15.4, ranged100 m, standard IEEE 802.15.4, Star, cluster, tree, mesh, topologies are examples of network topologies. Wireless sensor networks (WSN), consumption of low data, and power applications are among the network domains that can be used.

Bluetooth: IEEE 802.15.1, 10 m range, 1–3 Mbps data rates, 2.4 GHz bandwidth topology star. Wireless communication is one of the network regions that can be used for short-range control and monitoring.

Wi-Fi: IEEE 802.11, has a range of up to 5 kilometers. Data speeds range from 1 to 40 Mbps, with bandwidths of 2.4, 3.7, and 5 GHz. Network topologies include tree-topology, star-topology, and peer-to-peer (P2P)-topology. Mobile internet is a network region that is applicable.

WiMAX: IEEE 802.11, 15 km range, 75 Mbps data rate, 3.4 GHz, 2.5 GHz, and 2.3 GHz bandwidth Network topologies include star, tree, and peer-to-peer (P2P). Mobile internet is a network region that is applicable.

IrDA: Data rates were 115 kbps–1 Mbps and the range was 1 m. Almost all health-related appliances are ineligible. MICS: 2 m range, 500 kbps data rate, 400–405 MHz bandwidth.

MICS: MICS offers consumption of low power and data rate; however, it is not recommended for the management of health.

18.2.1 SENSORS FOR E-HEALTH

The following are a list of wearable sensors for tracking human activity[2]:

Heart rate sensors: Rate of the heart is used to determine the physical state and pressure of persons. This is a sensor for monitoring.

Electrocardiogram sensors: Small amplitude electrical impulses linked with heart failure and irregularity are detected or quantified by these sensors. An electrocardiogram sensor, when compared to cardiac rhythm, gives the most detailed information regarding the heart's state.

Electroencephalography sensors: Electroencephalography sensors are primarily utilized to collect data resulting from anesthesia/sedative in individuals who have been medically induced to lose consciousness (coma). Electrical activity is monitored by sensors, which have information about essential signals of brain damage.

Respiratory sensors: The rate of respiration is defined as the number of movements per minute or inhalation and exhalation per unit of time. The main purpose of respiratory rate monitoring is to see if your breathing is normal, irregular, abnormal, or nonexistent.

Blood pressure sensors: Blood pressure sensors detect diastolic, and systolic, blood pressure in humans. They're also useful for determining pulse rate.

Pulse oximetry sensors: This sort of sensor unit measures oxygen saturation level in blood and confirms consumption of oxygen as well as physical and psychological stress.

Electromyography sensors: When muscle cells are neurologically or electrically engaged, electromyography sensors detect the electric potentials they generate.

Temperature sensors: Temperature sensors are common sensors that detect conditions of human body. They constantly capture data, which is wirelessly communicated.

Perspiration sensors: Sweat concentration is measured by these sensors. They are not allowed to directly identify any aberrant state, but they can assist in determining the subject's objective and psychosomatic load.

Motion-tracking sensors: The progression of acceleration or changes in the spatial orientation of body portions can be used to detect exact physical activity.

Strain gauge sensors: Weight load, posture, walking ratio, and other factors are all resolved by these sensors. It can assess human activities and recognize the kind of harm caused by load distribution on the limbs with the help of an intelligent system.

Temperature sensors: Depending on the time during the day and the activity level of the bodily components, different sections of the human body have varying temperatures. Temperature sensors are useful for the above-mentioned tasks and in general for measuring discrete data.

Airflow sensors: It tracks a person's breathing pattern through their nostrils. It also measures continuous data and helps to detect early signs of hypoxia and apnea.

Galvanic response sensors: These sorts of sensors are effective for assessing the electrical transfer of skin and displaying changes in moisture levels. They function similarly to an ohmmeter.

Electrocardiography sensors: Electrocardiography is a technique for determining the electrical and the heart's muscular functionality. The electrocardiogram (ECG) is a continuous data measurement test that is one of the most prevalent and up-to-date medical examinations.

Electromyogram sensors: An electromyogram is used to compute electrical action made in the skeletal muscles by monitoring the muscle electrical movements during contraction and rest. Neuromuscular disease and low back pain are detected with electromyogram sensors.

18.3 INDICATORS THAT CAN BE COMPUTED IN HEALTHCARE MONITORING

The state of health of the human body can be determined by a variety of physical parameters and their specific functional relationships, as well as additional special techniques for evaluating and diagnosing certain diseases. However, to measure blood pressure, breathing rate, wrist pulse, heart rate, skin temperature, and other vital signs, criteria employed in the evaluation of health can differentiate vital signs anyplace on the body. Stress, discomfort, wound healing, angle position, clinical diagnosis, pressure, and emotional control can all be measured using body motions in the feet, throat, chest, face, and limbs. Gesture recognition, rehabilitation, athletic training, express identification, and other applications commonly involve motion or body sensors. Furthermore, metabolic rate-finding sensors are implanted in oral cavities and sweating areas to monitor pH, lactic acid, glucose, and body appearance.

18.3.1 VITAL SIGNS

Vital indicators are extremely significant in intensive care units. Each vital sign in real-time monitoring infers serious disorders such as heart attack, anesthesia, and hypertension, which are significant in high-risk category. Numerous vital indicators have been transformed because of improvements in wireless wearable sensing devices.

Pulse and heart rate: The heart is a vital organ in the human body because it supplies energy to all sophisticated self-organizing systems. As a result, any problems with cardiac function may result in death; continual inspection and monitoring are required to comprehend the heart's state, which is critical in the healthcare-monitoring system. In the field of healthcare, heart rate measurement is a crucial indicator. ECG and photo plethysmography (PPG) signals can both be utilized to determine this crucial value. An ECG is essential for protecting the heart from significant violations since it gives extensive information about the heart's working stages. Electrodes are used to collect data in ECGs, and they are usually placed on the wrists, throat, and ankles of the patient.

Breathing: Breathing problems can lead to serious problems like asthma, forgetfulness, chronic disease, and interruption of sleep. Masks with built-in sensor modules are attached to mouths and noses, and inhaled/exhaled airflow can be used to detect breathing rate or moisture rate. Variations in chest volume can cause a variety of movements in our bodies, which can be observed using wearable sensor nodes.

Saturation of blood oxygen: Blood oxygen saturation (SpO_2) monitoring, in addition to pulse oximetry results, is an expensive but necessary approach to quickly detect oxygen levels with PPG gear. PPG typically acquires blood vessel variation waveforms with two wavelengths (660 and 905 nm). When oxygen is linked to hemoglobin, the spectrum of absorbance of hemoglobin changes. Using oximetry data, it is possible to estimate the amount of oxygen passed through blood cells. This measurement aids in detecting the patient's state.

Glucose levels in the blood: In diabetic patients, blood glucose (BG) levels must be monitored. It should be emphasized that BG cannot be tested in regular circumstances, such as in a medical setting. Diabetes involves various physiological abnormalities, necessitating continual insulin injections to maintain diabetic parameters. The Medtronic Continuous Glucose

Monitoring (MCGM) device, which utilizes an adhesive patch and a needle to ascertain blood glucose levels and has sent data remotely to a wearable device for insulin management in the body.

Skin perspiration: Perspiration on the skin is a physical indicator that can be used to assess an individual's reaction to common conditions. Neurological responses to life-threatening situations cause sweating of skin, which affects the skin's conductivity and the amount of sweat generated by the sweat glands. Skin secretion is monitored as a vital physiological sign with various uses in aerobics, as perspiration and heart rate changes can assist disclose a player's mental tension in human behavior.

Capnography: Capnography is a low-cost method of estimating human aeration. It is used to avert clinical issues and guarantee patient safety by monitoring CO levels throughout the respiratory phase. Capnography is a noninvasive arterial oxygenation technique that is widely used. It primarily uses the external medical environment to monitor sleep apnea problem. On reciprocal/mutual devices, it becomes a common symptom, and it serves as a beginning point for life-saving discoveries.

Body temperature: In common, BT is the outcome of a sensation of equilibrium in the body's heat creation and heat loss. Because of the high temperature, blood temperature eliminates the majority of de-functionalization.

Blood pressure: The force exerted by blood clots on the inside of the artery is called blood pressure (BP). More crucially, it displays data on the circulation of blood and cellular oxygen delivery while the heart is in systolic and diastolic states. Pressure cuffs that inflate and a stethoscope are commonly used to measure blood pressure. Blood pressure is affected by a variety of physiological factors, including blood volume, cardiac output, vascular resistance, and artery elasticity. Many researchers believe that high-performance electrodes will be the solution to this dilemma. Furthermore, invisible biocompatible tattoo electrons will be simply worn to capture physical data on a constant basis.

18.4 TRANSPORTABLE DEVICES

With the advancement of current technical breakthroughs such as computer networks, micro-electromechanical systems, and wireless technologies, wearable devices are used in transportable sensors and health services. Body sensory control devices for wrists are split into wearable transportable devices, body garments, and heads based on current usage.

18.4.1 WRIST DEVICES

The primary goal of wrist-mounted devices is monitoring of physiological conditions, but their effectiveness is dependent on the hardware's compactness in turning raw data into readable statistics in real time; activity bracelets, smartwatches, and a few wrist gadgets are examples. As previously stated, blood pressure measurement is an important indicator for specific health concerns; however, typical blood pressure measurements are inaccurate and hard to handle when the person is moving. To address these challenges, some researchers have offered various ways. Hsu and Young[3] presented a special skin surface coupled configuration that could capture real-time BP waveforms and send them to a computer system or smartphone for analysis. A wrist-based PPG heart rate sensor and a bracelet-based photoplethysmogram heart rate sensor that can counter motion artifacts in daily activities are presently used in a range of appliances. Furthermore, smartwatches can be used to investigate tremor dysfunction. As a result, smartwatches can be employed in clinical settings and are well-received by patients.

18.4.2 DEVICES WORN ON THE HEAD

A head-mounted module that delivers sensory input is now included in certain wearable smart glasses, thanks to recent improvements. Sensors such as gyro meters, barometers, accelerometers, radio tags, magnetometers, and GPS are used in a few smart glasses. While driving or exercising, for example, these innovative smart glasses capture critical information about their owners' health and can present statistics about their behaviors on the display.

18.4.3 E-TEXTILES

In healthcare management, smart textiles or gadgets connected to garment material are commonly used to detect human physiological indicators, pressure, and body acceleration. Devices based on textile diagnostic use electrodes, textiles, and sensors. An e-textile is a textile that has electrodes and can monitor human behavior.[4]

18.5 ATTACHABLE DEVICES

It is increasingly simple to construct attachable or connected devices due to the recent advancements in micro-electromechanical systems, ICT, sensor technologies, and data analytics. These monitoring devices that can be attached can be employed in isolated medical management in healthcare devices. Attachable devices are unique in that they enable precise and constant sensing without interfering with a user's comfort, and usual movement, skin type adaptation, or litheness. For optical transparency and fabrication simplicity, polyethylene terephthalate and polyurethane are used; more crucially, flexible, attachable, soft substrate templates are the most significant sensor elements for attachable devices. Electrical circuits' tiny size is critical in the development of microelectromechanical and attachable devices systems which enables shrinking and cost reduction in electronic components. A fusion of smart attachable sensor devices is required for simultaneous health monitoring of physical signs that are closely related to physical status.

18.5.1 SKIN PATCHES

Wearable, transferable, and soft electronic devices are linked to skin surface and serve as a novel framework for mechanical advice and constant medical observation. Human wearable skin patches are utilized in cardiovascular, heat, and sweat sensors. The authors of Luo et al.[5] created improved functional materials with epidermal ECG and piezo resistive sensors for cuffless bold stress tests. Attached stretchable pulses were developed by Kim, Kim, Kwon, and Lee.[5] Sweat or bodily fluid is the primary measurement method because it contains vital biomarkers such as electrolytes, small molecules, and proteins. Biomarker inspection that is unbreakable is critical for monitoring human health in medication. More effort is essential to develop an epidermal biomonitoring system that with a single wearable sensor, it measures both skin and sweat fluid. A blood-associated sweat bio-sensing system with carbachol as a refreshment might be developed, according to the bulk of Heikenfeld's research.[6]

18.5.2 CONTACT LENS

The physical condition of the eyes and tears can be monitored with intelligent contact lenses. With the help of tear fluid's bio-electrical, optical, and chemical conductivity, a number of researchers have developed numerous kinds of contact lenses. March, Lazzaro, and Rastogi[7] used a photo fluorometer and a fluorescent contact lens comprised of liquid hydrogel nano spheres to monitor glucose levels.

18.5.3 IMPLANTABLE MEDICAL DEVICES

These devices have now merged with the advancement of micro-electromechanical systems technology, which has been aided by chemical, biological, and mechanical expertise. While these gadgets are inextricably linked to the human body, they should not be influenced negatively by it. These devices' wireless remote-based capability is critical not only for conveying patient-monitored information, but also for taking into account battery state and functional improvement. The majority of implanted devices have a power cell and bio-compatible resources that run along programmable rails. For cardiac patients, a pacemaker is a commonly used medical device that is placed in the body; the device detects abnormal heart-beats and restores regular beats when they are detected. Presently, most wearable health devices are strapped or taped to the human body.

18.5.4 INGESTIBLE PILLS

Ingestible pills developed are used to track when people take their medications. A little pill enters the stomach, has biochemical reactions with stomach acid, and sends respective signal wave to a device linked to the person's body. More development on this, such as tablets imbecile with real-time acquired data, will be required in the future.

18.6 APPLICATIONS OF WEARABLE SENSOR

In the fields of rehabilitation, identification of pathologies, health monitoring, home-based applications, ambient sensors, wearable and safety monitoring at early stage are extremely useful.

18.6.1 SAFETY MONITORING

Huge numbers of gadgets have been produced as a result of technological advancements, including home healthcare, safety monitor appliances for falls, Alarm-Net, alarm messages to alert an emergency team. Wearable sensors in some systems detect problems as separate occurrences distinct from normal events and physically convey an alert message to a medical response center. Patients/individuals can also be located using recent mobile technology to detect falls.

18.6.2 TREATMENT EFFICACY EVALUATION

One of the qualitative strategies for disease management is to assess treatment efficacy. It is possible to analyze the patient's condition by knowing what happens between outpatient appointments and prescription therapies. Additionally, understanding the ideal medication/treatment duration is vital for evaluating the effectiveness of new remedial treatments.

18.6.3 EARLY DETECTION OF DISORDERS

Wearable technology is becoming increasingly important for detecting early changes in patient's health that require clinical diagnosis. Identifying the patient's activities is the greatest technique to get an early detection. A decrease in activity is thought to indicate the likelihood of a worsening clinical condition in the individual with the advancement of sophisticated machine learning (ML) developments.

18.6.4 HOME REHABILITATION

Applications on home rehabilitation are growing with the advancement of wearable sensor technology. Wearable sensors are used in this study to allow for the implementation of rehabilitation examination plans that frequently influence a combination of sensing mechanics. A simple test to determine whether the patient has completed a well-organized exercise programme; activities can be adjusted by taking into account the patient's demands. Since wireless sensor-based health management technologies allow for the gathering of physiological and progress data, rehabilitation intervention in the home can be facilitated.

18.7 A BRIEF COMMENT ON SENSOR NETWORKS OF THE FUTURE

IoT is rapidly expanding in appliances ranging from private gadgets to industrial instruments where sensors of different categories are wirelessly interconnected to the internet. Simply said, the Internet of Things (IoT) uses sensor nodes to connect the physical and virtual worlds of humans and electronics. The primary goal is to design a system model that can automatically respond to and monitor environmental predictions like wildfires, hurricanes, air pollution, volcanic effects, seawater levels, and other controlled applications such as defect detection based on equipment, traffic control, construction safety, healthcare systems, armed applications, and other human-based applications. Specific specifications for wireless sensor networks could be defined based on this wide variety of use cases in various environmental and applications based on human need. The advance of micro-electromechanical systems can develop sophisticated, compact sensor nodes with limited computing and processing capabilities has sparked interest in wireless sensors around the world. Each sensor node has one or more sensors, as well as a small memory, processing unit, and transmission capabilities, as well as power management unit. The sensor's individual operation is mostly dependent on the power source. The sensor node is equipped with a variety of biological, magnetic, thermal, optical, and mechanical sensors for computing properties of environment. The present spectrum range is often subject to a licensing scheme. Customers must obtain a license from a local regulator to send transmissions in certain frequency bands. The majority of frequency bands were created for scientific, medical, and industrial applications; however, bands greatly range on different countries. 5 GHz, 2.4 MHz, 915 MHz, 868 MHz, 433 MHz are notable frequency bands. Higher frequency bands often have a wider bandwidth that allows for higher data throughput and more channels; lower radio frequency bands have a higher efficiency but are only obtainable inside a building.

Yang et al.[8] use a belt-type sensor to draw attention to target positioning issues. In this study, to deploy IoT sensor nodes, routing techniques that efficiently save energy are employed, as well as a decision mechanism to ensure security and dependability. For optical fiber transmission,[9] a line configuration generator and an online configuration reader were presented as part of a health monitoring system. The composition of clusters, content processing, and position management are all mentioned. The authors' main goal in Lee, Kwon, Song, Jeon, and Kim[11] was to depict the multi-model assignment of a received signal strength indicator (RSSI) in conjunction with medium access methods and signal modulation.

18.8 RESEARCH CHALLENGE

The fundamental goal of a wearable sensor health parameter monitoring system is to let people to live active lifestyles and be independent in their own familiar home like or home environment while maintaining noninvasive,[12] constant, seamless, and nonintrusive health and physical well-being observation. The rapid advancement of technology in recent decades has enabled the development and use of low-power, miniature actuators, sensors at low cost, powerful computers, and electronic modules, leading to noninvasive, and constant observing of health status at a very less cost.

Periodic health monitoring does provide comprehensive information on a person's health during the course of time. Wearable actuators and sensors, in combination with modern communication technology, have ushered in a new era of low cost remote health monitoring and healthcare services. Data analysis and monitoring, as well as predictive algorithms, may be included in the systems, which could lead to a more confident prognosis of some diseases, as well as early detection and treatment. If a potential health risk is recognized, using secure wireless platforms such as the internet or a cellular network, the system can alert and inform the individuals involved or health professionals, allowing for prompt medical assistance. Incorporating smart textiles technology into wearable healthcare systems, such as textile-based connection for sensors, could result in nonintrusive, more comfortable platforms for monitoring health. Furthermore, the substantial advancements were made in the last decade, there are still issues that need additional R&D to enhance the effectiveness of the cardiac health-monitoring system. First, bio-potential (EDA, ECG) monitoring devices frequently have a low SNR ratio (signal-to-noise ratio), which is mostly caused by noise created by the user's movement. High input impedance front end amplifiers or flexible electrodes can help to reduce motion artifact (MA). Empirical mode decomposition (EMD), time-frequency analysis, adaptive filtering, and independent component analysis (ICA) are examples of signal processing techniques and can also be used to improve SNR. Second, the security and privacy of the user's sensitive information regarding medical state is a major problem in wearable healthcare management system. More work is needed to build algorithms that will assure highly secure communication channels in short-range, low-power wireless devices. Third, in the case of long-term monitoring systems, less consumption of power and excellent efficiency in managing energy are essential. The system's power requirements can be met by utilizing more efficient batteries, low-power components, or techniques for energy harvesting. Battery life can also be extended by assuring that the sensors "sleep and wake up" in an appropriate manner with no compromising on the intended measurement frequency. Fourth, solutions must be inexpensive, simple to use, unobtrusive, and interoperable across many computing platforms to gain universal acceptance. To avoid losing the most crucial clinical data, just a small number of electrodes and sensors should be employed. As a result, greater R&D activities are required to improve the systems' usability. Finally, a quick overview of textile-based sensors, emphasizing their potential for monitoring physiological indications, is given. To ensure good sensor performance, more studies are required on the right embedding technique and sensory materials selection, as well as a robust sensor-skin contact. Furthermore, while making a smart health-monitoring module with textiles for long term, the sensors' durability and signal integrity should be improved throughout time and washing cycles.

18.9 CONCLUSIONS

It has been discussed and compared different wearable technologies involved; also their suitability for health-monitoring devices that can be worn is evaluated. In general, the systems collect and transmit many physiological indicators from the human body to a control node. The sensor data are analyzed, processed, and transmitted to a remote facility through the gateway node for healthcare personnel to be analyzed. More scientific research and technological development is required to provide data security and privacy, a dependable communication link, robust data compression techniques, and energy efficiency.

KEYWORDS

- **biomedical monitoring**
- **monitoring e-health**
- **wearable sensors**
- **intelligent sensors**
- **wireless sensor networks**

REFERENCES

1. Dementyev, A.; Hodges, S.; Taylor, S.; Smith, J. Power Consumption Analysis of Bluetooth Low Energy, ZigBee and ANT Sensor Nodes in a Cyclic Sleep Scenario. In

Proceedings of the 2013 IEEE International Wireless Symposium (IWS), Beijing, China, April 14–18, 2013; pp 1–4.

2. Amjadi, M.; Kyung, K.-U.; Park, I.; Metin, S. Stretchable, Skin-Mountable, and Wearable Strain Sensors and Their Potential Applications: A Review. *Adv. Funct. Mater.* **2016,** *26* (11), 1678–1698.
3. Hsu, Y.-P.; Young, D. J. Skin-Surface-Coupled Personal Health Monitoring System. In *Proceedings of the 2013 IEEE Sensors*, Baltimore, MD, USA, November 4–6, 2013, 2013; pp 1–4.
4. Liu, X.; Lillehoj, P. B. Embroidered Electrochemical Sensors for Biomolecular Detection. *Lab Chip* **2016,** *16*, 2093–2098.
5. Kim, J.; Kim, N.; Kwon, M.; Lee, J. Attachable Pulse Sensors Integrated with Inorganic Optoelectronic Devices for Monitoring Heart Rates at Various Body Locations. *ACS Appl. Mater. Interfaces* **2017,** *9*, 25700–25705.
6. Hauke, A.; Simmers, P.; Ojha, Y. R.; Cameron, B. D.; Ballweg, R.; Zhang, T.; et al. Complete Validation of a Continuous and Blood-Correlated Sweat Biosensing Device with Integrated Sweat Stimulation. *Lab Chip* **2018,** *18*, 3750–3759.
7. March, W.; Lazzaro, D.; Rastogi, S. Fluorescent Measurement in the Non-Invasive Contact Lens Glucose Sensor. *Diabetes Technology Therapeutics* **2006,** *8*, 312–317.
8. Yang, Y.; Ok, Y. S.; Kim, K. H.; Kwon, E. E.; Tsang, Y. F. Occurrences and Removal of Pharmaceuticals and Personal Care Products (PPCPs) in Drinking Water and Water/Sewage Treatment Plants: A Review. *Sci. Total Environ.* **2017,** *596*, 303–320.
9. Bandodkar, A. J.; Mohan, V.; López, C. S.; Ramírez, J.; Wang, J. Self-Healing Inks for Autonomous Repair of Printable Electrochemical Devices. *Adv. Electr. Mater.* **2015,** 1, 150–289.
10. Xu, J.; Wang, S.; Wang, G. N.; Zhu, C.; Luo, S.; Jin, L.; et al. Highly Stretchable Polymer Semiconductor Films through the Nanoconfinement Effect. *Science* **2017,** *355*, 59–64.
11. Lee, J.; Kwon, H.; Song, J.; Jeon, M.; Kim, J. Development of a Handheld Line Information Reader and Generator for Efficient Management of Optical Communication Lines. *Sensors* **2017,** *17*, 1950.
12. Renjith V. S.; Hency Jose P. S. A Noninvasive Approach Using Multi-Tier Deep Learning Classifier for the Detection and Classification of Breast Neoplasm Based on the Staging of Tumor Growth. In *2020 International Conference on Decision Aid Sciences and Application (DASA)*, 2020; pp 12–16. doi: 10.1109/DASA51403.2020.9317038.

CHAPTER 19

Design and Simulation of Cantilever Structured Flexible Polymer-Based Piezoelectric Pressure Sensors for Biomedical Applications

LOKESH SINGH PANWAR* and VARIJ PANWAR

Department of Electronics and Communication, Polymer Sensor and Actuator Lab, Graphic Era Deemed to be University, Dehradun, India

**Corresponding author. E-mail: lokesh31j@gmail.com*

ABSTRACT

Recent developments in electronics and smart materials allow the design of small size, flexible, lightweight and low-cost devices that are widely used as health monitoring devices, which are based on flexible and wearable sensor technologies that offer precise and efficient health monitoring of humans for better life prospects. This paper demonstrates an innovative approach to the design and simulation of flexible and self-powered pressure sensors for biomedical applications using piezoelectric material and also analyzes the sensing performance of it. The structure of this flexible pressure sensor is designed using a piezoelectric material based layer that is sandwiched between electrodes. PVDF (Polyvinylidene fluoride) is a piezoelectric polymer that is exploited as piezoelectric material for sensing layer in the proposed sensor. Due to the piezoelectric polymer-based sensing layer, this sensor is more flexible, self-powered, lightweight and cost-effective. This proposed sensor induces a self-sensing voltage when it sense a pressure, thus this sensor measures the applied pressure as induces electrical voltage. The

Intelligent Technologies for Sensors: Applications, Design, and Optimization for a Smart World.
S. Kannadhasan, R. Nagarajan, & Alagar Karthick (Eds.)

simulation of the proposed design shows the linear characteristic response and good sensitivity. The proposed device was designed with different thicknesses (40 μm, 60 μm, 80 μm, and 100 μm) and cross-sectional areas (0.06, 0.12, and 0.18) of the sensing layer, and simulation analysis revealed that altering the dimensions (thickness and cross-sectional area) of the sensing layer improved sensor sensitivity. The proposed pressure sensor using piezoelectric polymer is designed and optimized by COMSOL Multiphysics.

19.1 INTRODUCTION

With the continuous and rapid development of healthcare monitoring, soft robotics technology, and artificially electronic skins, the need for miniature sensor technology is attracting more and more attention.[1] To realize this, research organizations have started an integrated effort toward biomedical sciences and electronics technologies to measure and analyze a wide range of human physiological parameters. Advancement in electronics, particularly sensor technology, microcomputers, and digital signal processing has greatly affected our day-to-day lives, particularly the health care domain.[2] The front end of most of the medical equipment is sensors that are sophisticated components that translate one form of energy into another. The evolution in healthcare devices has led to a smart, small size, flexible,[3] multiparameter, remote-controlled, and noninvasive sensor that converts biological qualitative inputs such as various human pressures, temperature, and pH body fluids into quantitative measurable electrical output. These sensors are used as health care indicators, which also regulate several human body pressures including intraocular pressure and blood pressure.[4] This research paper discussed the wearable, flexible, cost-effective polymer-based piezoelectric pressure sensor.

The flexible pressure sensors are used for accurate and precise measurement of pressure in different fields such as healthcare devices,[5] automotive industry, and robotics industry.[6] There are various sensing mechanisms used in pressure sensors that are capacitive, piezoresistive, and piezoelectric-sensing mechanism.[7] Each sensing mechanism has unique merits and demerits. Capacitive pressure sensors are effective for high sensitivity and dynamic range, but it suffers nonlinear output. Piezoresistive sensing techniques generally capable of lower sensitivity and temperature compensation are required for an accurate and efficient measurement.[8] Also external power is required for both capacitive and piezoresistive sensing operation, whereas piezoelectric pressure sensor that requires no external power for operation also offers the high sensitivity, high durability, high linearity, and low noise device.[9] Therefore, for accurate and precise blood pressure measurement of the human body, a noninvasive, versatile, and cost-effective piezoelectric pressure sensor has been discussed in this paper. The noninvasive method is significant and largely acceptable by receivers, has almost no effect on the human body, and is easy to operate and sterilize. Thus, the piezoelectric pressure sensors meet the design specifications for bio-related pressure-measuring devices.

This paper demonstrates an innovative, flexible, self-powered, and highly sensitive piezoelectric sensor for accurate and precise blood pressure measurement. The proposed sensor is a cantilever-structured flexible pressure sensor (FPS) using Polyvinylidene fluoride (PVDF) piezoelectric polymer that is sandwiched between two electrodes and polydimethylsiloxane (PDMS) polymer-based cover layer supports whole device and provides flexibility to the proposed sensor. In this paper, pressure equivalent to human blood pressure range (50–200 mmHg) is applied to this proposed piezoelectric pressure sensor with the help of COMSOL Multiphysics, and the proposed sensor generates a sensing voltage when it senses a different blood pressure value.

19.2 STRUCTURAL AND MATERIALS OF PROPOSED SENSOR

The proposed flexible pressure sensor using piezoelectric polymer is a cantilever-structured sensor shown in Figure 19.1. In this design, the PVDF piezoelectric polymer material is used as a sensing layer with thickness ranging from 50 to 100 µm sandwiched between two silver electrodes that are covered by top and bottom PDMS layers with 2 mm thickness; PDMS is a biocompatible, elastic transparent material that provides more flexibility to the proposed sensor and is widely used in wearable devices. The proposed flexible piezoelectric pressure sensor is cantilever structured due to its high sensitivity to measure pressure or stress.[10,11] The proposed sensor is an ultra-thin skin-attachable sensor to monitor real-time arterial pressure.

The ultrathin geometry of the sensor provides sufficient flexibility to allow the proposed sensor to be transferred onto the microscopic arbitral curvature of body skin and generates sensing voltage by the arterial pressure. For the design and simulation of the proposed sensor, the COMSOL Multiphysics is used. COMSOL Multiphysics is a dominating tool for designing and simulating various structural and material-based devices.

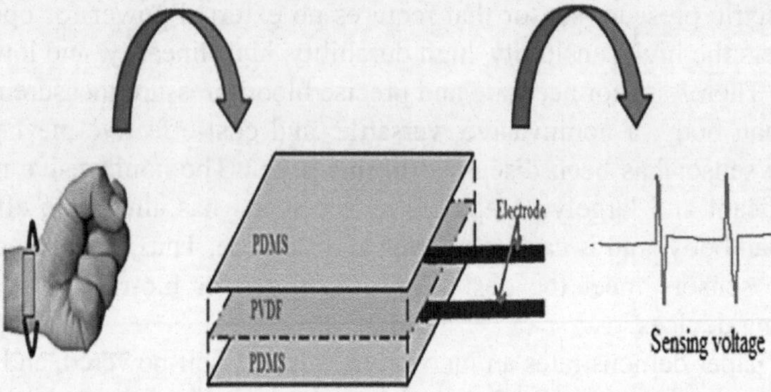

FIGURE 19.1 Structure of the piezoelectric sensor using PVDF polymer.

The working mechanism of the proposed sensor is based on the periodical deflection/deformation on the wrest wall of human skin due to the corresponding artery pressure; the PVDF-based piezoelectric sensing layer is mechanically stretched which is covered by PDMS thin layers, and corresponding voltage is generated, as shown in Figure 19.2.

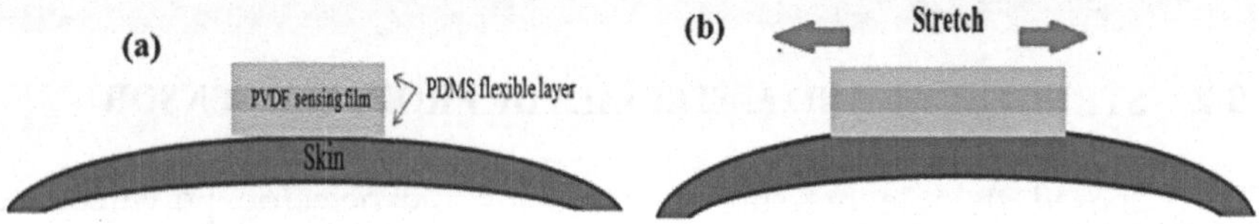

FIGURE 19.2 Proposed sensor mechanism (a) at rest state and (b) at deformation state.

In the proposed flexible pressure sensor, a piezoelectric material is employed as the sensing layer. When external mechanical pressure, force, stress, or strain is applied to a specific type of crystal material, an electric voltage is generated; this phenomenon is known as piezoelectric effect or piezoelectricity,[11] and these crystal materials are called piezoelectric. In this sensor PVDF material is used as piezoelectric material. The PVDF is a semicrystalline, highly pure organic polymer. Among various piezoelectric materials, PVDF has attracted the interest of researchers due to its excellent properties such as biocompatible, high flexibility, high mechanical strength, good piezoelectricity, high thermal stability, high chemical resistance, low permittivity, and outstanding membrane-forming properties.[12] PVDF is synthesized by the polymerization process of 1,1-difluoroethylene (CH_2=CF_2) monomers,[13] as shown in Figure 19.3, and this polymerization occurs by suspension or emulsion at a temperature from 10 to 150°C and pressure of 10–300 atm.

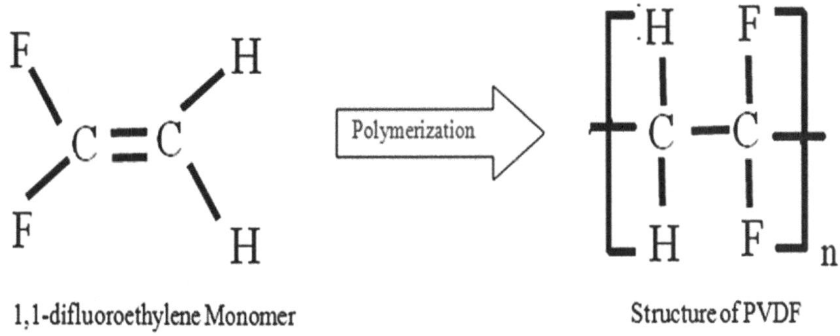

FIGURE 19.3 Polymerization of PVDF.

PVDF is a semicrystalline polymer and has four different crystalline phases, α, β, γ, and δ-phases.[14] The different phases of PVDF material differ by their chain conformations.[15] The electric dipole moments in the α-phase of PVDF are oriented in opposite directions, then net polarization is zero, which shows that the α-phase of PVDF has high thermal stability. The β-phase is the only one that exhibits a spontaneous polarization, the C–F bonds are polarized, all electric dipoles are aligned in the same direction, the highest electronic dipole moment occurs, and thus piezoelectricity occurs in the β-phase of the PVDF, as shown in Figure 19.4. The β-phase of PVDF is the desired phase to its piezoelectric sensing property.[16–18] The β-phase of a PVDF crystallite consists of hydrogen atoms which have opposing fluorine atoms in all-trans-linkage, which gives the highest dipole moment per monomer, therefore the β-phase of PVDF is generally preferred for piezoelectric materials to measure vibration or pressure.[19]

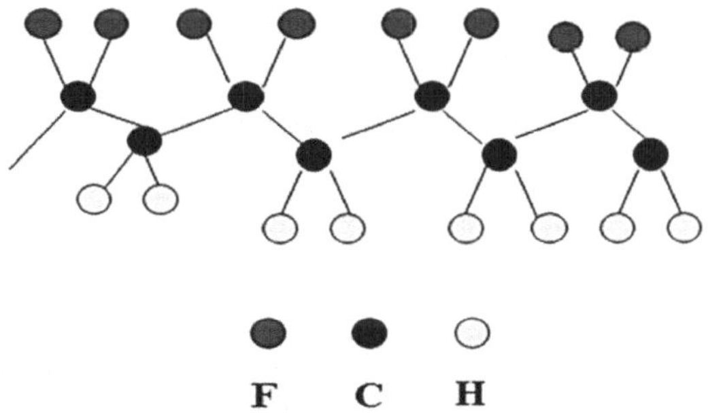

FIGURE 19.4 β-Phase of PVDF.

PVDF piezoelectric material is usually anisotropic dielectric material,[20,21] and it has a unique atomic structure in which molecules are ionically bonded and positive and negative ions make a pair called unit cells; there is equilibrium between the positive and negative charges, which remains neutral on the polar axis, no output voltage is generated that is shown in Figure 19.5.

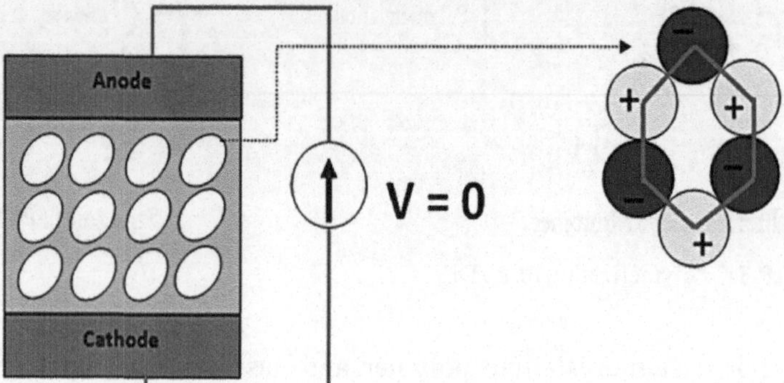

FIGURE 19.5 Piezoelectric material crystalline structure without applied pressure.

At the point, when mechanical force, stress, or pressure are applied to the PVDF piezoelectric sensing layer, the geometry of the atomic structure of PVDF is changed due to the total net movement of both the positive and negative ions to each other; electric dipole is produced and polarization occurs, as shown in Figure 19.6. Finally, both the positive and negative charge carries are split by dissolving the neutrality of the PVDF molecule and enhancing a surface charge density near to the electrodes.[22] Consequently, both positive and negative charges are collected at the opposite end of the material; an electric voltage is generated. Thus, this PVDF converts from a dielectric material to a charged piezoelectric material. The induced self-generating output voltage is directly proportional to the amount of stress or pressure applied to piezoelectric sensing layer.

The advantages of piezoelectric materials are high sensitivity and fast response time, so it is useful for the detection of vibration or dynamic pressure change.[23] Since piezoelectric materials generate a self-sensing voltage, it is a good candidate for low power consumption; therefore, it is also called a self-power sensor or active sensor.[24,25]

When force F is applied to the PVDF-based piezoelectric material thin film whose length is L, width is W, and thickness is t, it results in electric dipole or polarization; both the positive and negative charge carriers get separated by terminating the neutrality of the molecule,[26] the charges collected on the surface of the electrode,[27] total induced surface charge Q collected in the electrode is

$$Q = d \cdot F \quad \text{Coulomb} \tag{19.1}$$

where F = applied force; N, d = charge sensitivity, or piezoelectric constant of crystal; C/N.

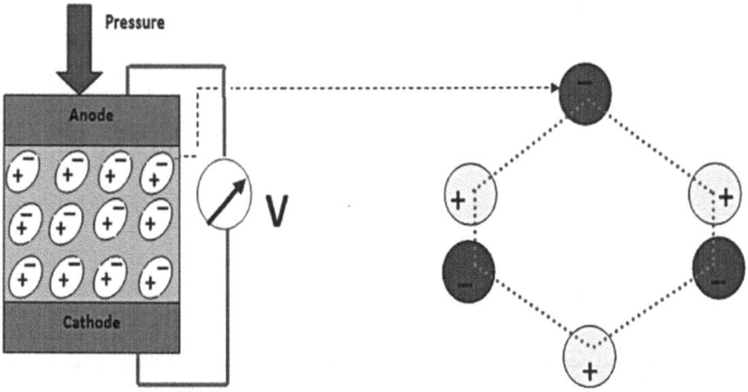

FIGURE 19.6 Piezoelectric material crystalline structure with applied pressure.

Since polarization occurs, the positive and negative charges are collected in the opposite electrode of piezoelectric materials; capacitance (C_p) is produced between two opposite electrodes. Output voltage (V_0) is induced due to the formation of capacitance between the electrodes.

The induced output voltage is

$$V_0 = \frac{Q}{C_p} = \frac{Q}{\epsilon_0 \epsilon_r \frac{A}{t}} = \frac{dFt}{\epsilon_0 \epsilon_r A} = \frac{dtP}{\epsilon_r \epsilon_0} = g\,t\,P = g\,t\frac{F}{A} \tag{19.2}$$

where A is the cross-sectional area of the piezoelectric (PVDF) thin film which is W·L (m^2); t is the thickness of the PVDF film before applying force F (m); ϵ_0 is the dielectric constant of free space; ϵ_r is the dielectric constant of material (PVDF); P is the applied pressure on piezoelectric film which is $\frac{F}{A}$; g is the voltage sensitivity of the piezoelectric materiel which is $\frac{d}{\epsilon_r \epsilon_0}$ (V m/N).

According to Equation (19.2), the induced output voltage depends on piezoelectric material properties, and sensor dimensions for the applied input force, pressure, or stress. The piezoelectric material properties are piezoelectric constant (d), voltage sensitivity (g), and the relative dielectric constant ($\in_r$); therefore, induced output voltage (V_0) is high if piezoelectric material has high piezoelectric constant (d), high voltage sensitivity (g), and the low relative dielectric constant ($\in_r$). Also, the output sensing voltage (V_0) is increased with the sensor dimensions: thickness (t) and the cross-sectional area (A) of the sensor.[28]

19.3 SIMULATION RESULTS AND ANALYSIS

The proposed PVDF-based piezoelectric pressure sensor is performed with various investigations such as self-generating sensing voltage, linearity, and sensitivity analysis, which plays a major role in the performance of the proposed sensor.

19.3.1 SENSING VOLTAGE ANALYSIS

19.3.1.1 EFFECT OF THICKNESS (T) OF SENSING LAYER ON SENSING VOLTAGE OF PIEZOELECTRIC SENSOR

The PVDF piezoelectric polymer-based pressure sensor produces self-generating sensing voltage by applied distinct pressure. The output sensing voltage for different applied pressures is listed in Table 19.1 and Figure 19.7 shows induced sensing voltage after applying distinct pressure for different thicknesses of the sensing layer of the proposed flexible sensor. The described pressure sensor is utilized in a limited pressure range of 60–200 mmHg, indicating that increasing the thickness (t) of the sensing layer of the proposed sensor improves the output self-generating voltage. The graph plot of output voltage versus applied pressure for different thicknesses of the sensing layer is shown in Figure 19.8, which also shows that the output sensing voltage rises linearly with the thickness of the sensor's sensing layer. When thickness of sensing membrane t = 40 μm, linearity of sensor is very less; hence, linearity is increased as thickness of sensing layer increases.

19.3.1.2 EFFECT OF CROSS-SECTIONAL AREA (A) OF SENSING LAYER ON SENSING VOLTAGE OF PIEZOELECTRIC SENSOR

Figure 19.9 shows the simulation of output self-generating voltage at the different cross-sectional areas of the sensing layer and it is listed in

TABLE 19.1 Output Sensing Voltage (V) Versus Applied Pressure (P) at Different Thicknesses of Sensing Layer.

Applied pressure (mmHg)	Output sensing voltage (V)			
	t = 40 μm	t = 60 μm	t = 80 μm	t = 100 μm
60	0.002	0.043	0.054	0.061
80	0.003	0.053	0.072	0.081
100	0.0042	0.072	0.09	0.101
120	0.0045	0.086	0.108	0.122
140	0.005	0.101	0.126	0.142
160	0.006	0.115	0.144	0.162
180	0.0071	0.13	0.162	0.182
200	0.0073	0.144	0.179	0.203

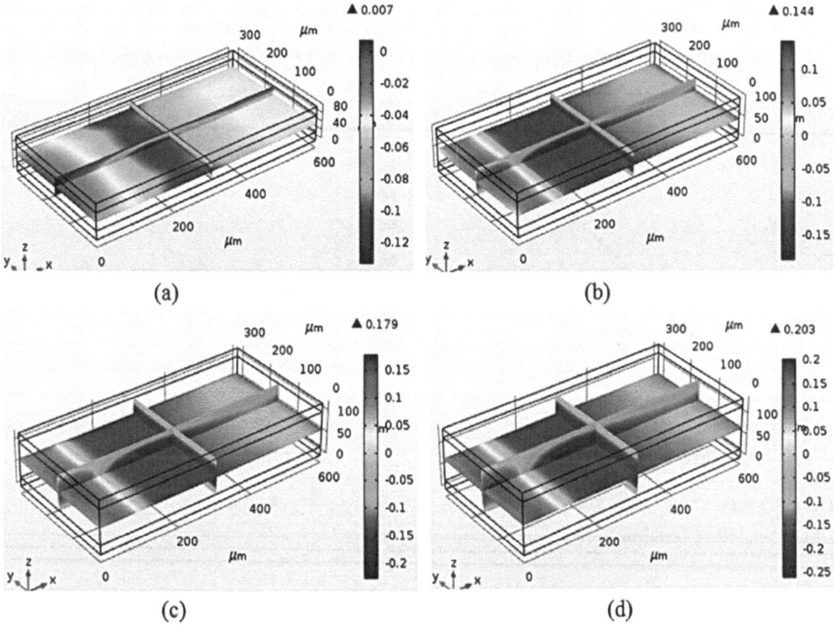

FIGURE 19.7 Simulation of the output voltages at the different thickness (t) of the sensing layer (a) t = 40 μm, (b) t = 60 μm, (c) t = 80 μm, and (d) t = 100 μm.

Table 19.2, which indicates that output self-generating voltage is increased by reducing a cross-sectional area (A) of the sensing layer of the proposed sensor. Figure 19.10 represents the graph plot of output voltage versus applied pressure for the different cross-sectional areas of the sensing layer, which also indicates that output sensing voltage is linearly increased with reducing the cross-sectional area of the sensing layer of the sensor, which also indicates that output sensing voltage is linearly increased with reducing the cross-sectional area of the sensing layer of the sensor.

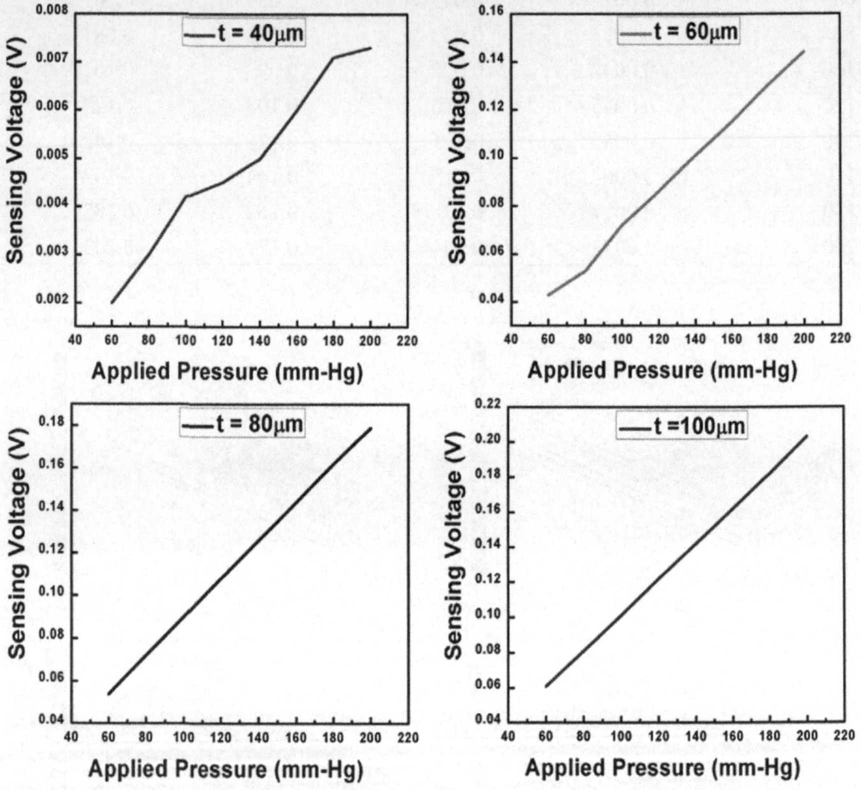

FIGURE 19.8 Graph between applied pressure (mmHg) and output voltage (V) at different thicknesses (t) of sensing layer.

19.3.2 SENSING VOLTAGE ANALYSIS

19.3.2.1 LINEARITY ANALYSIS

The linearity is the important desirable parameters that impact sensor accuracy. The nonlinearity starts with the line from zero output pressure and extends to the rated output pressure; this line is known as an end-point straight line. The maximum deviation from "the end-point straight line" represents nonlinearity. Figures 19.8 and 19.10 show the linear relationship between output sensing voltage and input applied pressure; therefore, the proposed piezoelectric pressure sensor is a more accurate sensor.

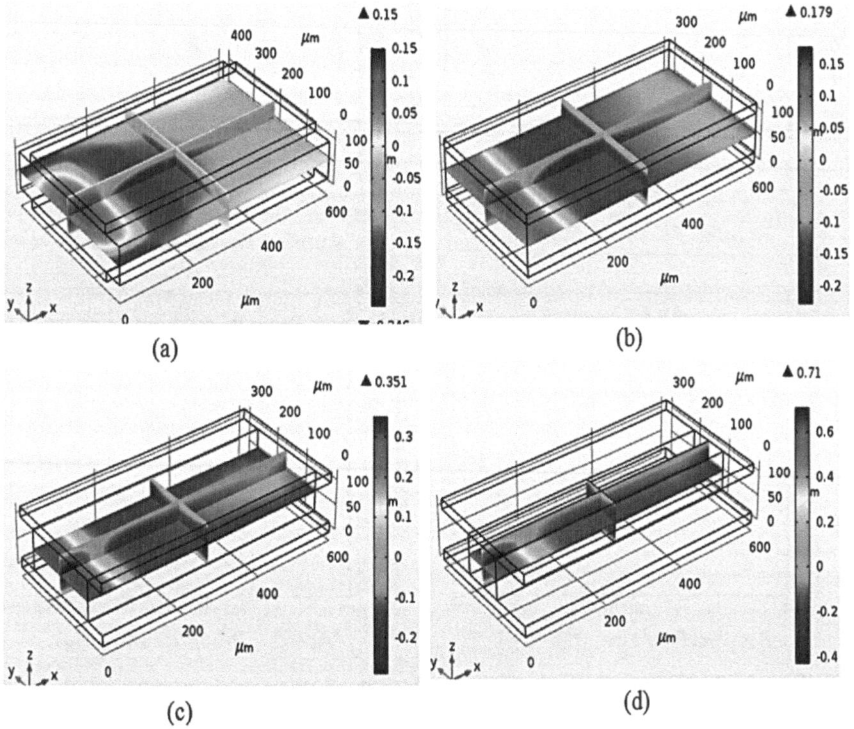

FIGURE 19.9 Simulation of the sensing voltages at the different cross-sectional areas (A) of the sensing layer (a) A = 0.24 mm², (b) A = 0.18 mm², (c) A= 0.12mm², and (d) A= 0.6 mm².

TABLE 19.2 Output Sensing Voltage (V) Versus Applied Pressure (P) at Different Cross-Sectional Areas of Sensing Layer.

Applied pressure (mmHg)	Output sensing voltage (V)			
	A = 0.24 mm²	A = 0.18 mm²	A = 0.12 mm²	A = 0.06 mm²
60	0.04	0.054	0.105	0.213
80	0.06	0.072	0.141	0.284
100	0.075	0.09	0.176	0.355
120	0.09	0.108	0.211	0.426
140	0.105	0.126	0.246	0.497
160	0.12	0.144	0.281	0.568
180	0.135	0.162	0.316	0.639
200	0.15	0.179	0.351	0.71

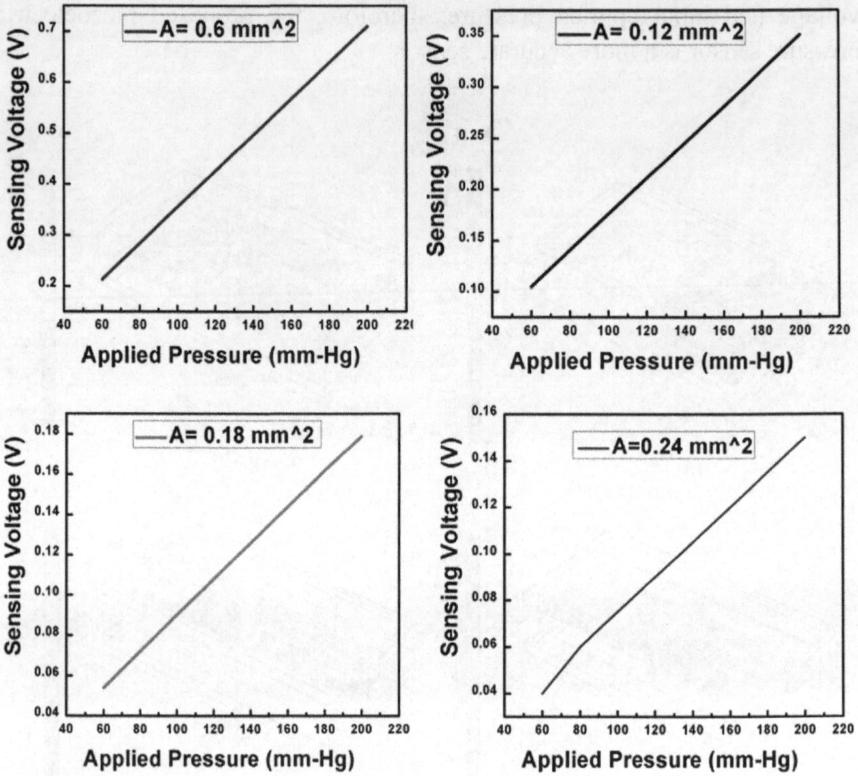

FIGURE 19.10 Graph between applied pressure (mmHg) and output sensing voltage (V) at different cross-sectional areas of sensing layer.

19.3.2.2 SENSITIVITY ANALYSIS

The sensitivity of the sensor defined as a small change in input comes about in large changes in output. Therefore, sensitivity is the minimum input (pressure) that will produce a detectable output (sensing voltage) change.

Sensitivity (S) of a proposed sensor is expressed by

$$S = \frac{\Delta V}{\Delta P} = \frac{V_{max} - V_{min}}{P_{max} - P_{min}}$$

The sensitivity of the proposed sensor is 0.9 mV/mmHg for the sensing layer thickness 80 μm and cross-section area is 0.18 mm^2 for 500 mmHg pressure range. The sensitivity of the proposed piezoelectric sensor can be changed by altering the thickness and cross-sectional area of the sensing layer, as shown in Tables 19.3 and 19.4, respectively. Figure 19.11 shows the graph plot of the effect of the thickness (t) and cross-sectional area (A) of the sensing layer on the sensitivity of the sensor, which indicates that the sensitivity of the proposed sensor can be enhanced by increasing thickness and reducing the cross-sectional area of sensing layer of the piezoelectric sensor.

TABLE 19.3 Thickness (t) of Sensing Layer Versus Sensitivity (S) of Proposed Sensor for Cross-Sectional Area (A) = 0.18 mm^2.

Thickness (t) (mm)	Sensitivity (s) (mV/mmHg)
40	0.037
60	0.72
80	0.90
100	1.014

TABLE 19.4 Cross-Sectional Areas (A) of Sensing Layer Versus Sensitivity (S) of Proposed Sensor for Thickness (t) = 80 μm.

Cross-Sectional Area (A) (mm2)	Sensitivity (s) (mV/mmHg)
0.24	0.78
0.18	0.89
0.12	1.75
0.06	3.55

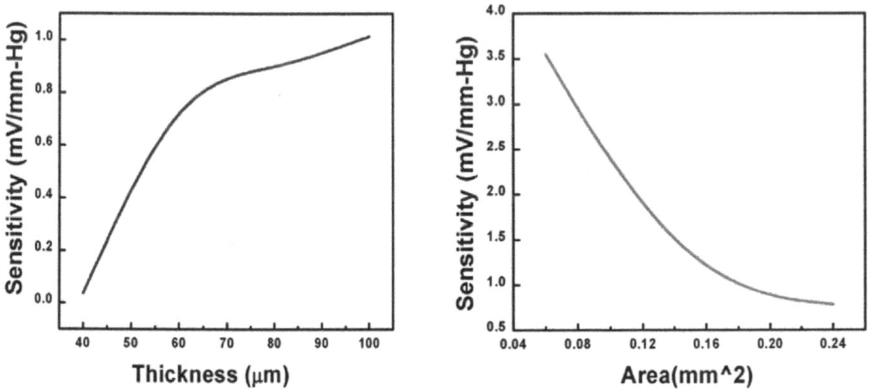

FIGURE 19.11 (a) Graph between the sensitivity of the sensor and thickness (t) of the sensing layer of the sensor and (b) graph between the sensitivity of the sensor and the cross-sectional area (A) of the sensing layer of the sensor.

19.4 CONCLUSION

This research paper explores a cantilever-structured PVDF-based piezoelectric pressure sensor for biomedical application using COMSOL Multiphysics. The analysis of the simulation results shows that the proposed PVDF piezoelectric polymer-based blood pressure sensor senses a blood pressure as a self-generating voltage at output for small pressure range from 50 to 200 mmHg. The entire sensor is designed for different thicknesses (40, 60, 80, and 100 μm) and the different cross-sectional areas (0.06, 0.12, and 0.18 mm^2) of the sensing layer. This designed sensor demonstrates a linear relationship between output sensing voltage and applied pressure. The analysis of the simulation also shows the output self-generating voltage and the sensitivity of the proposed sensor is enhanced by increasing the thickness and reducing the cross-sectional area of the sensing layer of this sensor. Therefore, this paper illustrated a more flexible, self-powered, highly sensitive, and efficient piezoelectric pressure sensor using PVDF polymer is designed and simulated.

KEYWORDS

- **polymer**
- **piezoelectric materials**
- **PVDF**
- **flexible sensor**
- **sensitivity**

REFERENCES

1. Xu, K.; Lu, Y.; Takei, K. Multifunctional Skin-Inspired Flexible Sensor Systems for Wearable Electronics. *Adv. Mater.* **2019,** *4*, 1800628. doi: 10.1002/admt.201800628.
2. Al-khafajiy, M.; Baker, T.; Chalmers, C.; et al. Remote Health Monitoring of Elderly through Wearable Sensors. *Multimed. Tools Appl.* **2019,** *78*, 24681–24706. doi: 10.1007/s11042-018-7134-7.
3. Dagdeviren, C.; Joe, P.; Tuzman, O. L.; K-I., P.; Lee, K. J.; Shi, Y.; Huang, Y.; Rogers, J. A. Recent Progress in Flexible and Stretchable Piezoelectric Devices for Mechanical Energy Harvesting, Sensing and Actuation. *Extreme Mech. Lett.* **2016,** *9*, 269–281. doi: 10.1016/j.eml.2016.05.015.

4. Zang, Y.; Zhang, F.; Di, C.-A.; Zhu, D. Advances of Flexible Pressure Sensors toward Artificial Intelligence and Health Care Applications. *Mater. Horiz.* **2015,** *2*. doi: 10.1039/C4MH00147H.
5. Li, G.; Wen, D. Wearable Biochemical Sensors for Human Health Monitoring: Sensing Materials and Manufacturing Technologies. *J. Mater. Chem. B* **2020,** *8*, 3423–3436. doi: 10.1039/C9TB02474C.
6. Song, P.; Ma, Z.; Ma, J.; Yang, L.; Wei, J.; Zhao, Y.; Zhang, M.; Yang, F.; Wang, X. Recent Progress of Miniature MEMS Pressure Sensors. *Micromachines (Basel)* **2020,** *11* (1). doi: 10.3390/mi11010056.
7. Shin, D.-M.; Hong, S. W.; Hwang, Y.-H. Recent Advances in Organic Piezoelectric Biomaterials for Energy and Biomedical Applications. *Nanomaterials* **2020,** *10* (1), 123. doi: 10.3390/nano10010123.
8. Seo, Y.; Kim, D.; Hall, N. A. Piezoelectric Pressure Sensors for Hypersonic Flow Measurements. *J. Microelectromech. Syst.* **2019,** *28*, 271–278. doi: 10.1109/JMEMS.2019.2899266.
9. Horowitz, S.; Nishida, T.; Cattafesta, L.; Sheplak, M. Development of a Micromachined Piezoelectric Microphone for Aeroacoustics Applications. *J. Acoust. Soc. Am.* **2007,** *122* (6), 3428–3436. doi: 10.1121/1.2785040.
10. Panwar, L. S.; Panwar, V.; Panwar, S. S. Performance Enhancement of PZT Wearable Blood Pressure Sensor Using Cantilever Structure. In *2020 International Conference on Advances in Computing, Communication & Materials (ICACCM)*, 2020; pp 148–153. doi: 10.1109/ICACCM50413.2020.9212895.
11. Panwar, L. S.; Kala, S.; Panwar, V.; Panwar, S. S.; Sharma, S. Design of MEMS Piezoelectric Blood Pressure Sensor. In *2017 3rd International Conference on Advances in Computing,Communication & Automation (ICACCA) (Fall)*, 2017; pp 1–7. doi: 10.1109/ICACCAF.2017.8344698.
12. Wan, C.; Bowen, C. R. Multiscale-Structuring of Polyvinylidene Fluoride for Energy Harvesting: The Impact of Molecular-, Micro- and Macro-Structure. *J. Mater. Chem. A* **2017,** 5 (7), 3091–3128. doi: 10.1039/C6TA09590A.
13. Li, T.; Ma, J.; Es-Souni, M.; Woias, P. Advanced Piezoelectrics: Materials, Devices, and Their Applications. *Smart Mater. Res.* **2012**. doi: 10.1155/2012/259275.
14. Safaei, M.; Sodano, H. A.; Anton, S. R. A Review of Energy Harvesting Using Piezoelectric Materials: State-of-the-Art a Decade Later (2008–2018). *Smart Mater. Struct.* **2019,** *28*. doi: 10.1088/1361-665X/ab36e4.
15. Schlaberg, H. I.; Duffy, J. S. Piezoelectric Polymer Composite Arrays for Ultrasonic Medical Imaging Applications. *Sens. Actuators A: Phys.* **1994,** *44* (2), 111–117. doi: 10.1016/0924-4247(94)00791-8.
16. Gomes, J.; Serrado, N. J.; Sencadas, V.; Lanceros-Mendez, S. Influence of the β Phase Content and Degree of Crystallinity on the Piezo- and Ferroelectric Properties of Poly(Vinylidene Fluoride). *Smart Mater. Struct.* **2010,** *19*. doi: 10.1088/0964-1726/19/6/065010.
17. Ruan, L.; Yao, X.; Chang, Y.; Zhou, L.; Qin, G. Properties and Applications of the β-Phase Poly(Vinylidene Fluoride). *Polymers* **2018,** *10* (3). doi: 10.3390/polym10030228.
18. Xia, W,; Zhang, Z. PVDF-Based Dielectric Polymers and Their Applications in Electronic Materials. *IET Nanodielectr.* **2018,** *17*, 17–31. doi: 10.1049/iet-nde.2018.0001.
19. Caliò, R.; Rongala, U. B.; Camboni, D.; Milazzo, M.; Stefanini, C.; De Petris, G.; Oddo, C. M. Piezoelectric Energy Harvesting Solutions. *Sensors* **2014,** *14*, 4755–4790. doi: 10.3390/s140304755.

20. Xu, F.; Li, X.; Shi, Y.; Li, L.; Wang, W.; et al., Recent Developments for Flexible Pressure Sensors: A Review. *Micromachines* **2018,** *9* (11). doi: 10.3390/mi9110580.
21. Sappati, K. K.; Bhadra, S. Piezoelectric Polymer and Paper Substrates: A Review. *Sensors (Basel)* **2018,** *18* (11), 3605. doi: 10.3390/s18113605.
22. Kang, M. G.; Jung, W. S.; Kang, C. Y.; Yoon, S. J. Recent Progress on PZT Based Piezoelectric Energy Harvesting Technologies. *Actuators* **2016,** *5* (1). doi: 10.3390/act5010005.
23. Wu, N.; Chen, S.; Lin, S. Li, W.; Xu, Z.; Yua, F.; Huan, L.; Hu, B.; Zhou, J. Theoretical Study and Structural Optimization of a Flexible Piezoelectric-Based Pressure Sensor. *J. Mater. Chem. A* **2018,** *6*, 5065–5070. doi: 10.1039/C8TA00688A.
24. Nag, A.; Mukhopadhyay, S. C.; Kosel, J. Wearable Flexible Sensors: A Review. *IEEE Sens. J.* **2017,** *17* (13), 3960. doi: 10.1109/JSEN.2017.2705700.
25. Sundar, U.; Banerjee, S.; et al. Piezoelectric and Dielectric Properties of PZT-Epoxy Composite Thick Film. *Acad. J. Polym. Sci.* **2018,** *1* (5). doi: 10.19080/AJOP.2018.01.555574.
26. Wei, H.; Wang, H.; Xia, Y.; Cui, D.; Shi, Y.; Dong, M.; Liu, C.; Ding, T.; Zhang, J.; Ma, Y.; Wang, N.; Wang, Z.; Sun, Y.; Wei, R.; Guo, Z. An Overview of Lead-Free Piezoelectric Materials and Device. *J. Mater. Chem. C* **2018,** *6*, 12446–12467. doi: 10.1039/C8TC04515A.
27. Jaffe, B.; Cook, W. R.; Jaffe, H. L. *Piezoelectric Ceramics*; Academic Press: London, 1971.
28. Panwar, L. S.; Panwar, V.; Kala, S.; Sharma. S. Modelling of Different MEMS Pressure Sensors using COMSOL Multiphysics. *Int. J. Curr. Eng. Technol.* **2017,** *7* (1), 243–247.

CHAPTER 20

Design and Optimization of Low-Cost RF Energy-Harvesting Circuit

PRADEEP CHINDHI[1*], H. P. RAJANI[2], GEETA KALKHAMBKAR[3], and RAJASHRI KHANAI[2]

[1]*Department of Electrical Engineering, SGMCOE, Mahagaon 416503, Maharashtra, India*

[2]*Department of Electronics and Communication Engineering, KLE'S Dr. MSSCET, Belgaum, Karnataka, India*

[3]*Department of Electronics and Telecommunication Engineering, SGMCOE, Mahagaon, Maharashtra, India*

[*]*Corresponding author. E-mail: pradeepchindhi.4003@gmail.com*

ABSTRACT

In this chapter, a low-cost Radio Frequency (RF) rectifier, capable of harvesting RF energy is presented. The conversion efficiency is maximized by optimizing the impedance matching circuit elements and load for an input RF power of -10 dBm at 3.45GHz. A half-wave rectifier Schottky diode (HSMS2850) was selected for RF to DC conversion. Theoretical analysis and Harmonic Balance Advanced Design System (ADS) simulation is performed. The simulation result shows a conversion efficiency of 12.70% for an input power of -10 dBm for unoptimized impedance matching circuit elements. The maximum conversion efficiency of 54.33% is achieved for an input power of -10 dBm in the case of optimized impedance matching circuit elements. The output voltage Vs load and conversion efficiency Vs RF input power is also presented. The conversion efficiency and output voltage are observed for an input RF power sweep from -20 dBm to 20 dBm and load resistance of 1 Ω to 2 kΩ.

Intelligent Technologies for Sensors: Applications, Design, and Optimization for a Smart World.
S. Kannadhasan, R. Nagarajan, & Alagar Karthick (Eds.)

20.1 INTRODUCTION

In past and present, most of the wireless senor networks (WSN) are powered by batteries. The major concern with WSN/IoT devices is typically battery life and its substitution. Recent advances in ultralow power sensors and exponential growth in wireless sensor/IoT networks, development in the wireless technology (GSM900/1800, 3G, 4G, Wi-Fi, and 5G)[7] and rising implications of energy costs and carbon footprints, the need to adopt low-cost, green energy-harvesting approaches are of paramount importance. To address this, the practicability of harvesting low-power density RF energy becomes another method. The technology can be positioned in low-power devices such as smart farming, autonomous driving, security surveillance, smart cities, industrial IoT, and many more.[5,8] The IoT sensors' marketplace is projected to grow from USD 8.4 billion in 2021 to USD 29.6 million by 2026; it is expected to grow at a compound annual growth rate (CAGR) of 28.6% during 2021–2026 (https://www.globenewswire.com/news-release/2021/03/15/2192487/0/en/). The RF energy-harvesting system consists of antenna, impedance matching circuit, rectifier/voltage multiplier, power management unit/storage device. In RF energy-harvesting system, antenna is used to receive available radio frequency signal, impedance matching circuit to transfer maximum RF input available power to the load, rectifier to convert extracted RF power into DC voltage, a power management unit, which decides whether to store the harvested power obtained from the RF energy harvester or to use directly. Figure 20.1 shows the block diagram representation of the RF energy-harvesting system. The chapter is put together as follows: theoretical analysis with impedance matching circuit is presented in Section 20.2. Section 20.3 presents the advanced design system (ADS) harmonic balance (HB) simulation for unoptimized and optimized rectifier circuit. Section 20.4 wraps up the chapter.

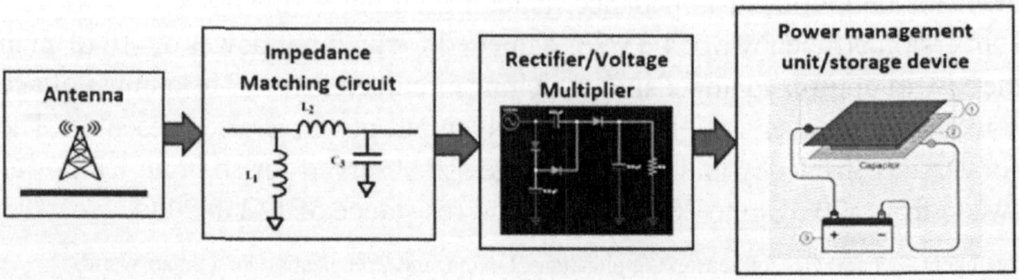

FIGURE. 20.1 Block diagram representation of RF energy harvesting.

In this chapter, a low-cost RF rectifier, capable of harvesting radio frequency (RF) power, is presented. The conversion efficiency is maximized by optimizing impedance matching circuit elements and load for an input RF power of −10 dBm at 3.45 GHz. A half-wave Schottky diode (HSMS2850) rectifier was selected for RF to DC conversion. Theoretical analysis and harmonic balance advanced design system (ADS) simulation is performed. The simulation results show a conversion efficiency of 12.70% for an input power of −10 dBm for unoptimized impedance matching circuit elements. The maximum conversion efficiency of 54.33% is achieved for an input power of −10 dBm in case of optimized impedance matching circuit elements. The output voltage versus load and conversion efficiency versus RF input power details is also presented. The conversion efficiency and output voltage are measured for input RF power sweep from −20 dBm to 20 dBm and load resistance of 1 Ω–2 kΩ.

20.2 RF HARVESTER AND IMPEDANCE MATCHING CIRCUIT

The rectenna efficiency is inspected by equation (20.1)

$$\eta = \frac{P_{DC}}{P_{RF}} = \frac{\frac{V_{DC}^2}{R_L}}{P_{RF}} \tag{20.1}$$

where V_{DC} = Output DC voltage, R_L = Load resistance P_{RF} = Input RF power

The input RF power (P_{RF}) received by the reviving antenna is calculated by using expression (20.2)

$$P_{RF} = P_t G_t G_r \left(\frac{c}{4\pi f}\right)^2 \left(\frac{1}{R}\right)^n e^{-\alpha R} \tag{20.2}$$

where P_t = transmitter antenna power, G_t = gain of reciving antenna, G_r = gain of transmitting antenna, c = speed of light, n = path loss exponent, n = 2 in free space, f = RF signal frequency, α = effective decay coefficient. In air it is 0.001, R = distance from the RF source.

The details of impedance matching circuit analysis and selection of circuit element is given in Ref. [1–4,6].

The required values of impedance matching circuit elements are given by expressions (20.3) and (20.4)

$$C_m = \frac{1}{w_r \times R_s}\sqrt{\frac{Rs}{R_{in} - R_s}} \tag{20.3}$$

$$L_m = \frac{R_{in}}{w_r} \frac{1}{w_r R_{in} C_{in} + \frac{1}{\sqrt{\frac{R_s}{R_{in} - R_s}}}} \tag{20.4}$$

Voltage gain and quality factor of rectenna is computed by using (20.5) and (20.6).

Voltage gain,

$$G = \frac{v_{in}}{v_s} = \frac{1}{2}\sqrt{\frac{R_{in}}{R_s}} \tag{20.5}$$

$$Q = \sqrt{\frac{R_{in}}{R_s} - 1} \tag{20.6}$$

20.3 SIMULATION RESULT ANALYSIS

Figure 20.2 shows the ADS HB circuit simulation schematic of the proposed unoptimized RF energy harvester. A high-pass L matching circuit is composed of capacitor (C1) and inductor (L1) with Schottky diode half-wave rectifier and filter capacitor (C2). The initial values of capacitor (C1) and inductor (L1) are obtained by using equations (20.3) and (20.4). In ADS HB simulation input RF power sweep of −20–dBm, filter capacitor C2= 1 nF, and HSMS-2580 Schottky diode is selected. The details of SPICE parameters for the HSMS2850 is mentioned in Table 20.1.

TABLE 20.1 HSMS2850 SPICE Parameters.

Parameters	Units	Value
B_v	V	3.8
C_{j0} (C_{in})	pF	0.18
E_G	Ev	0.69
I_{BV}	A	3E−4
I_S	A	3E−6
N	No unit	1.06
R_S	Ω	25
$P_R(V_J)$	V	0.35
$P_T(XTI)$	No unit	2
M	No unit	0.5

The obtained simulated results for the unoptimized RF energy harvester are shown in Figures 20.3–20.6. Figure 20.3 shows conversion efficiency versus RFpower relationship, at 0 dBm input RFpower the conversion efficiency is 12.70% and Figure 20.4 shows maximum conversion efficiency versus Rload relationship, at 1 kΩ load the maximum conversion efficiency is 2.07%. In Figure 20.5, Vout at 2 kΩ is 62.11 mV peak and in Figure 20.6, 504.0 V Vout at 0 dBm input RFpower is observed. Table 20.2 shows the impedance matching circuit values for unoptimized and optimized RF rectifier circuits.

TABLE 20.2 Impedance Matching Circuit Parameters.

Circuit parameters	Unoptimized circuit values	Optimized circuit values
Capacitor (C1)	0.14774 pF	0.32 pF
Inductor (L1)	6.5674 nH	4.00 nH
Input power sweep	−20 dBm–20 dBm	
Load sweep	1 Ω–2 kΩ	

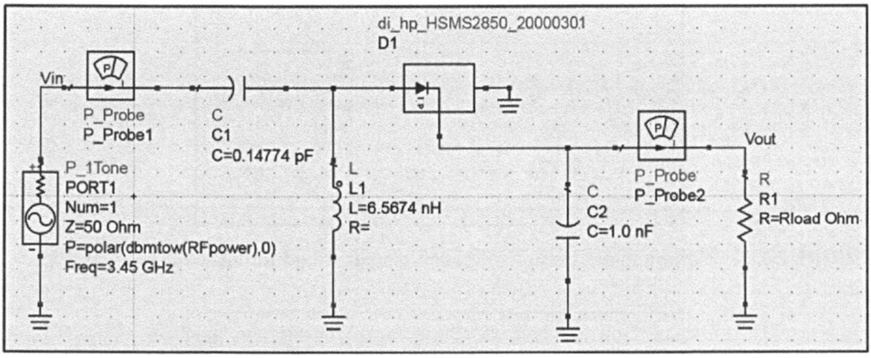

FIGURE 20.2 Unoptimized single-diode (HSMS2850) rectifier in ADS.

Figure 20.7 shows the ADS HB circuit simulation schematic of the proposed optimized RF energy harvester. Figures 20.8 and 20.9 depict 54.33% conversion efficiency at 0 dBm input RFpower and 27.358% efficiency at 1.101 kΩ load. In Figures 20.10 and 20.11, Vout at 2 kΩ is 227.4 mV peak and 1.042 V Vout at 0 dBm input RFpower is observed. It is clear that optimized C1 and L1 gives higher conversion efficiency and output voltage.

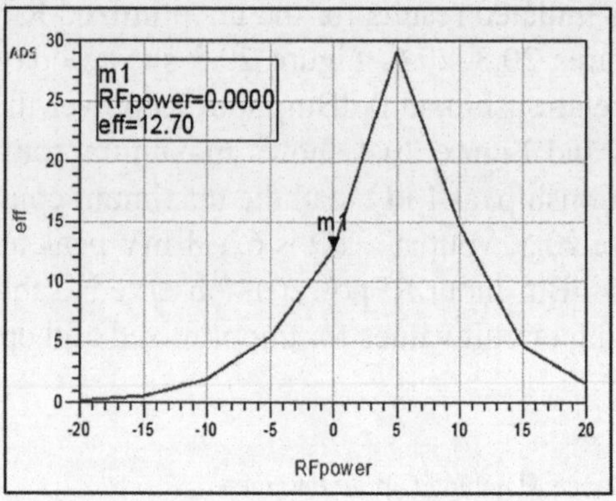

FIGURE 20.3 Conversion efficiency versus RFpower for unoptimized rectifier.

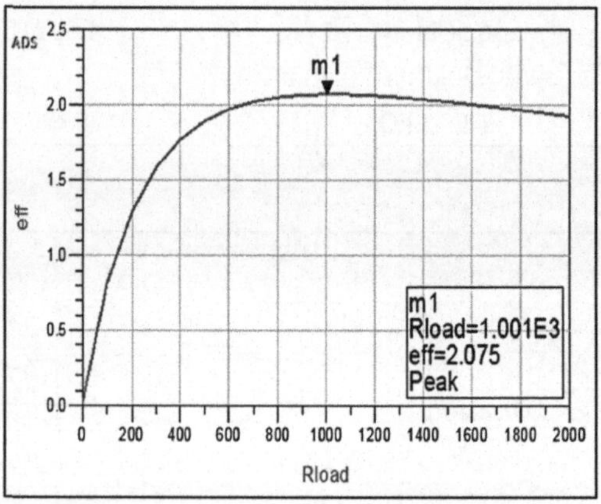

FIGURE 20.4 Maximum conversion efficiency versus Rload for unoptimized rectifier.

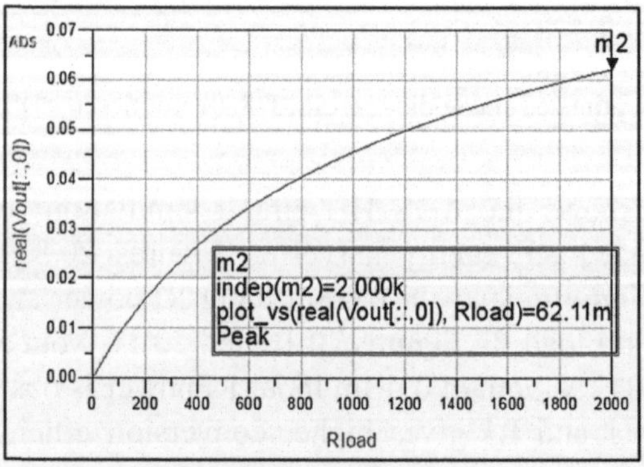

FIGURE 20.5 Vout versus Rload for unoptimized rectifier.

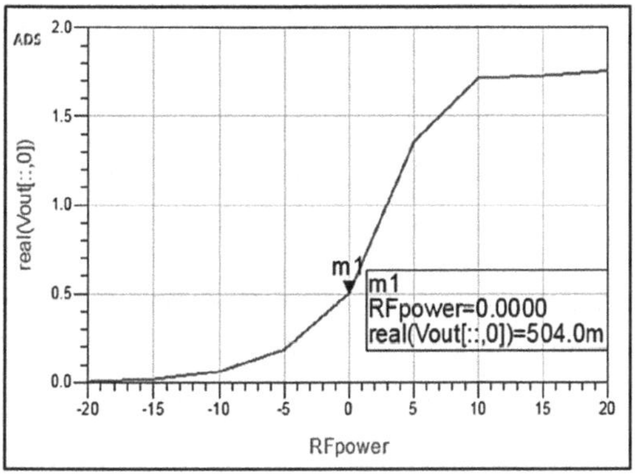

FIGURE 20.6 Vout versus RFpower for unoptimized rectifier.

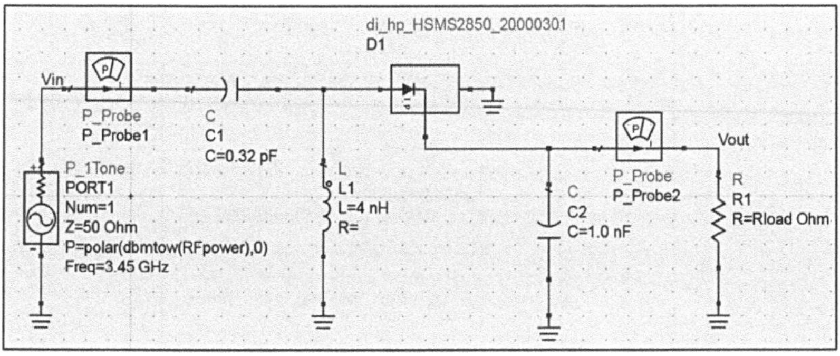

FIGURE 20.7 Optimized single-diode (HSMS2850) rectifier in ADS.

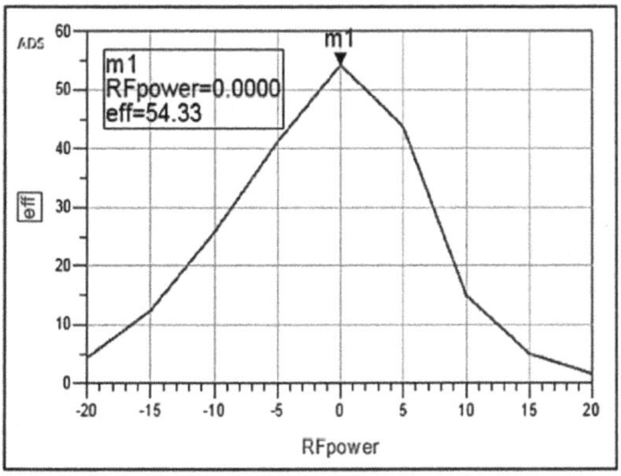

FIGURE 20.8 Conversion efficiency versus RFpower for optimized rectifier.

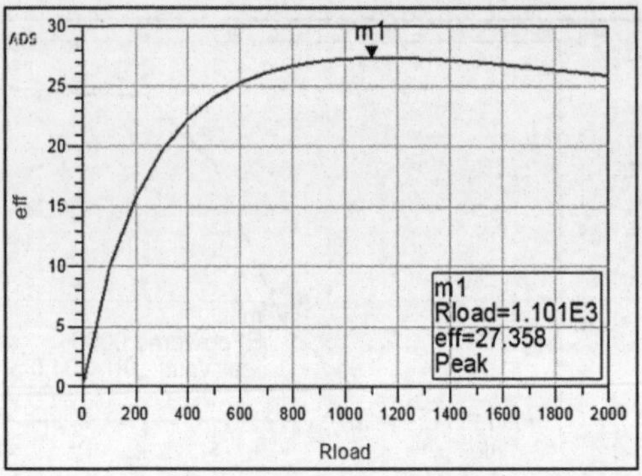

FIGURE. 20.9 Maximum conversion efficiency versus Rload for optimized rectifier.

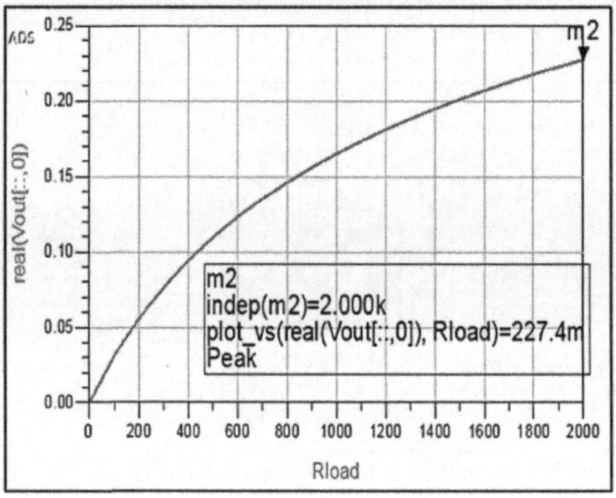

FIGURE 20.10 Vout versus Rload for optimized rectifier.

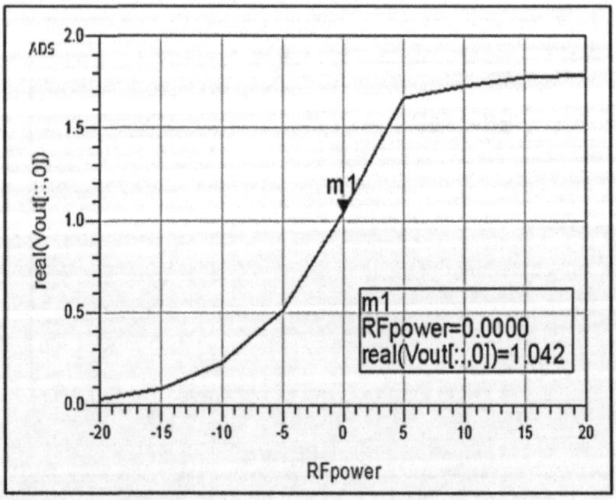

FIGURE 20.11 Vout versus RFpower for optimized rectifier.

20.4 CONCLUSION

In this chapter, we have proposed a low-cost RF energy-harvesting circuit. To boost the conversion efficiency at low RF input power level a high-pass L matching circuit is implemented. A theoretical detail to find out a suitable value of impedance matching circuit has been presented. The ADS HB simulation at a frequency of 3.45 GHz shows 54.33 conversion efficiency at an input RF power of −20–20 dBm respectively. The simulation results obtained for optimized RF energy harvester can be successfully used for powering WSN/IoT sensor nodes.

KEYWORDS

- **radio frequency**
- **energy harvesting**
- **rectenna**
- **HSMS2850**
- **efficiency**

REFERENCES

1. Nimo, A.; Grgić, D.; Reindl, L. M. Optimization of Passive Low Power Wireless Electromagnetic Energy Harvesters. *Sensors* **2012,** *12,* 13636–13663. doi:10.3390/s121013636. ISSN 1424-8220, www.mdpi.com/journal/sensors.
2. Mohan, A.; Mondal, S. An Impedance Matching Strategy for Micro-Scale RF Energy Harvesting Systems. *IEEE Transactions on Circuits and Systems II: Express Briefs.* doi:10.1109/TCSII.2020.3036850.
3. Bahhar, C.; Aidi, M.; Mejri, F.; Aguili, T. Design and Optimization of High-Efficiency Rectenna for RF Energy Harvesting. *Photonics & Electromagnetics Research Symposium | Spring (PIERS | SPRING)*, Rome, Italy, 17–20 June 2019.
4. Anchustegui-Echearte, I.; Jimenez-Lopez, D.; Gasulla, M.; Giuppi, F.; Georgiadis, A. A High-Efficiency Matching Technique for Low Power Levels in RF Harvesting. In *PIERS Proceedings, Stockholm*, Sweden, Aug. 12–15, 2013.
5. Internet of Things. *Wireless Sensor Networks, White Paper*; International Electrotechnical Commission: Geneva, Switzerland, 2014.
6. Jordana, J.; Reverter, F.; Gasulla, M. Power Efficiency Maximization of an RF Energy Harvester by Fine-Tuning an L-Matching Network and the Load. *Proc. Eng.* **2015,** *120,* 655–658. doi:10.1016/j.proeng.2015.08.698.

7. Alsharif, M. H.; Kelechi, A. H.; Albreem, M. A.; Chaudhry, S. A.; Sultan Zia, M.; Kim, S. Sixth Generation (6G) Wireless Networks: Vision, Research Activities, Challenges and Potential Solutions. *Symmetry* **2020,** *12*, 676. doi:10.3390/sym12040676. www.mdpi.com/journal/symmetry.
8. Alharbi, N.; Soh, B. Roles and Challenges of Network Sensors in Smart Cities. *IOP Conf. Series: Earth Environ. Sci.* **2019,** *322*, 012002. doi:10.1088/1755-1315/322/1/012002.

CHAPTER 21

The Future Network 2030: A Simplified Intelligent Transportation System

VELMANI RAMASAMY[1*] and MĂDĂLIN-DORIN POP[2]

[1]*Computer Science and Engineering Department, Adithya Institute of Technology, Kurumbapalayam, Coimbatore 641107, Tamil Nadu, India*

[2]*Computer and Information Technology Department, Politehnica University of Timişoara, România*

[*]*Corresponding author. E-mail: velmanir@gmail.com*

ABSTRACT

The continuous growth in the human population and the advancements in technology increase the number of vehicles required in the transportation system. To fulfill the current economy and globalization needs, the world is looking at efficient intelligent transportation systems (ITS) in real-time road traffic monitoring and management. The already existing transportation system for vehicle monitoring and the road traffic congestion avoiding mechanisms on the roadways cannot fulfill the requirements in 2030. Therefore, the ITS with the Internet of Things (IoT) is required in future network (FN) 2030. It plays a vital role in road infrastructure monitoring systems and simplifies traffic mobility in crowded urban areas. Here, the FN 2030 includes the future of sensor networks and vehicular networks, which helps to collect the real-time traffic information from the ad-hoc and the sensor node. In addition, it helps to send the real-time collected data to the traffic management centers (TMCs) to control the timings of the traffic light. In wireless sensor networks (WSNs), the size of the sensor node is very small

Intelligent Technologies for Sensors: Applications, Design, and Optimization for a Smart World.
S. Kannadhasan, R. Nagarajan, & Alagar Karthick (Eds.)

and they are used to operate with low-power batteries. Therefore, the data processing capacity and the communication range are limited due to the nature of the hardware configuration and they can be used for short-range wireless communication. In WSNs, the sensor nodes are logically connected to make efficient sensor network communication and is connected with an internet connection to transfer the real-time data to the remote location. In this paper, architecture is proposed to overcome the existing design issues and the limitations of ITS in 2030. In addition, it will make effective communication and meets the general requirements in 2030.

21.1 INTRODUCTION

The intelligent transportation system (ITS) has captured business in the international market size of USD 29.69 Bn in 2020, and is expected to hit in the market size from USD 47.89 Bn to USD 60 Bn by 2030. The market size is growing at a CAGR from 6.3% to 10%. Therefore, the existing transportation infrastructure for vehicle monitoring and the road traffic congestion-avoiding mechanisms on the roadways cannot fulfill the requirements of 2030. Hence, the advancement in ITS with the Internet of Things (IoT) is required in FN 2030.[2,43]

Generally, the ITS is the integration of several cutting-edge information and communication technologies, which helps to manage the road traffic and congestion, as well as transforms the way of consumers, government, and business deals on transportation.[52] ITS plays a vital role on various services like intelligent traffic monitoring and management, security and surveillance, weather monitoring system, fleet management, environment protection, parking management system, emergency vehicle notification, smart ticketing and passenger information solution, mobility services, etc. The FN 2030 includes the future of sensor networks and vehicular networks that help to collect the real-time traffic information from the ad-hoc and the sensor node.[40,51]

Most of the governments planned to invest heavily for the smart transport infrastructure projects. It is aimed to develop smart cities that increase the adoption of smart transportation modes to speed up vehicle mobility of traffic and moderate the congestion. The most crucial aspect of any smart city development project is smart mobility.[11] It is intended to coordinate products and human-centric transportation with information and communication technology in order to provide well-organized vehicle mobility in that particular geographic area. The latest breakthroughs in digital technology and the IoT are critical components in establishing smart ITS that are more effective, dependable, intelligent, and environmentally friendly.

Many nations are now constructing ITS to achieve smart mobility, and many of these countries are supporting the ITS strategy by integrating it into the cooperative-ITS (C-ITS) system. Vehicle drivers, road consumers, private and public agencies, and service providers are all part of the smart mobility technique and they follow the obliging pattern on vehicle-to-vehicle (V2V), vehicle-to-infrastructure (V2I), and vehicle-to-everything (V2X) technologies.[12] The Future ITS: 2030 architecture is proposed in this study to address existing design challenges and ITS constraints in terms of effective communication, and to meet the general requirements in 2030.

IoT-based sensor and actuator network monitor the real-time intelligent ambiances to improve the quality of transportation, education, medical facilities, etc. IoT collects the real-time data from various sources of information and establishes coordination according to the real-time environments. Further, smart ITS ensures driver safety and improved traffic mobility in the crowded areas. In general, traffic management centers (TMCs) collect the real-time monitored information from the cameras and the sensors via the IoT backbone and take timing control choices on traffic lights for traffic mobility. In addition, vehicle-to-IoT (V2IoT) dynamic traffic lights control advantages to evaluate the metrics such as travel time, road safety, noise emissions, and CO_2 emissions. Weather condition (e.g., foggy, rainy) plays a major role on road safety and road visibility. To mitigate these kind of issues, vehicle ad-hoc networks (VANET) infrastructure added an extra flavor in ITS that alerts the driver through pictorial and audible aids by using an optical sensor.

The smart city projects researchers introduced various advancements that combined many technologies like IoT services, which encapsulate the ITS in many real-time components such as ambience, buildings, and energy. Moreover, it accesses the resources dynamically and intelligently to take decision through local administrations.[45] Even, IoTs in a smart city have confidentiality and safety perspective; it is essential to have an overview between the chances, benefits, costs, and concerns. Further, it should fulfill the general requirements and respect the confidentiality and safety guidelines given by the general data protection regulation (GDPR) of the European Union Commission.[40]

Furthermore, the green Internet of Things (G-IoT) play an important role to provide a socio-economic ecological system for the efficient use of environmental assets, energy conservation, and trash administration, among other things, in order to protect the environment. In addition, the Internet of Everything (IoE) and the Internet of Services (IoS) have created service-oriented architectures, platforms (server, storage, and communication), next-generation blockchain technology, on-demand distributed computing, cipher-physical systems, and semantic technologies. The new technologies must be aware of the 5G wireless communication system, as well as support the blockchain in terms of security and scalability.[5] The proposed design for Future ITS: 2030 takes into account current difficulties as well as FN 2030's particular requirements.

21.2 RELATED WORK

Urbanization is growing prosperity for the humans in developing countries, which increases the demand in the public and private transportation. The UN estimates that more than 55% of population residing in the urban areas and that it will be raise to 68% in 2025. Road traffic and congestion is straining the transport system, which is driving the need to adopt ITS. Developing regions like Latin America, Asia Pacific, and Africa have lack of infrastructure in city planning; therefore, the transportation system is facing high traffic congestion. Furthermore, the traffic management sector have a leading share on the global ITS market. Even semi-autonomous vehicles are in the commercial market; truck platooning and fully autonomous taxis pilot projects have been undertaken for the share of the global ITS market and that has gained popularity around the world.

The international ITS market is sectored into seven broad categories (Figure 21.1), which are region, major players, transport mode, system, deployment mode, type, and application; furthermore, it is categorized into sub-categories.[45] The international business market is grouped into North America, European Union, Asia Pacific, Middle East, Latin America, and Africa regions. At present, the dominant players in the international ITS market are Siemens AG, IBM Corporation, Lanner Electronics, Hitachi Ltd, Kapsch TrafficCom, Nuance Communications Incorporation, Cubic Corporation, Denso Corporation, WS Atkins PLC, Tom Tom, Thales Group, Clever Devices Ltd., EFKON GmbH, Inc., Garmin International Inc., Iteris, Inc., and Telenav.

The global transportation can be categorized into three modes such as roadways, railways, and airways. Now, the ITS system is divided into region/zone-based traffic control and monitoring, data collection and dissemination, commercial vehicle, road weather, travel information, bus information and cashless ticketing systems, advanced public transportation, automated traffic enforcement, integrated corridor management, and electronic toll collection systems.[37] The ITS deployment mode can be divided into on-premise and cloud.

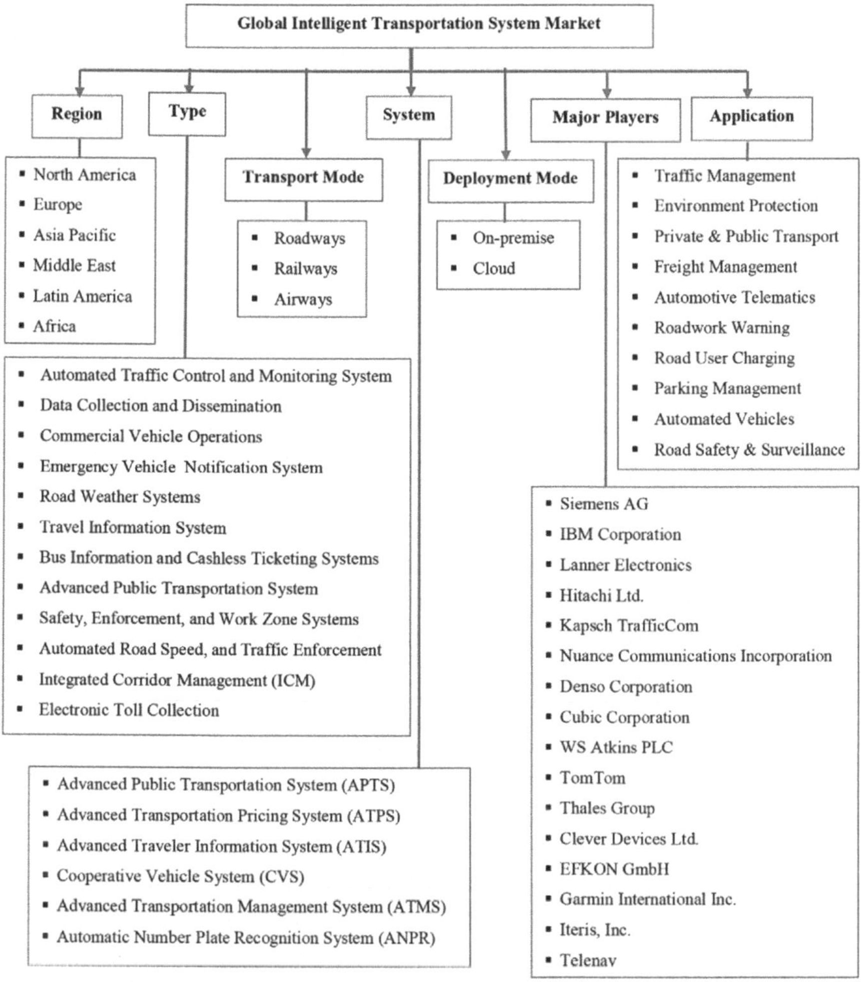

FIGURE 21.1 Global intelligent transportation system market.

Further, the system is divided into six major types such as:

1) **Advanced Public Transportation System (APTS):** The U.S. Department of Transportation's Federal Transit Administration launched this program in 1991. APTS aims to do the research in ITS, and develop more efficient and effective solutions to current

transportation challenges in an advanced navigation communication, and vehicle technologies are to get benefit on public transportation. The applications of APTS are onboard DVR/NVR, passenger counting and information system, passenger entertainment, wayside signaling control, and train control and supervision.[1]

2) **Advanced Transportation Pricing System (ATPS):** It includes motor vehicle miles traveled (VMT) fee systems, e-Toll systems, automatic gate machine, fee-based express lanes, automatic fare collection, ticket vending, congestion pricing, and back-end systems.[2]
3) **Advanced Traveler Information System (ATIS):** It also a type of ITS that integrates the advanced computer communication and the information technologies to deliver real-time data to the users of a system. Further, it assists travelers with planning perception, analysis and decision-making regarding roadside service system, traffic regulation, environmental monitoring, path and location assistance, traveler information system, hazardous conditions and security advisory, varied speed information, warning and caution messages, and parking system.[30,32]
4) **Cooperative Vehicle System (CVS):** It operate ITS services on V2V communication in dedicated short-range communications (DSRC) either single or multiple path small-range to medium-range of wireless safety frequencies bands. AITS in ITS is specifically considered for automation purpose.[46] A set of protocols and standards are used to probe the vehicle information's, reports and the hazard warnings by transferring the real-time information among suitably equipped vehicles as they are in travel. In AITS, a geographical information system (GIS) performs a huge amount of data handling, processing, logical association, analysis, and storage. Further, it will provide an effective traveler information in graphical display to the users.
5) **Advanced Transportation Management System (ATMS):** The Texas Department of Transportation (TxDOT) developed a critical tool called an ATMS in transportation management centers (TMCs) across Texas. It comprises an integrated traffic management and rescue console to offer various futures such as smooth and uninterrupted traffic flow, improved transportation system efficiency, smart mobility, enhanced road safety, alert for abnormal road and weather conditions, and reduction in journey time and inconvenience. The integrated solution of multiple technologies in ATMS improves the flow of vehicle traffic. A set of logically incorporated roadside equipment

is attached to afford a high-class highway solution to ensure a safe and highly secured journey.[48] The major equipment's of ATMS are Emergency Calling Box (ECB), Video Incident Detection System (VIDS), Variable Message Sign (VMS), Meteorological data station (MDS), Mobile Radio Communication System (MRCS), Automatic traffic counter cum classifier (ATCC), and control room.

6) **Automatic Number Plate (or) License Plate Recognition System (ANPR/ALPR):** It comprises of optical character recognition (OCR) that is coupled with artificial intelligence (AI)-based sophisticated software algorithms and related hardware. The Frame Grabber Device provide a wide range of facilities including on-board programming region of significance, image scaling, lookup table, and a pixel decimation for high-resolution and high-speed digital imagining for machine vision and scientific applications. With the intervention of deep learning technologies in ITS, the system will be able detect and differentiate the passing vehicles according to the category, weight, license plate, brand, model, and color.[39] By linking the ANPR with vehicle information such as make/(or) manufacture, model, insurance, and color, data is particularly important when detecting replaced or stolen license plates.

The major applications of ITS are traffic management, environment protection, private and public transport, freight management, automotive telematics, road user charging, parking management, automated vehicles, road safety and surveillance, etc. Since, the ITS is moving toward human-centric transportation system, the application of ITS is critical to enable smart urban mobility. The major stakeholders of FN 2030 are private and public agencies, industry players, and academic and research institutions (Figure 21.2). Further, active participation and close collaboration of the stakeholders in FN 2030 will accelerate the advancements in ITS development for various transportation modes.[11]

21.3 ITS—CURRENT LIMITATIONS AND POSSIBLE IMPROVEMENTS

The current limitations of IoT services in the integration with evolving ITS represents one of the recent research trends in the ITS field. The real-time traffic data acquisition from sensors networks and IoT equipment together with establishing and maintaining their real-time communication shows limitations regarding the current *network infrastructure's competency and scalability*.[31,40,49,54] The implementation of the wireline network concept overcomes this limitation by addressing issues related to network and mobility management as a result of creating a centralized network architecture.[44] Current concerns are also related to the spectral efficiency in 5G that can reach the boundary in the future because of the advances in massive multiple-input and multiple-output (MIMO) antenna architectures, network densification, and millimetre-wave transmission as well as by a set of legacy multiplexing techniques inherited from 4G.[19] The improvements on a large scale for spectral efficiency will be difficult for the 6G infrastructure due to the Shannon limit constraint.[19]

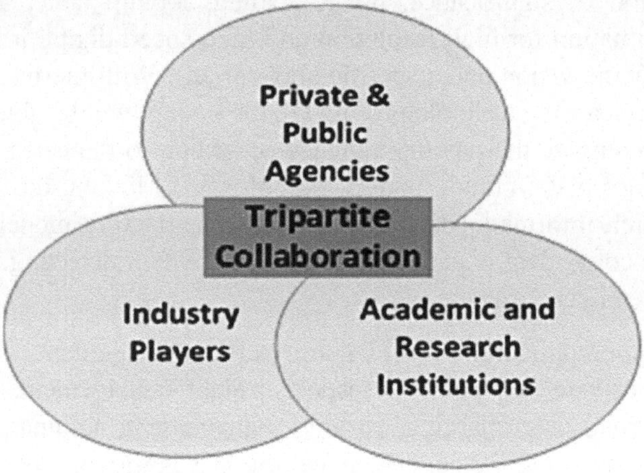

FIGURE 21.2 Tripartite collaboration.

Existing ITS have many limitations and the *lack of flexibility* that leads to reduce the interaction between the vehicles and various flavors of IoT-based applications. Furthermore, they are using old technologies that cannot provide technology integration support with new technologies (or) they cannot ensure backward compatibility.[40,49] Therefore, upcoming 6G networks must be clearly defined to solve many existing issues through cell-less architecture, decentralized resource allocation, and three-dimensional super-connectivity.[14,19]

The employment of wireless solutions facing various difficulties due to the *limited battery power* furthermore they could not be able to ensure highly intensive full functionalities on real-time applications.[4,40] Consequently, this limitation relates also to limited processing power and memory storage.[26] The traffic monitoring systems are directly influenced by this limitation, even if we refer to wireless video cameras systems that require good power supplies or if we talk about data aggregation need with different types of sensors such as microwave radars, inductive loops, pneumatic road tubes, etc.[28,38] In order to meet this limitation, the Internet of Energy (IoE) concept emerged. IoE combines the visions of IoT and smart grid (SG) and aims to provide intelligent monitoring and management of energy distribution, efficient use of energy, identifies possible sources of energy losses, promotes the use of low-cost low-power devices, and the reduction of electro-magnetic pollution.[9,25,42] IoE communication infrastructure supports big data transmission and processing[42] and requires optimized versions of Internet Protocols. In this regard, internet engineering task force (IETF) is leading the efforts in creating several working groups to tailor the networking constrained devices with low-energy requirements to SG needs[9,42]:

- IPv6 over low-power wireless personal area networks (6LoWPAN) for the network layer.
- Routing over low-power and lossy networks (ROLL) for the routing of datagrams.
- Constrained RESTful environment (CoRE) for the application layer, in conjunction with world wide web Consortium (W3C)'s definition for the efficient XML interchange specification (EXI).

Currently, energy-harvesting methods are being implemented by the 5G infrastructure to meet many challenges caused by its coexistence with the communications and the performance degradation during the conversion of harvested signals to electric current.[21] Dang et al.[19] suggest that the planning of the 6G communication technology shall include a human-centric and sustainable vision for 2030. This vision considers that the 6G infrastructure shall apply intelligent harvesting and green-based methodologies to accomplish the energy efficiency gain. Good sources in the energy harvesting process can be the ambient radio-frequency signals, micro-vibrations, and sunlight.[19,50]

Current IoT has a *limitation in self-recovering mechanisms* to handle the transition from an error state to the normal operational mode for ITS systems.[40] Most of the existing systems require human intervention in the diagnosis and fixing of the identified faults process.[40,49] The solution provided by TCP/IP, by default, consists of an acknowledge (ACK)-based end-to-end (E2E) reliability mechanism. Unfortunately, this E2E mechanism is not appropriate for sensor networks due to their high loss rates as a result of signal attenuation and path loss arising from low power radios and channel contention from dense sensor deployment.[26] Such type of protocol sends out control messages after packets reached their destination and starts each recovery attempt right from the beginning source, wasting too much of the energy.[15]

Hop-by-hop protocols solve these issues through their characteristics regarding the immediately sending back of the data packets to the upstream node to restore the lost packets, the traffic congestion avoidance, and the management of the transmission rates according to the buffer size of a neighbor node or the link load between two nodes.[15] Another solution for self-recovering and self-stabilization mechanisms for ITS can implement blockchain-based systems to store relevant state data managed by a "master" process implemented using a long-living smart contract.[49] The "master" process runs periodically, investigates the system state, and in case of an erroneous state, starts the recovering procedure.

The *limited re-usability* leads to additional costs and waste of many devices when it is decided the migration to new technologies and the old system is retired.[40] The IoT development needs a management system that reuses existing hardware platforms and assigns them to other existing domains or to new ones.[40,54]

Other important sources of limitation of the current networks relate to *security, secrecy,* and *privacy*.[6,53] The lack of policies and regulations reduces the type of data to be stored and used further in adopting road traffic improvement strategies.[40] Moreover, there are still missing specific and clear policies for creating a global framework for implementing at a large-scale ITS based on technologies like autonomous driving, V2V, V2I, and V2X. The main challenges here include the design of identity management systems to serve many entities, the trustworthiness, and control of distributed platforms, and the secure and trusted interaction with real-world objects and entities through sensors and actuator network infrastructures.[36]

Regarding the privacy topic, European Union's General Data Protection Regulation (GDPR) introduced strict rules regarding digital privacy and trust. A big challenge here is to be compliant with these defined privacy rules, and at the same time to address the legitimate need of networks provided to be aware of what happens in their network.[16] The current 5G infrastructure still provide data encryption solutions based on Rivest–Shamir–Adleman (RSA) public key, but the latest technologies like Big Data and artificial intelligence (AI) technologies tend to lower the security level of these algorithms.[19] Further, improvements must be addressed to mitigate these kind of issues and 6G infrastructure shall provide high security, secrecy, and privacy as key features.

The actual sensors networks have limitations to *ensure the retrieved road traffic data on authentication, accuracy, and precision* during the monitoring process. The limitation of the cellular bandwidth of wireless technology reduces the video stream quality in the case of wireless video-based traffic monitoring systems.[40] Consequently, the identification process of the vehicles from the traffic lanes becomes more difficult and leads to wrong vehicle types classification by TMCs or wrong traffic volumes identification that directly influences the traffic lights timings, which leads to difficult situation on traffic congestion. Referring to a video camera used for autonomous driving vehicles, data accuracy is essential in the identification process of traffic lights, traffic signs, obstacles, bicyclists, and pedestrians.

Currently, many solutions try to provide the trade-off between accuracy, precision, and coverage for the given limited resources of sensor devices (Puzis et al., 2013).[16,33] Bluetooth-based data acquisition systems provide a feasible identification process of the travel mode due to Bluetooth-enabled devices available for bicyclists and pedestrians.[8] FN 2030 will provide solutions to improve the data acquisition accuracy through 6G infrastructure and will enable wider bandwidths, higher frequencies, and massive antenna arrays; furthermore, it has the capability to achieve accurate positioning.[7,20]

21.4 ITS ARCHITECTURE IN THE CONTEXT OF FN 2030

21.4.1 REQUIREMENTS FOR FN 2030

Besides its benefits, the FN 2030 will meet many challenges related to the migration of current IoT applications, especially from providing the desired and full-functionality perspectives. This chapter further provides a set of requirements to be fulfilled to ensure the success of this migration to FN 2030 as were already defined by the existing scientific literature[13,27,29,34,55] and takes over much of the discussion from previous work by Pop et al.[40] Additionally, this chapter also defines requirements that consider the integration of IoE with FN 2030.

FN 2030 requirements are defined from the current limitations in the existing technologies and the advancements in latest technologies.[4] Therefore, the FNs are characterized with respect to the existing mobile-oriented environment. Here, the machine-to-machine (M2M) communication offers the physical telecommunication environment required to exchange data between two interconnected cellular connectivity-based end devices,[4,34] and this perspective gives rise to the first requirement.[27]

Requirement 1: *A mobile host shall be treated as a basic user.*

Considering the same challenge of a mobile-oriented environment, another issue to be addressed is network security assurance. The robustness of the FN system will consist of the efficiency of data transmission management considering also the instability of wireless connections.[13,27]

Requirement 2: *The system shall prevent data loss because of network traffic congestion or unstable connection.*

Cloud computing, as an IoS enabler, will ensure a part of security needs from the starting of service development for data handling and throughout its entire life-cycle.[34]

Requirement 3: *The confidentiality of information, data integrity, and authenticity shall be ensured through privacy-by-design.*

Apart from the security ensured through the previous two requirements, data transmission security is responsible also for managing the cooperation between different technologies that are using the same frequency band.[13,55] In this regard, the following requirement shall be fulfilled.

Requirement 4: *The system shall ensure freedom from interference between existing communication technologies.*

Mobile hosts will surround future communication world. Therefore, power consumption will increase dramatically and, from the G-IoT perspective, will result in the need for energy-saving.[13,35] In addition, the energy consumption, green energy sources, and the network management protocols will define the internal network states tailored to FN to make internet–work communication between the devices (mobile or fixed host).[29,35]

Requirement 5: *The mobile hosts shall enter into an idle/sleep mode until it initiates communication with other devices or receives a request from other devices.*

Usually, the communication network protocols will convert real-time data in to the packets and an Internet Protocol (IP) address will be attached before going to be transmitted over a network. The IP address consists of two important information: a unique identifier (ID) to identify the packet destination identity and a locator (LOC) to define the packet destination location. A good current solution based on ID/LOC separation reduces the possible scalability issues but, for FN, it is not enough and a new approach is needed for mobile hosts.[27,56] For these reasons, requirements 6 and 7 shall be fulfilled.[27]

Requirement 6: *ID shall be allocated to the host itself instead of to its interface.*

Requirement 7: *A mobile host shall not have assigned a static LOC.*

Talking about mobile environments, mobility support and ultra-low delays are mandatory. Current solutions introduced some protocols using patch-on transmission control protocol/Internet Protocol (TCP/IP) as a feature for mobility support but brought disadvantages in terms of performance (e.g., overuse of proxy agent, triangle routing, etc.).[27,55]

The following requirements describe how FN 2030 will consider the mentioned challenges:

Requirement 8: *The mobility support shall be built-in.*

Requirement 9: *A decentralized network architecture shall be used to increase routing performance and reduce failures or possible attacks.*

The continuous growth in ICT new technologies will challenge the network architecture requirements of FN 2030. Therefore, the FN 2030 network architecture must be designed to fulfill the future requirements, and that must be adaptable for communication networks in 2030. Again, it must support the application development in future trend (or) high frequency bandwidth requirements on high-resolution video streaming.[4] Due to these reasons, the current IP routing will be inefficient in FN 2030 and a special data delivery mechanism must be introduced to fulfill the future requirements in 2030.[27]

Requirement 10: *The routing system shall be capable to adapt its routing mechanism according to different network architectures.*

IoE integration with FN 2030 will be mandatory considering the trend in developing energy efficiency policies and the increasing interest in the usage of green sources of energy. Consequently, FN 2030 shall define clear requirements to cover this environmental-friendly approach.[19,21,26] The high number of requirements for this category compared to the previous topics proves again the need for energy savings and that the future is toward a green technology direction.

Requirement 11: *The FN 2030 shall define energy-efficient transport layer protocols.*

Requirement 12: *The placement location of the sensors networks shall provide the access to green energy sources.*

Requirement 13: *The sensors' network architectures shall provide energy harvesting mechanisms.*

21.4.2 CHALLENGES IN MIGRATION OF CURRENT ITS TO FN 2030-BASED PLATFORMS

To satisfy the FN 2030 requirements and considering the current limitations on ITS, many challenges will arise in implementing the Future ITS: 2030. Further, this chapter will present an overview of these challenges as they were depicted in relevant research papers and was already discussed by

Pop et al.[40] Some of them are identified as short-term challenges that need to be solved, such as migration to the 5G technologies, which will also be considered in the context of FN 2030.[10,41,47]

The biggest problem identified in ICT point will be the *overload of wireless networks* in 2030. It is due to the number of wireless devices to be utilized in the IoT network for traffic monitoring and control. In such cases, an adaptive routing protocol must be introduced in the real-time *resource allocation and prioritization.* Furthermore, it must be capable of handling the *system for storage and management of big road traffic data volumes* (Masek et al., 2016).[41,44]

Vehicle-to-vehicle (V2V) communication will be the subject of other challenges. New certificate management systems shall be designed to *ensure the privacy, safety, and security of the infrastructure communication networks*.[22,23,31,47] V2V messages can be the target of transmission interferences or different attacks like taking vehicle control or manipulating the data sent to the other mobile hosts which can lead to accidents.

The complex sensors networks used for road traffic monitoring embedded in the road infrastructure, traffic lights systems, and V2V communication can use different technologies. For this reason, the Future ITS: 2030 shall provide *a common platform that allows the deployment for various technologies and architectures*.[41] The future autonomous driving will request the existence of autonomous networks and efficient integration between them and ITS. Moreover, these systems shall be reliable and capable of *self-recovering from failure states*.[49]

Cost reduction is another problem, which arised due to the use of low-cost sensors, deployment services, and reducing energy consumption.[10,35] *Orientation to eco-friendly solutions* is another biggest challenge and can be solved by cost saving. Here, the system could be developed by using a software and hardware solution, reducing consumption, building a renewable energy system and on demand data transmission, etc.[10,35]

Further, Future ITS: 2030 will meet specific challenges in IoT like *scalability, robustness, energy consumption, bandwidth, connectivity, latency, and security.* These are all the critical challenges due to the exponential growth of the ITS data in real-time data transmission and processing.[10,31,41,44,49]

Energy efficiency-needed protocols and *green-based sources* usage will be other challenges in the migration of current ITS to FN 2030. Both built-in vehicle sensor and road infrastructure sensor networks will need these energy protocols, which can lead to delays in data transmission in case of V2X communication due to the needed transition of the destination sensor from the sleep mode to normal, operating mode. Also, energy harvesting and ensuring a possible green source of energy for each type of sensor would be difficult. Many of the energy efficiency and harvesting challenges will be solved by the 6G infrastructure according to the vision provided by Dang et al.[19]

Many of the mentioned challenges are related and can be solved partially by *appropriate policies and regulation*s frameworks.[36] This will have an important role in data storage and usage due to permissions allowed legally to FN 2030 systems.

21.4.3 ARCHITECTURE OF FUTURE ITS: 2030

This section mainly concentrates on describing the proposed architecture for Future ITS in 2030 as Pop et al.[40] previously proposed it. The ITS: 2030 architecture shows the novelty through its tailoring to an architecture that additionally incorporates the concepts of IoE,[9,42] blockchain-based solutions,[49] and 6G technology.[14,19,21]

Figure 21.3 illustrates a refined version of the Future ITS architecture built by Pop et al.[40] that refers to the current limitations of IoT in ITS. In the proposed architecture, the future requirements and the identified challenges are included (e.g., IoE, blockchain, 6G concepts integration, etc.,) in the FN 2030. The three-layered architecture[29] is tailored to the FN 2030 architecture.[18] Here, both of them is being shown in parallel for a better overview. In the design process, our proposal was taken into account with other architectures which are related to future IoTs (Masek et al., 2016).[13,22,24,29,34,35]

The *Perception Layer* is the correspondent to the *Autonomous Network Architectures* FN 2030 cluster. This includes the sensor networks and actuators that are responsible for relevant data collection with an important role in system self-awareness assurance, self-configuration, and autonomous operation. Additionally, this cluster includes the energy transfer and harvesting units specific to the integration process of Future ITS with IoE.

Network Layer is responsible for ensuring high-quality communication and data transmission between the perception layer and application. Here, many services fulfill the security and privacy requirements through current network protocols (e.g., IPv4 and IPv6) or specific algorithms designed for access, control, authorization, and encryption. Also, the gateway networks are included for interoperability assurance between networks. Moreover, the 5G technology has its place here through concepts like tactile internet, nano-networks, and flying networks.[55] New advancements show that the 6G technology also will have its place at this layer, especially for the possible provided solutions such as augmented and virtual reality.[14,19,21] *Core + Access Networks* is the FN 2030 corresponding cluster and has as objectives the multi-operator resource sharing and the "always best connected and managed" vision achieved.[18] This cluster also provides solutions for energy distribution through 6LoWAN.[9,42]

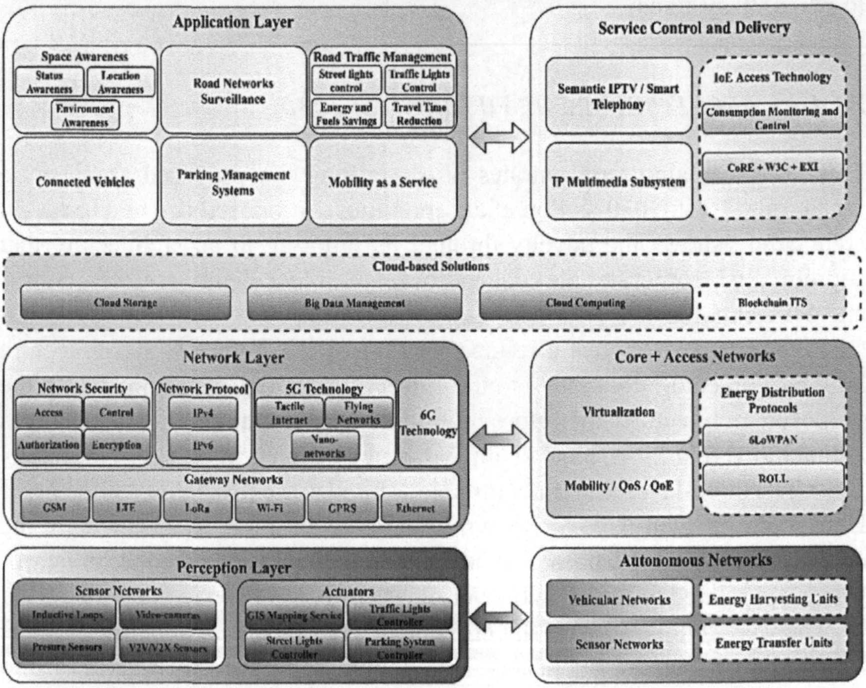

FIGURE 21.3 Future ITS: 2030—refined architecture (adapted after Pop et al.[40]).

Application Layer is the highest level and is responsible for applications related to spatial awareness, road network surveillance, road traffic management, connected vehicles, parking management systems, and mobility-as-a-service (MaaS) feature. This is the correspondent of the *Service Control and Delivery* layer from FN 2030 architecture, which uses semantic Internet Protocol Television (IPTV) or IP multimedia subsystem and aims to design a common framework for integrating personalized, context-sensitive, and semantically rich multimedia applications.[18] At this level, the IoE Access Technology ensures the management of the ITS-related sensors networks and integrates specific services to meet the networking constrained devices with low-energy requirements (e.g., the mixture of CoRE with W3C's definition of the EXI).[9,42]

Cloud-based solutions layer was included as a buffer area between the network and application layers. Its purpose is to provide a common platform responsible for storage, computing, and big data management. In this manner, the collected real-time road traffic data are processed before being used by the application.[24] At this level, blockchain ITS specific solutions can be provided for ensuring the relevant road traffic data privacy, secrecy, and security.[49]

21.5 CONCLUSION

The information and communication technology advancements in ITS led to growth of FN: 2030, and that is very essential, complicated, advanced, and automatic. Therefore, all the countries have to achieve the goal of a FN: 2030 ITS with the vehicle policies on smart mobility. To achieve smart transportation and mobility, inclusive policy proclamation, pilot project instigation, standards creation, and interagency alliance factors must be considered. FN: 2030 provides complete information about ITS in 2030, further it can be used in the transportation networks (i.e., bus, airport, and railway stations), hospitals, government and private offices, stadiums, and tourist places. Due to the rapid growth of IoT and computer and wireless communication technologies and cloud-based solutions to the ITS in recent years, the FN: 2030 will enable reliable and quality road infrastructure monitoring systems in 2030. The FN: 2030 will simplify the congestion and traffic flow in crowded urban areas; furthermore, it will fulfill the future requirements by 2030.

KEYWORDS

- **Future Network 2030**
- **intelligent transportation system**
- **Future ITS: 2030**
- **Internet of Things**
- **wireless sensors networks**

REFERENCES

1. Adeleke, O. O.; Jimoh, Y. A.; Akinpelu, M. A. Development of an Advanced Public Transportation System for Captive Commuters on Urban Arterials in Ilorin, Nigeria. *Alexandr. Eng. J.* **2013,** *52* (3), 447–454. doi: 10.1016/j.aej.2013.04.004.
2. Prescient & Strategic Intelligence Private Limited. *Advanced Transportation Pricing Systems Market*. Prescient & Strategic Intelligence Private Limited, January 26, 2016. https://www.psmarketresearch.com/market-analysis/advanced-transportation-pricing-systems-market.
3. Aindralabs. *What Is Intelligent Transportation System (ITS): Applications and Examples*, August 6, 2020. https://www.aindralabs.com/what-is-intelligent-transportation-system-its-applications-and-examples/.
4. Al Mtawa, Y.; Haque, A.; Bitar, B. The Mammoth Internet: Are We Ready? *IEEE Access* **2019,** *7*, 132894–132908. doi: 10.1109/ACCESS.2019.2941110.
5. Albreem, M. A.; Sheikh, A. M.; Alsharif, M. H.; Jusoh, M.; Mohd Yasin, M. N. Green Internet of Things (GIoT): Applications, Practices, Awareness, and Challenges. *IEEE Access* **2021,** *9*, 38833–38858. doi: 10.1109/access.2021.3061697.
6. Badii, C.; Bellini, P.; Difino, A.; Nesi, P. Smart City IoT Platform Respecting GDPR Privacy and Security Aspects. *IEEE Access* **2020,** *8*, 23601–23623. doi: 10.1109/ACCESS.2020.2968741.
7. Barakat, B.; Taha, A.; Samson, R.; Steponenaite, A.; Ansari, S.; Langdon, P. M.; Wassell, I. J.; Abbasi, Q. H.; Imran, M. A.; Keates, S. 6G Opportunities Arising from Internet of Things Use Cases: A Review Paper. *Fut. Internet* **2021,** *13* (6), 159. doi: 10.3390/fi13060159.
8. Bathaee, N.; Mohseni, A.; Park, S.; Porter, J. D.; Kim, D. S. A Cluster Analysis Approach for Differentiating Transportation Modes Using Bluetooth Sensor Data. *J. Intell. Transport. Syst.* **2018,** *22* (4), 353–364. doi: 10.1080/15472450.2018.1457444.
9. Bui, N.; Castellani, A.; Casari, P.; Zorzi, M. The Internet of Energy: A Web-Enabled Smart Grid System. *IEEE Network* **2012,** *26* (4), 39–45. doi: 10.1109/MNET.2012.6246751.
10. Chettri, L.; Bera, R. A Comprehensive Survey on Internet of Things (IoT) toward 5G Wireless Systems. *IEEE Internet Things J.* **2020,** *7* (1), 16–32. doi: 10.1109/JIOT.2019.2948888.
11. Chin, K. K.; Grace, O. N. G. *Smart Mobility 2030—ITS Strategic Plan for Singapore*, November 2015. https://www.esci-ksp.org/. https://esci-ksp.org/wp/wp-content/uploads/2012/06/J15Nov_p04Chin_SmartMobility2030.pdf.
12. Choosakun, A.; Chaiittipornwong, Y.; Yeom, C. Development of the Cooperative Intelligent Transport System in Thailand: A Prospective Approach. *Infrastructures* **2021,** *6* (3), 36. doi: 10.3390/infrastructures6030036.
13. Chopra, K.; Gupta, K.; Lambora, A. Future internet: The Internet of Things—A Literature Review. In *2019 International Conference on Machine Learning, Big Data, Cloud and Parallel Computing (COMITCon)*, 2019; pp 135–139. doi: 10.1109/COMITCon.2019.8862269.
14. Chowdhury, M. Z.; Shahjalal, Md.; Ahmed, S.; Jang, Y. M. 6G Wireless Communication Systems: Applications, Requirements, Technologies, Challenges, and Research Directions. *IEEE Open J. Commun. Soc.* **2020,** *1*, 957–975. doi: 10.1109/OJCOMS.2020.3010270.
15. Chuang, P.-J.; Yu, Y.-C.; Lin, C.-S. Reliable and Congestion-Controlling Transport in Wireless Sensor Networks. *J. Chin. Inst. Eng.* **2013,** *36* (1), 2–16. doi: 10.1080/02533839.2012.726021.

16. Clemm, A.; Zhani, M. F.; Boutaba, R. Network Management 2030: Operations and Control of Network 2030 Services. *J. Netw. Syst. Manage.* **2020,** *28* (4), 721–750. doi: 10.1007/s10922-020-09517-0.
17. ATTAP. *Connected Vehicles (CV)*. ATTAP, May 2, 2017. https://attap.umd.edu/cvs-and-avs/connected-vehicles-cvs/.
18. DAI-Labor. *Research: Network and Mobility*. n.d. Retrieved January 27, 2020. https://dai-labor.de/en/research/.
19. Dang, S.; Amin, O.; Shihada, B.; Alouini, M.-S. What Should 6G Be? *Nat. Electron.* **2020,** *3* (1), 20–29. doi: 10.1038/s41928-019-0355-6.
20. De Lima, C.; Belot, D.; Berkvens, R.; Bourdoux, A.; Dardari, D.; Guillaud, M.; Isomursu, M.; Lohan, E.-S.; Miao, Y.; Barreto, A. N.; Aziz, M. R. K.; Saloranta, J.; Sanguanpuak, T.; Sarieddeen, H.; Seco-Granados, G.; Suutala, J.; Svensson, T.; Valkama, M.; Van Liempd, B.; Wymeersch, H. Convergent Communication, Sensing and Localization in 6G Systems: An Overview of Technologies, Opportunities and Challenges. *IEEE Access* **2021,** *9*, 26902–26925. doi: 10.1109/ACCESS.2021.3053486.
21. Giordani, M.; Polese, M.; Mezzavilla, M.; Rangan, S.; Zorzi, M. Toward 6G Networks: Use Cases and Technologies. *IEEE Commun. Mag.* **2020,** *58* (3), 55–61. doi: 10.1109/MCOM.001.1900411.
22. Guerrero-ibanez, J. A.; Zeadally, S.; Contreras-Castillo, J. Integration Challenges of Intelligent Transportation Systems with Connected Vehicle, Cloud Computing, and Internet of Things Technologies. *IEEE Wireless Commun.* **2015,** *22* (6), 122–128. doi: 10.1109/MWC.2015.7368833.
23. Hsu, I. Y.-Y.; Wódczak, M.; White, R. G.; Zhang, T.; Russell Hsing, T. Challenges, Approaches, and Solutions in Intelligent Transportation Systems. In *2010 Second International Conference on Ubiquitous and Future Networks (ICUFN)* 2010; pp 366–371. doi: 10.1109/ICUFN.2010.5547180.
24. Jabraeil Jamali, M. A.; Bahrami, B.; Heidari, A.; Allahverdizadeh, P.; Norouzi, F. IoT Architecture. In *Towards the Internet of Things*; Jabraeil Jamali, M. A., Bahrami, B.; Heidari, A.; Allahverdizadeh, P.; Norouzi, F., Eds.; Springer International Publishing: Cham, 2020; pp 9–31. doi: 10.1007/978-3-030-18468-1_2.
25. Jaradat, M.; Jarrah, M.; Bousselham, A.; Jararweh, Y.; Al-Ayyoub, M. The Internet of Energy: Smart Sensor Networks and Big Data Management for Smart Grid. *Proc. Comput. Sci.* **2015,** *56*, 592–597. doi: 10.1016/j.procs.2015.07.250.
26. Jones, J.; Atiquzzaman, M. Transport Protocols for Wireless Sensor Networks: State-of-the-Art and Future Directions. *Int. J. Distrib. Sens. Netw.* **2007,** *3* (1), 119–133. doi: 10.1080/15501320601069861.
27. Jung, H.; Koh, S.-J. Problem Statements and Requirements for Mobile Oriented Future Internet. *ICTC* **2011,** *2011*, 610–611. doi: 10.1109/ICTC.2011.6082696.
28. Kafi, M. A.; Challal, Y.; Djenouri, D.; Bouabdallah, A.; Khelladi, L.; Badache, N. A Study of Wireless Sensor Network Architectures and Projects for Traffic Light Monitoring. *Proc. Comput. Sci.* **2012,** *10*, 543–552. doi: 10.1016/j.procs.2012.06.069.
29. Kaur, K. A Survey on Internet of Things—Architecture, Applications, and Future Trends. *2018 First International Conference on Secure Cyber Computing and Communication (ICSCCC),* 2018; pp 581–583. doi: 10.1109/ICSCCC.2018.8703341.
30. Kem, O.; Balbo, F.; Zimmermann, A. Traveler-Oriented Advanced Traveler Information System Based on Dynamic Discovery of Resources: Potentials and Challenges. *Transport. Res. Proc.* **2017,** *22*, 635–644. doi: 10.1016/j.trpro.2017.03.059.

31. Koeneman, C. *The Benefits and Challenges of IoT in Public Transportation*, April 25, 2017. https://www.metro-magazine.com/10002967/the-benefits-and-challenges-of-iot-in-public-transportation.
32. Kumar, P.; Singh, V.; Reddy, D. Advanced Traveler Information System for Hyderabad City. *IEEE Trans. Intell. Transport. Syst.* **2005,** *6* (1), 26–37. doi: 10.1109/tits.2004.838179.
33. Li, X.; Ouyang, Y. Reliable Traffic Sensor Deployment under Probabilistic Disruptions and Generalized Surveillance Effectiveness Measures. In *Transportation Research Board 90th Annual MeetingTransportation Research Board*, 2011. https://trid.trb.org/view/1091313.
34. Lovrek, I.; Caric, A.; Lucic, D. Future Network and Future Internet: A Survey of Regulatory Perspective. In *2014 22nd International Conference on Software, Telecommunications and Computer Networks (SoftCOM)*, 2014; pp 186–191. doi: 10.1109/SOFTCOM.2014.7039093.
35. Maksimovic, M. Greening the Future: Green Internet of Things (G-iot) as a Key Technological Enabler of Sustainable Development. In *Internet of Things and Big Data Analytics Toward Next-Generation Intelligence*; Dey, N., Hassanien, A. E., Bhatt, C., Ashour, A. S.; Satapathy, S. C. Eds.; Springer International Publishing: Cham, 2018; Vol. 30, pp 283–313. doi: 10.1007/978-3-319-60435-0_12.
36. Misuraca, G.; Broster, D.; Centeno, C. Digital Europe 2030: Designing Scenarios for ICT in Future Governance and Policy Making. *Govern. Inf. Q.* **2012,** *29*, S121–S131. doi: 10.1016/j.giq.2011.08.006.
37. Mordor Intelligence. *Intelligent Transport Systems Market | Growth, Trends, and Forecast (2021–2026)*, n.d. https://www.mordorintelligence.com/. Retrieved August 12, 2021. https://www.mordorintelligence.com/industry-reports/intelligent-transport-systems-market.
38. Park, S.-J.; Sivakumar, R. Energy Efficient Correlated Data Aggregation for Wireless Sensor Networks. *Int. J. Distrib. Sens. Netw.* **2008,** *4* (1), 12–26. doi: 10.1080/15501320701774592.
39. Patel, C.; Shah, D.; Patel, A. Automatic Number Plate Recognition System (ANPR): A Survey. *Int. J. Comput. Appl.* **2013,** *69* (9), 21–33. doi: 10.5120/11871-7665.
40. Pop, M.-D.; Pandey, J.; Ramasamy, V. Future Networks 2030: Challenges in Intelligent Transportation Systems. In *2020 8th International Conference on Reliability, Infocom Technologies and Optimization (Trends and Future Directions) (ICRITO)*, 2020; pp 898–902. doi: 10.1109/ICRITO48877.2020.9197951.
41. Puri, S.; Rai, R. S.; Saxena, K. Barricades in Network Transformation from 4G to 5G in India. *2018 7th International Conference on Reliability, Infocom Technologies and Optimization (Trends and Future Directions) (ICRITO)*, 2018; pp 695–702. doi: 10.1109/ICRITO.2018.8748303.
42. Rana, M. Architecture of the Internet of Energy Network: An Application to Smart Grid Communications. *IEEE Access* **2017,** *5*, 4704–4710. doi: 10.1109/ACCESS.2017.2683503.
43. Research, P. *Intelligent Transportation System Market Size to Hit US$ 47.89 Bn by 2030*. GlobeNewswire News Room, June 1, 2021. https://www.globenewswire.com/en/news-release/2021/06/01/2239977/0/en/Intelligent-Transportation-System-Market-Size-to-Hit-US-47-89-Bn-by-2030.html.
44. Sallai, G. Chapters of Future Internet Research. In *2013 IEEE 4th International Conference on Cognitive Infocommunications (CogInfoCom)*, 2013; pp 161–166. doi: 10.1109/CogInfoCom.2013.6719233.

45. *Smart City Solutions | Trafiksol.* (n.d.). Trafiksol. Retrieved August 12, 2021. https://www.trafiksol.com/smart-city-solutions/.
46. Sepulcre, M.; Gozalvez, J.; Hernandez, J. Cooperative Vehicle-to-Vehicle Active Safety Testing under Challenging Conditions. *Transport. Res. C: Emerg. Technol.* **2013,** *26,* 233–255. doi: 10.1016/j.trc.2012.10.003.
47. Thayananthan, V.; Algarni, A. Internet of Things Based 5G Infrastructure for Securing Transportation Facilities in Smart Cities. *Int. J. Comput. Appl.* **2019,** *181* (46), 26–33. doi: 10.5120/ijca2019918614.
48. TraficSol. *Advanced Traffic Management Systems | Intelligent Traffic Management | Trafiksol.* n.d. https://www.trafiksol.com/. Retrieved August 12, 2021. https://www.trafiksol.com/advanced-traffic-management-system/.
49. Tseng, L.; Wong, L. Towards a Sustainable Ecosystem of Intelligent Transportation Systems. In *2019 IEEE International Conference on Pervasive Computing and Communications Workshops (PerCom Workshops)*, 2019; pp 403–406. doi: 10.1109/PERCOMW.2019.8730669.
50. Ulukus, S.; Yener, A.; Erkip, E.; Simeone, O.; Zorzi, M.; Grover, P.; Huang, K. Energy Harvesting Wireless Communications: A Review of Recent Advances. *IEEE J. Select. Areas Commun.* **2015,** *33* (3), 360–381. doi: 10.1109/JSAC.2015.2391531.
51. Ramasamy, V. Design of ITS Through IoT. *Int. J. Eng. Adv. Technol.* **2019,** *8* (5C), 1559–1561. doi: 10.35940/ijeat.e1229.0585c19.
52. Vimal, V.; Khosla, A.; Prabhakar, P.; Arora, S.; Ashok, A. Comparison of Adaptive Filtering Scheme for Sustainable and Efficient Communication in Smart City. *Sustain. Energy Technol. Assess.* **2021,** *47*, 101472. doi: 10.1016/j.seta.2021.101472.
53. Weinberg, B. D.; Milne, G. R.; Andonova, Y. G.; Hajjat, F. M. Internet of Things: Convenience vs. Privacy and Secrecy. *Bus. Horizons* **2015,** *58* (6), 615–624. doi: 10.1016/j.bushor.2015.06.005.
54. Xu, L.; Mcardle, G. Internet of too Many Things in Smart Transport: The Problem, the Side Effects and the Solution. *IEEE Access* **2018,** *6*, 62840–62848. doi: 10.1109/ACCESS.2018.2877175.
55. Yastrebova, A.; Kirichek, R.; Koucheryavy, Y.; Borodin, A.; Koucheryavy, A. Future Networks 2030: Architecture & Requirements. In *2018 10th International Congress on Ultra Modern Telecommunications and Control Systems and Workshops (ICUMT)*, 2018; pp 1–8. doi: 10.1109/ICUMT.2018.8631208.
56. You, T. Toward the Future of Internet Architecture for IoE: Precedent Research on Evolving the Identifier and Locator Separation Schemes. In *2016 International Conference on Information and Communication Technology Convergence (ICTC)*, 2016; pp 436–439. doi: 10.1109/ICTC.2016.7763513.

CHAPTER 22

The Future of Web Crowdfunding: An Ethereum Blockchain Approach

SUBASH CHANDRA BOSE JAGANATHAN[1*], MURUGESH VEERASAMY[2], AZATH MUBARAKALI[3], CHANDRAMOHAN DHASARATHAN[4], VELMANI RAMASAMY5, R. KALPANA[6], and NINOSLAV MARINA[7]

[1]*School of Computing Science and Engineering, VIT Bhopal University, Sehore, MP, India*

[2]*Department of CSE, Koneru Lakshmaiah Education Foundation, Vaddeswaram, AP, India*

[3]*College of Computer Science, King Khalid University, Abha, Saudi Arabia*

[4]*Computer Science and Engineering Department, Thapar Institute of Engineering and Technology, Patiala, Punjab, India*

[5]*Professor and Head, Computer Science and Engineering Department, Adithya Institute of Technology, Kurumbapalayam, Coimbatore, Tamil Nadu, India*

[6]*Dept. of Computer Science and Engineering, Madanapalle Institute of Technology and Science, Angallu, Andhra Pradesh, India*

[7]*University of Information Science and Technology, "St. Paul the Apostle", Ohrid, North Macedonia*

[*]*Corresponding author. E-mail: jsubashme@gmail.com*

ABSTRACT

One of the most efficient methods of crowdfunding in today's world is using websites such as GoFundMe or KickStarter or Indiegogo. Those are websites where people are creating a project, presenting it with a whitepaper and

Intelligent Technologies for Sensors: Applications, Design, and Optimization for a Smart World.
S. Kannadhasan, R. Nagarajan, & Alagar Karthick (Eds.)

people who like the idea or for an example, there are also people who are looking to get some money to do something like a wish or a dream come true, and for their medical treatments. The downside of this is that people who are creating the ideas or "projects" they are obliged to pay 20% or more in commissions to these platforms. Here comes the new era of cryptocurrency where decentralization is the main point of its existence. By cutting the middle man, such as the banks, these type of websites, people can use this technology to send peer to peer transactions to each other without being tracked by a third party or being in some way commissioned for some service. The technology developed by Vitalik Buterin and his team, who are developers of the Ethereum cryptocurrency, they made this coin and a platform to code with it. The language for coding in the Ethereum network called Solidity. With some or none coding skills, anyone can make their own "token" which is an expressed currency depending on the needs of the creator, could be any service or platform that can accept anyone's token. The momentum of this movement is that anyone can create a project, for example a crowdfunding that will be expressed with the currency of the token, which will be given from the creator (the one who is having the idea) and the price also. Therefore, a creator is generating the token and is selling it proportional with the price of the Ethereum (ETH). When people are buying this token they are actually supporting the creator who receives the ETH from the supporters and they in return get Token which in the future if the project is successful they can use that token for the services of the platform or to sell it on an online cryptocurrency exchange. The presented work shows the process of creating a personal ICO (Initial Coin Offering). The goal is to create a powerful anonymous way to fund projects and to receive a product from the developers, which in this case would be a Token named ServiceCoin (SRV), for which they would have to buy it using Ethereum cryptocurrency in order to receive the tokens. We will go thru the process of installing a cryptocurrency wallet using MetaMask (Google chrome extension wallet), then we will use Brackets text editor in order to sort our code. After we will use Remix IDE to compile that code and at the end we will use MyEtherWallet.com service to publish our ICO so everyone can verify and see the actual tokens.

22.1 INTRODUCTION

In May 2010, Brad Damphousse and Andrew Ballester created a company named GoFundMe. The idea behind the company was to give the people a platform, a software, a community that will have the power to fund projects such as ideas in business, in medical help, wishes, and travels, and if the people, users liked that idea they would fund it with few dollars from their pocket. The power of many is more than a power of the few.

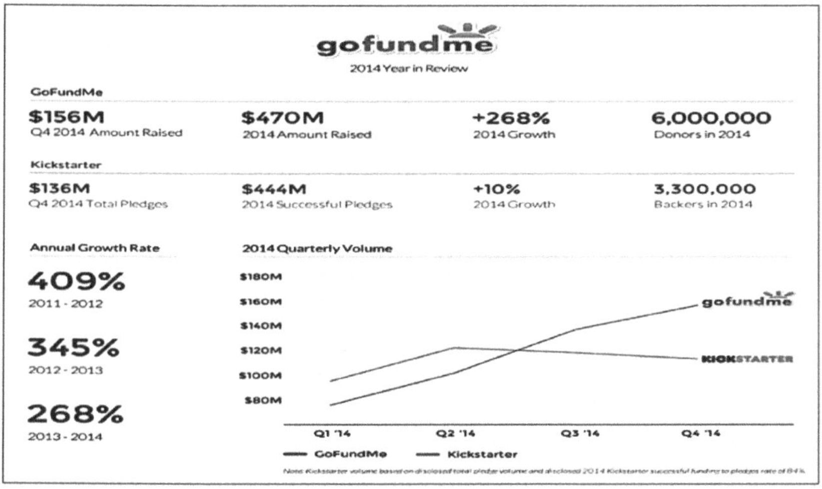

FIGURE 22.1 GoFundMe vs Kickstarter (platform that we will use as an example in this article).

GoFundMe has now a different fee for projects, 5% from total funds gathered, 3% for the payment processing, and few added for money transfers to project owners approximately ~11%. Another platform worth mentioning is Indiegogo, founded in 2008 by Danae Ringelmann as shown in Figure 22.1. The idea behind Indiegogo is somewhat unique in a way that when someone helps, funds, invests in a project he is going to receive a gift back from the creators of the project. Let us say there is a product that is being developed and after the project is finished, all of the investors get a prototype or the whole product as a gift.

The idea is the same in 2009; Perry Chen, Charles Adler, and Yancey Strickler created the Kickstarter platform that is a crowdfunding platform. As mentioned, this platform is a big success. A lot of projects and ideas get to life because of this platform (Figure 22.2). The idea behind this was to give everyone an equal chance to present their idea in front of a greater audience. However, downside is that this platform is taking commissions from the successful projects that got the required funding 5% from the total funds plus payment processing commissions. The total cost can go up to 20%. This is the process and idea behind crowd funds, as shown in Figure 22.3.

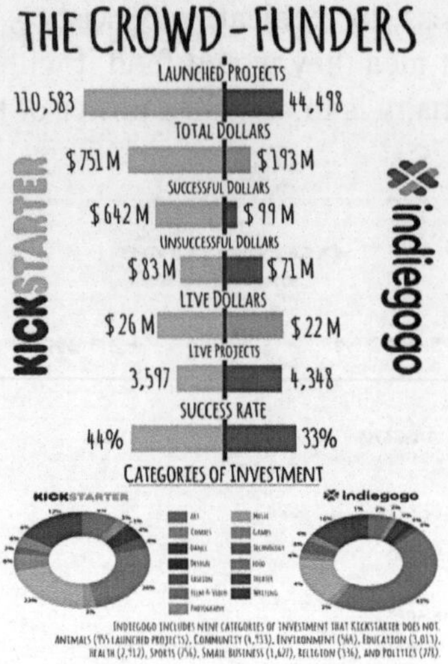

FIGURE 22.2 IndieGogo vs KickStarter efficiency chart.

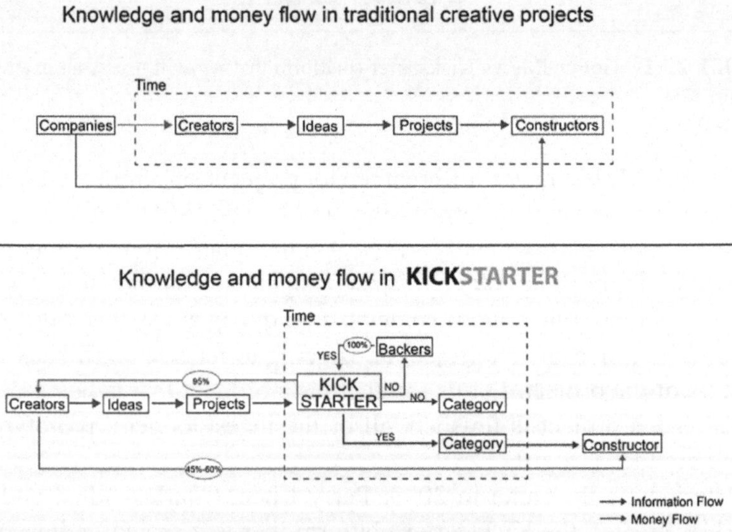

FIGURE 22.3 The idea behind KickStarter "shortening time."

An interesting research paper by Yi Zhang "An Empirical Study into the Field of Crowdfunding" published at Lund University[29] has a great data about the differences between KickStarter and Indiegogo. It states the differences in the model of doing business and how they are different from each other. We can represent that data; it has been shown in Table 22.1.

TABLE 22.1 Direct Comparison between KickStarter and Indiegogo.

	KickStarter	**Indiegogo**
Model	"Fund or return"	"Fund or return" or "keep all"
Commission	5% of successful project + 5% payment processing	4% for successful or 9% for partial 3% credit card pay
Pay methods	Credit/debit card	Paypal
Types	13 types, 36 subtypes	3 types, 24 subtypes
Allowance	1. Charity or cause funding project	No
	2. "Fund my life" projects	
	3. Other prohibited contents	
Partners	No	16
Law restriction	Only US campaigns	Global

Now, the idea behind cryptocurrencies is to cut the intermediary such as KickStarter, so anyone can start their own project, build a whitepaper around it, and present it online or wherever he/she wishes. Popular successful ICO's that today are worth millions are as follows.

The process is simple for having the ICO; you build the token and give out the published address of the "Contract" after generating the bytecode that will be explained further in Figure 22.4 (the published code for your ICO, where people send their funds in Ethereum (ETH) to get their Tokens (shares/currency) from your project. Therefore, after sending out the contract address, now it depends on the creator's innovation to market his project. Therefore, the process would be:

So what we are going to do in our article? We are going to present the first three steps of creating an ICO: Design and plan, publish the smart contract with the address for sales, and explain the idea in Figure 22.5. The fourth step is usually a real-life scenario where you as a creator pay to the ETH network some ETH to pay the miners to publish your contract, which would be from 1 to $10 depending on ETH price on the crypto market.

There are a lot of differences, but the most important ones between ERC20 (Tokens) and Crowdfunding such as Kickstarter for example are the liquidity of the funds inside, meaning the Tokens are immediate and you receive them in the same moment and in the Crowdfunding you wait for the full release of the project, not knowing it would be successful or not (Figure 22.6).

Table 22.2 has illustrated main difference between ICO and normal Crowdfunding on company-platforms. The main difference and results we can see in the actual history that happened in 2017 in this case. How the cryptocurrency changed the way, people go for funding their projects and ideas. For comparison, we can see Figure 22.7.

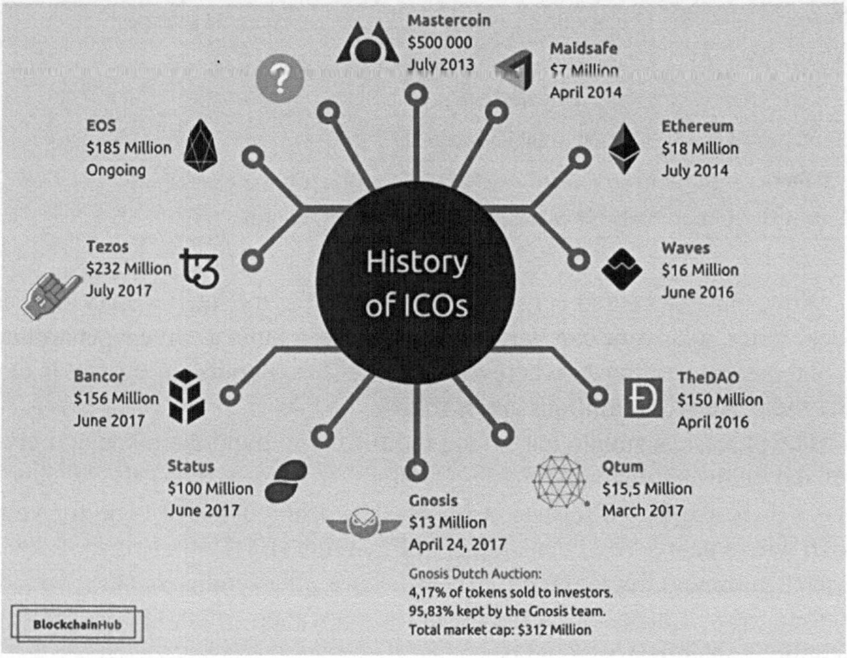

FIGURE 22.4 Report of raised funds for ICO projects.

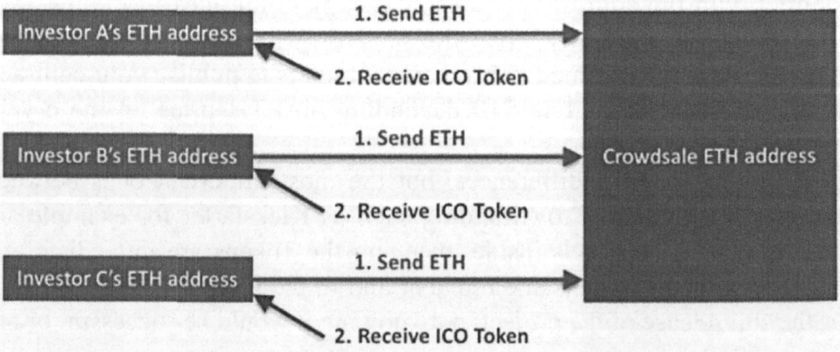

FIGURE 22.5 The process of investing to the given ICO crowd sale project.

FIGURE 22.6 The real-life process of creating an actual ICO.

TABLE 22.2 ICO and normal Crowdfunding on Company-Platforms.

	Selection	Coaching	Funds	Liquidity
Crowdfunding	X	X	Small investors	5–10 years
ICO	X	X	Small investors	Immediate

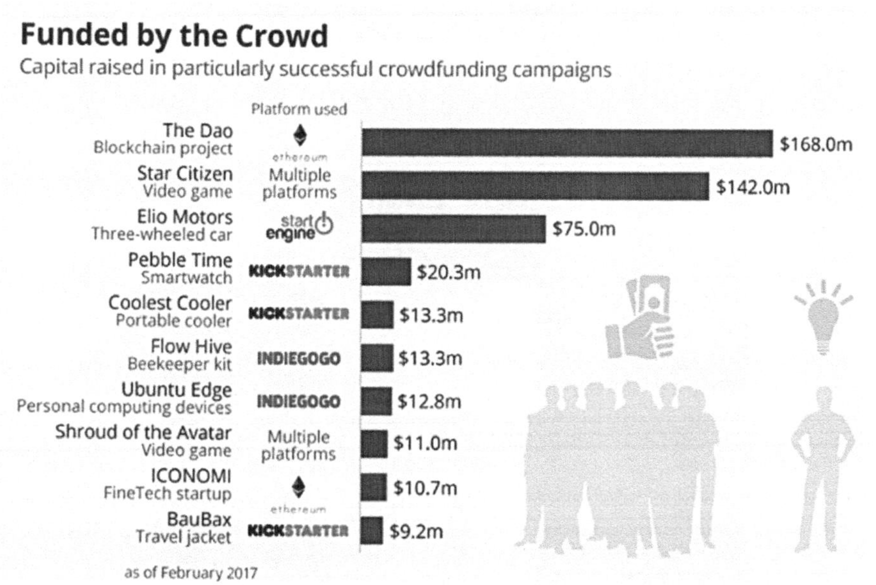

FIGURE 22.7 The crowdfunding campaigns (The Dao-ICO ERC20 ETH Token).

This article will show the benefits of the newest innovation available to every person in the world who has access to the internet no matter the device of use. This idea and article will demonstrate how anyone can fund an idea avoiding many activities included in the today's process to create a crowdfund. A blockchain is a completely disseminated, distributed programming system that utilizes cryptography to safely have applications, store information, and effectively move computerized instruments of significant worth that speak to certifiable cash as shown in Figure 22.8. Another thing to mention is the mining proof-of-work algorithm Equihash. Ethereum is based on this algorithm to process transactions and to publish the contracts. The algorithm is famous for being ASIC resistant, which means that only GPU (graphic units) can process the transactions and processes. Cryptography is the craft of correspondence by means of cryptic message as described in Figure 22.9. Working with Ethereum, cryptic methods are being used to create safe environment for everyone using the AI system that can control and make sure that there is no third party getting involved into any transaction or generation of public and private keys. With the power given, there is no reason why we will not see market speculation, frauds, skepticism on this topic, given the people's lack of knowledge of this new technology.[9] Therefore, we are going to create a Ethereum ERC20 Token and name it ServiceCoin (SRV). We are going to publish this token as an ICO (initial coin offering). The plan behind this is to create a crowdfund to gather funds for the development of the same project, for example, to develop the website, to develop the infrastructure, and to get the right human resources.

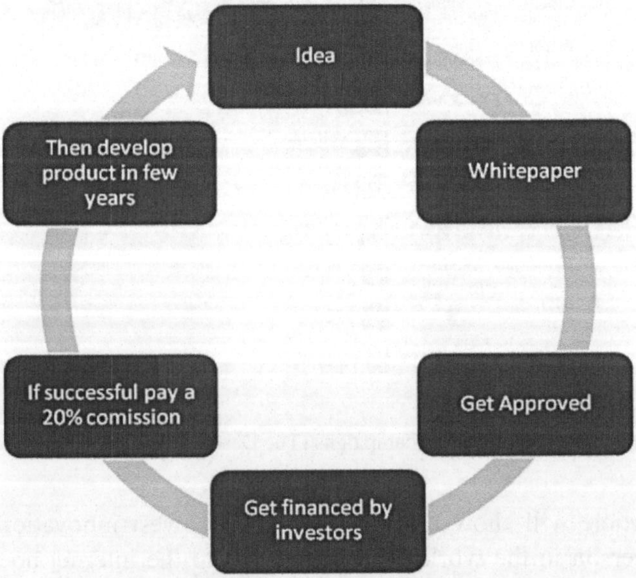

FIGURE 22.8 The normal crowdfunding company platform process (Kickstarter as ex.).

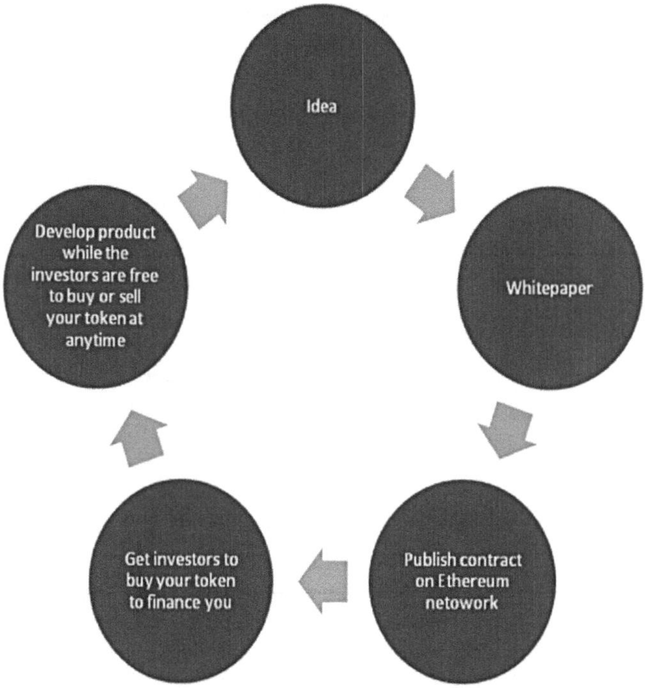

FIGURE 22.9 The ERC20 Token ICO process that cuts the middle man.

By creating this, no one is obligated to invest, the matter of investment is a personal choice and there are a lot of factors included in the process, such as marketing, the way the project is developed, the whitepaper, reliable team—all of these things attract new investors as shown in Figure 22.10. This would be an example of how would ICO start for a website such as Freelancer.com and Upwork.com.. All of the projects or job offerings could be funded with our coin named ServiceCoin. By that rule, SRV will receive some value in the market, and can be easily listed on cryptocurrency exchanges so anyone could reach and buy tokens. The advantage of doing this would be the anonymity the buyer of job poster would receive, also the intermediary process of payment processors, banks, and debit cards. That is the reason why peer-to-peer payments are the future of the economy. The amount of money raised by crowdfunding is from CrowdBerkeley—a project that developed a database from Californian University—Business School Berkley's Haas. The used database had previously been used many times and gives outstanding results. The database shows us the amount of money from four crowdfunding platforms such as KickStarter, FundRazer, RocketHub, and Indiegogo. The data shows the raising of crowdfunding. The period has been shown from Q2 2009 to Q1 2014.[28]

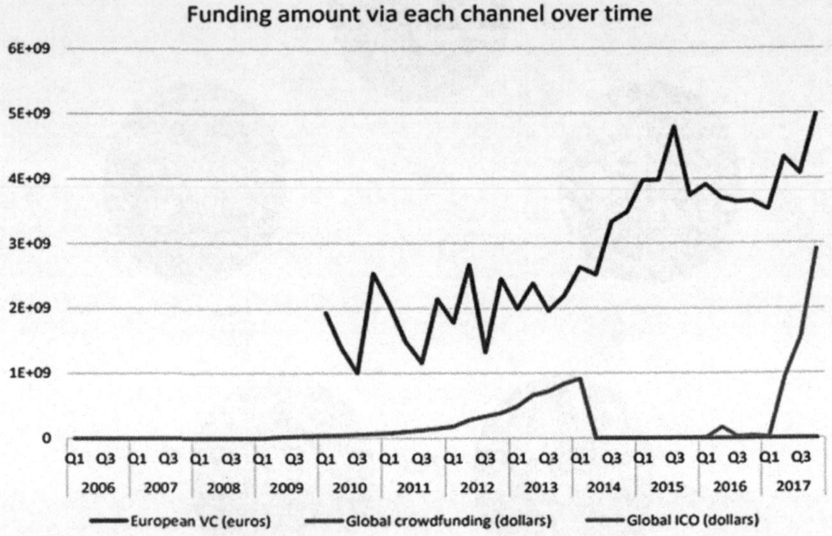

FIGURE 22.10 Funding raised for a given period compared to VC, crowdfunding. ICO.

22.2 AN INITIATIVE FOR CROWDFUNDING

The name of our token is going to be ServiceCoin. This is going to be a decentralized currency that anyone who owns Ethereum can buy when the idea is finished.

We are going to use Metamask as our wallet because of the simple interface; we will use the coding language Solidity that is built solely for this type of projects in the Ethereum network, and we are going to run our code in Remix IDE, which will compile our code to an understandable code for the Ethereum network (Figure 22.11) examinee's the system controlling the information on digital types of value. A cryptocurrency such as Ethereum creates a system of many computers that work together. Every computer included has a history of all the things happening in the network such as transactions, storing values, and remembering every wallet value that contains some Ethereum inside. A transaction is a file that states, "Angela sends Y Ethereum to Alex" and it has been signed by Mark's private key. Next thing after sign, a transaction is being created in the network pool, being sent from one user to another (Figure 22.12) method that is called p2p-technology.[26]

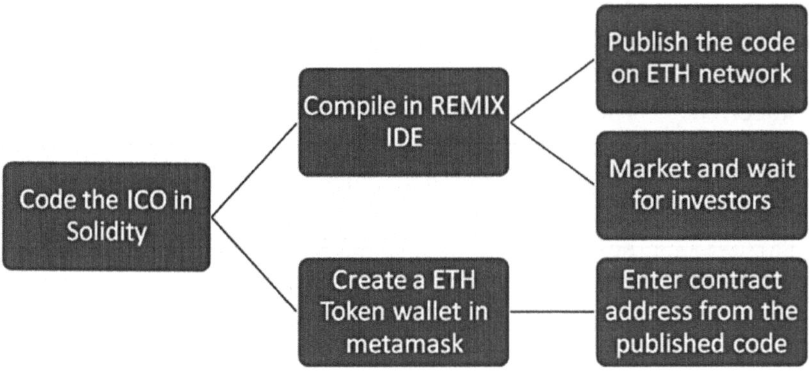

FIGURE 22.11 The Ethereum network process.

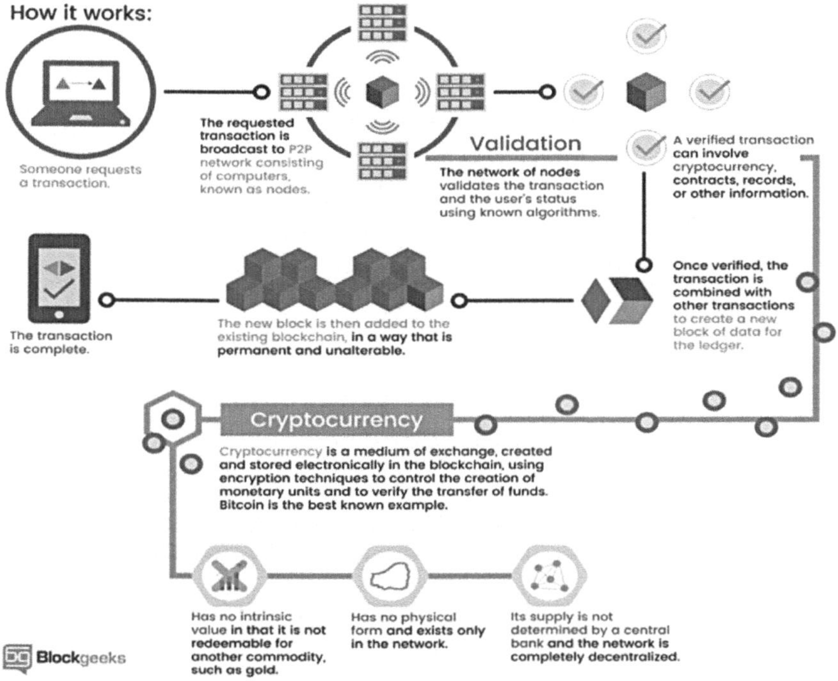

FIGURE 22.12 Functionality of cryptocurrency and its workflow.

22.2.1 PURPOSE

Ideologist crypto-enthusiast Vitalik Buterin has grown to be the second largest name in the cryptocurrency game. As Bitcoin, not only the network does not just confirm transactions and hold value. This means Ethereum could be used to build projects and ideas implemented in published contracts. This property gives Ethereum the power to be one of the biggest names in cryptocurrency.[12] The purpose of this coin would be to decentralize the market for online work, since many people around the world are limited to work because of the country law, meaning that they cannot get their funds through a certain payment processor, or such as legally offer a job to someone willing to help with your project. This would be the perfect solution since anyone at any age can do this all, also be anonymous in the process, such as where the funds are coming. The future is coming with peer-to-peer payments. Let us see some statistics in Figure 22.13.

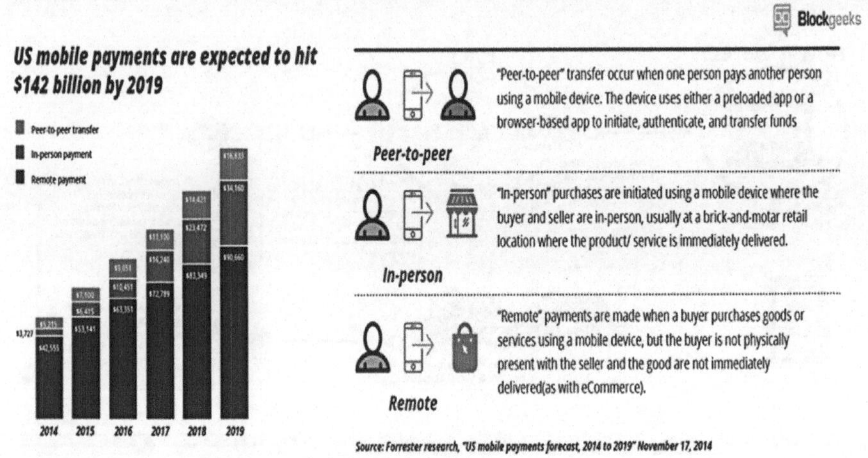

FIGURE 22.13 US mobile payments prediction by 2019.

22.2.2 ADVANTAGES

Tokens inside Ethereum network are able to speak to other volatile exchangeable great cryptocurrencies, dependability focuses, in-app purchases, and so forth. Tokens actualize few essential highlights in a norm manner, and additionally imply that the token is going to be immediately perfectly incorporated inside the Ethereum wallet.[13] Our task will demonstrate the points of interest that are primarily incorporated into the shown pros:

a) **Cannot be reversed:** After being published or sent a transaction, no one can interact with the contract or the transaction ID. That means that anything happening in the network will forever stay there in the network. Therefore, this is one thing that could not be forged since the network always knows how much ETH everyone has in the wallet and every time there is a transaction, the new values are stored.

b) **Anonymous:** No one really knows anyone's names since every Ethereum wallet has unique 30 character name. There is no personality or description besides that. This means that only the string of characters and numbers is given.

c) **Great speed and global:** The transactions are generated after 2–3 min till the network catches up on them. Anyone can use this technology since there are no geopolitical restrictions that prohibit usage.

d) **Safe:** Crypto assets are safe; cryptography behind it using private and public keys makes the network impossible to break into. The system is built not to be intervene with.

e) **Without restrictions:** You don't need any permission from anyone, after creating a wallet nothing can stop you to send some Ethereum into another wallet, there is no third party watching what are you doing.

22.2.3 USE

On the top, we can see Figure 22.14. shows the use of an ERC2 ICO Ethereum used in many sectors; we can still say that it is in early stages. It's an ecosystem based on usage and reward. By usage, we mean that we can transfer assets from one place to another and by reward we mean that for the transfer to happen someone is receiving a reward for doing that; in this case are the so-called miners that are processing the transactions.

FIGURE 22.14 Example of ERC2 ICO.

22.2.4 ERC20

The published contracts usually are creating ERC-20 tokens. They are also built for propose; trade of tokens, and to safe the changes to the given tokens. They are written in a coding language named "Solidity" based on (IFTTT) If-This-Then-That logic.[8] So what happens when an agreement makes a coin? Now is where ERC-20 comes into light. After a ERC-20 token is created it is built to be transferred from wallet A to wallet B. Also it has the power to exchange one token for another. In the past ERC-20 tokens, developers used another code—let us take an example, some developers used (totalAmount) while others used (totalNumber) in their code. With the standards, the token can be now traded on exchanges that approve the token or be stored in some wallet. ERC-20 gives us the power to generate a new token easily, and that is the reason ERC-20 Ethereum was the most used network for generating ICO's in 2017. "ERC" is a term for "Ethereum Request for Comments." To be more precise, ERC is an official topic on top of giving updates to the Ethereum Blockchain. "20" in "ERC20" means the unique ID. In the few years, ERC-20 is one of the biggest reasons why cryptocurrency got a lot of attention. The protocol gave us these days the standards, which were missing before, and gave the acceleration of development of today's DApps (Distributed Applications) in a standard platform.

22.3 INITIAL COIN OFFERING SUCCESSES

With the dispatch of many new digital currencies day by day, few of them are destined to have the boom like Bitcoin, yet some will crash, be a terrible project, or be destroyed by the market. Is there a recipe for being a successful ICO? There are various approaches to quantify "achievement"; however, we have assembled this rundown arranged by increment in worth. These four successful ICO have happened.[11]

a) NEO, before known as Antshare. It is an open-source blockchain developed by Da HongFei and Erik Zhang. It is known as China's Ethereum since the technology is similar because you can create your own tokens and contracts on a similar way. The success behind NEO is speculated to come from the big names behind it that supported its development. Their ICO was one of the biggest booms on the start that made for the ICO buyers a 294,000% Return on Investment (ROI).[11]

b) Ethereum is the second biggest name in cryptocurrency; Vitalik Buterin and his team created something extraordinary. It is nothing like Bitcoin. It has a remarkable innovation that has the ability to develop an "intelligent contract"—a network that gives a new point of view of how cryptocurrencies could be used. Different from the boost of Bitcoin in 2011, it took 4 years for Ethereum to get that type of hype. It is known that early investors in Ethereum made approximately return on investment of 230,000%. Early entry was $0.31, an Ether token now sits at a whopping $213, at the second place after Bitcoin. The reason for that is it gives a powerful innovation to build apps on the network. Ethereum had the attention of many corporate companies because of the usage that could be made to fund or develop projects on their network.[11]

c) Stratis, the "secure based cryptocurrency" is a debatable coin that exicted the enthusiast that could give another point of view of sending cash worldwide. Giving the ability to its users to have a secure way to send money, Stratis made most governments cringe since it could be traced on any way. The team behind this coin is underground. They are using Tor system and the powerful encryption, Stratis is untraceable not even giving out transaction IDs. The token in November 2016, if you invested $0.001 today could be worth $0.64, or Return on investment of 64,000%.[12] While being popular, Stratis got a real attention from the market since businesses always need this kind of finances and Stratis looks like the right choice to invest.

d) **Ark.** The technology behind Ark was to develop multiple coins on one blockchain network. Everything would be decentralized with no one controlling the network with 15 different developers from 11 countries. With the blast from other coins, any news blog can connect them with making a huge profit, and Ark is that type of coin. Advisors were really excited about this token and the investors got 35,400% return on investment if we look at today's cost of $3.54 per coin.[11]

22.4 ETHEREUM

Ethereum is an open-source programming network that is dependent on a blockchain innovation that empowers developers to create decentralized applications.

22.4.1 DEFINITION

Just like Bitcoin, Ethereum is an open-source blockchain network. They are some critical points in the differences between Bitcoin and Ethereum; the most important difference is the reason of existence since Bitcoin was developed as a peer-to-peer network and to store value, while Ethereum is built to do the same thing plus to create other decentralized apps that could be used in many more ways. Bitcoin offers just transfers of money and to hold value; it has stated that Bitcoin is like the gold for cryptocurrency. Bitcoin is used for transferring funds, and probably in the future would be used to send big just big amounts of cash due to the transaction costs, Ethereum is used for the programming innovation while building apps on the decentralize network. In the ETH blockchain, miners are not working to acquire bitcoin; rather they are working on getting Ethereum. Like other cryptocurrencies have, Ether is used to pay for the expenses and maintaining the network alive to process the transactions etc.[25]

22.4.2 WHY ETHEREUM?

Ethereum makes the developers innovative and creative while building the apps. DAAP stands for a decentralized application (which is run on the ethereum network). Bitcoin is basically a Daap since it gives its users the power to transfer money by peer-to-peer method. Decentralized apps on the blockchain can give users the power to not just send money; they are not limited to just basic scalability. In Figure 22.15, we can see what are the benefits of using cryptocurrency networks and what is the advantage of using these types of networks, and in Figure 22.15, we see Ethereum network.

22.4.3 FEATURES

ETH is used also for creating other cryptocurrencies. By using the ERC20 Token technology, developers can develop their own tokens/coins that would be different depending on their needs or projects; these types of models are usually developed and looking to raise funds by ICO. With this funding strategy, the developers can set their token name, the maximum amount, or the minimum amount for the ICO to be successful (there are cases where the funding is not successful and the ETH sent are given back to the sender addresses). Many billions are raised by this method and one of the most famous projects is EOS, which is an ERC20 token.[25] While Dapps are running on the blockchain network, they usually profit from these possessions: its unchangeable—no one can intervene with new updates or information.

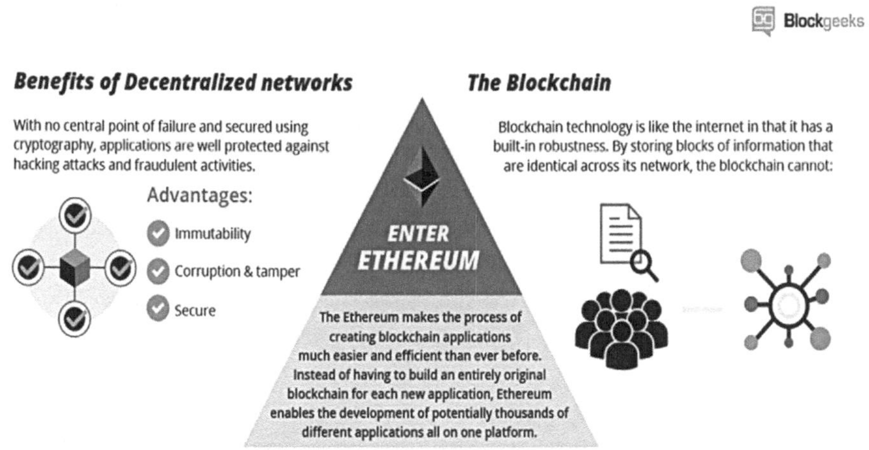

FIGURE 22.15 Benefits of decentralized networks with blockchain technology.

Designed with care—the apps are created to be dependent on the network, so controlling it from outside is not possible. Safe—by using verified cryptography, the apps are hacking-proofed from any attacks or brute force or DDosing the network that is impossible because of the PC's around the world connected to the network. No time off—apps are never turned off since the blockchain networking is running 24/7 with no resting because all of the miners that are active they are working all the time. In Figure 22.16, we can see what kind of apps are developed or are going to be developed in the future.

22.5 DEVELOPMENT PROCESS

In this section, we will discuss the actual process of developing the coin. Most important steps to pay attention to, are creating the wallet with Metamask, getting to run the code, and Remix IDE compiling to MyEtherWallet publishing the contract that we will own.[23]

22.5.1 METAMASK WALLET

By definition MetaMask is a software, which enables you to create an Ethereum wallet to store your ETH, also connects you to the blockchain so you can be up to date. In addition, it gives you the power to run your own contracts without downloading the Ethereum Hub software. Therefore, it is a tool that can be used to store tokens.

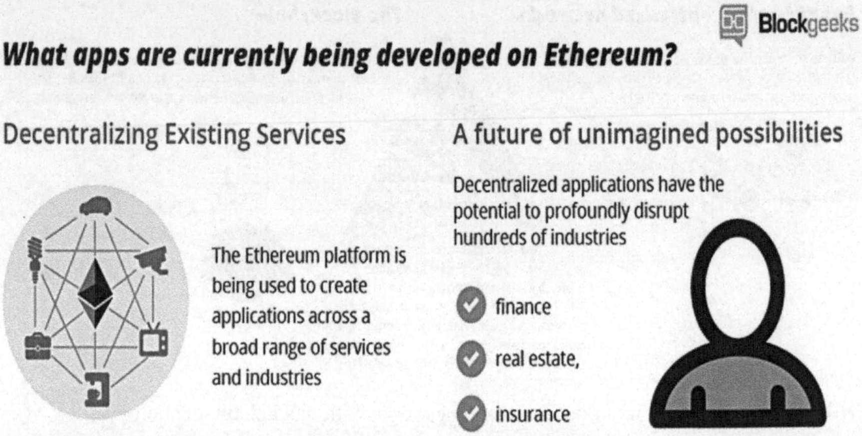

FIGURE 22.16 Presentation of services developed on Ethereum.

In the background, Metamask ensures the values entered has been stored and converts them to a language that the network can understand with an easy user-friendly UI.[6] First, you need to install the chrome extension and following the steps you can see as in Figure 22.17.

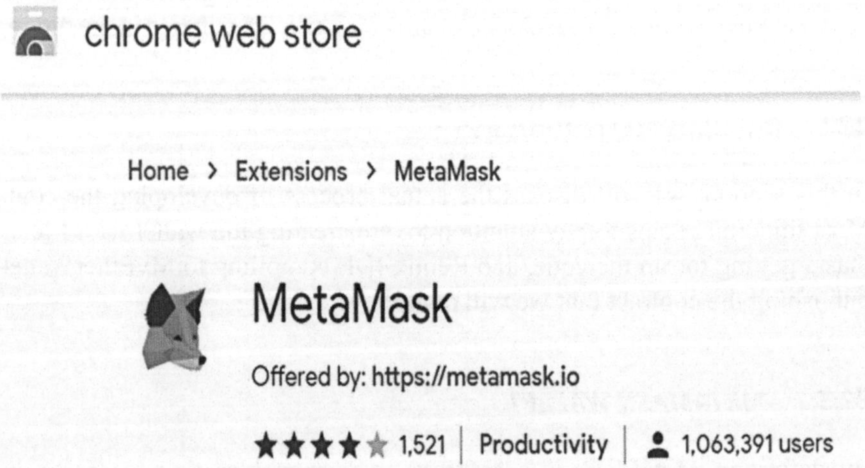

FIGURE 22.17 Metamask on the chrome web store.

After installing run the extension and sign up by entering your password, it is really fast and easy. In addition, after logging in you can see your ETH wallet and balance. It should look something like this Figure 22.18.

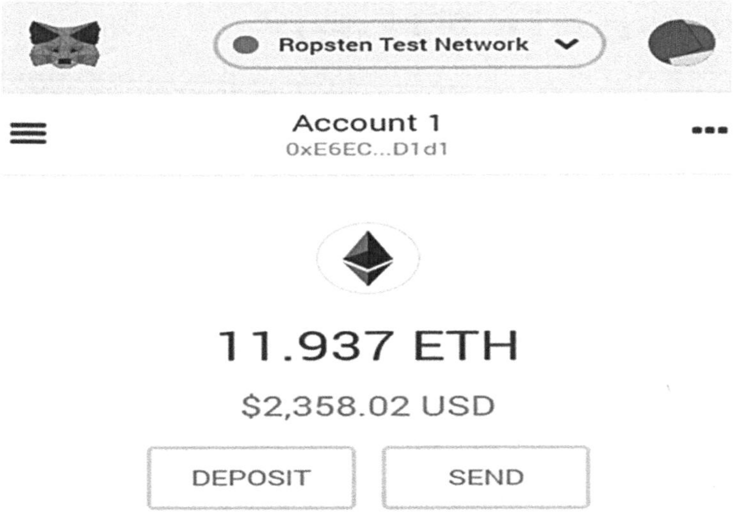

FIGURE 22.18 View after creating your account on MetaMask.

You can note at the top it states Ropsten Test Network; we will explain the meaning in the Ropsten section. Ropsten network is a test network, so the balance is actually not a real one; only for testing purposes we have the option to receive ETH from ropsten faucet that is included in the extension. By pressing on DEPOSIT, we will have an option to get ETH from the faucet. By pressing get faucet after the link transfer you should be done by now. The next step follows as we start to build our coin.[20] In Figure 22.19, we can see how to get test ETH to our wallet.

22.5.2 BRACKETS.IO

Brackets is a lightweight, yet groundbreaking, present-day editor. We mix visual instruments into the supervisor so you get the perfect measure of assistance when you need it without impeding your innovative procedure. You will appreciate composing code in Brackets.[5] I lean toward Brackets word processor with the simple Live Preview it has. Get a constant association with your program. Make changes to CSS and HTML and you will right away observe those progressions on screen. It is the intensity of a code editorial manager with the comfort of in-program dev devices. Now we download Brackets since we will need to edit our code.

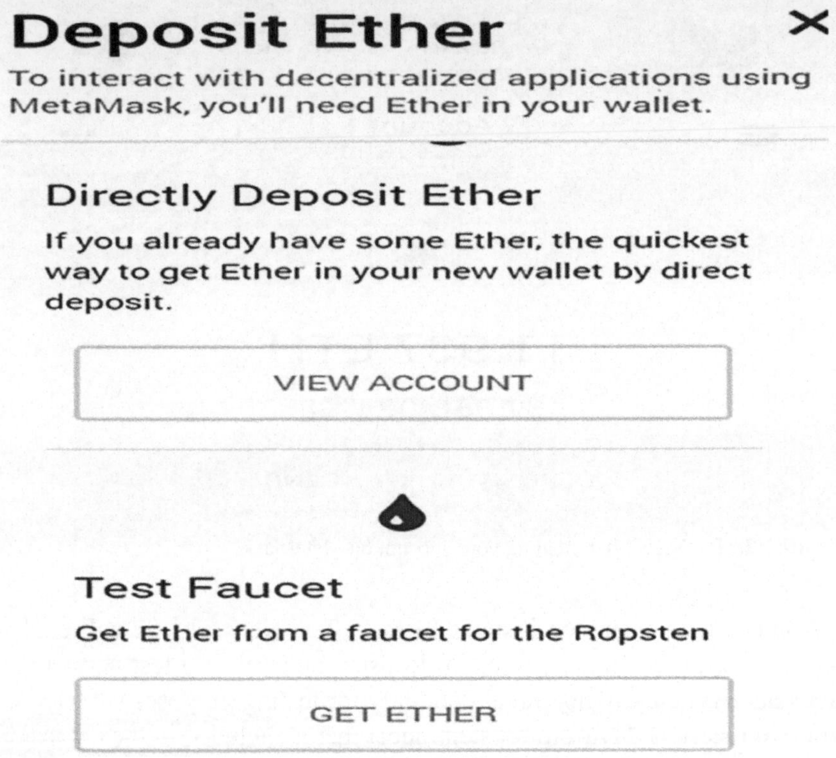

FIGURE 22.19 Receive free test ETH from faucet.

22.5.3 SOLIDITY

Solidity is a coding language, used to create smart contracts. Python, JavaScript, and C++ inspired it. Solidity is a programming language intended for creating keen gets that keep running on the ethereum network. Using the language Solidity, developers can develop decentralized applications, which are self-implementing eco-system typed smart contracts, making a nonrepeatable and definitive history of transactions. Solidity has built using ECMAScript coding structure to make it easier for developers, which have experience already with ECMAScript. Solidity is used to create more complex variables for contracts that have hierarchical mappings. So now after mentioning both Solidity and Brackets we need to head to Moritz Neto's article on how to issue your own token using the Ethereum network.[21] To see an example for creating our ICO, we are going to start from there.

Pseudo Code for Creating ICO

```
StandardToken is Token {
    function transfer(address _to, uint256 _value) returns (bool success) {
        if (balances[msg.sender] >= _value && _value > 0) {
            balances[msg.sender] -= _value;
            balances[_to] += _value;
            Transfer(msg.sender, _to, _value);
            return true;
        } else { return false; }
    }
transferFrom(address _from, address _to, uint256 _value) returns
(bool success)
{
        if (balances[_from] >= _value && allowed[_from][msg.sender]
        >= _value && _value > 0) {
            balances[_to] += _value;
            balances[_from] -= _value;
            allowed[_from][msg.sender] -= _value;
            Transfer(_from, _to, _value);
            return true;
        } else { return false; }
}
approve(address _spender, uint256 _value) returns (bool success) {
        allowed[msg.sender][_spender] = _value;
        Approval(msg.sender, _spender, _value);
        return true;
}
    if(!_spender.call(bytes4(bytes32(sha3("receiveApproval(address,uint25
    6,address,bytes)"))), msg.sender, _value, this, _extraData)) { throw; }
```

The pseudo-code ICO needs few prerequisite variables to fit our need for our coin such as coin name, supply of coins (in our case we need to update the supply to 1000,000; therefore, we need 1 and 24 zeroes written as a variable). So now, after changing the code to our needs, I also added comments to describe the functions and meaning of the structure. By having the basic functions and not adding, the ones that are not needed for example a mint function could destroy our eco system with the ability to resupply our coin that would make the investors fearful to get into our project.

Our most important functions are TotalSupply that we have to be unique, since it is stating how much actually tokens we are going to have and it will not be changeable.

Another important function is ServiceCoin, since it contains all our unique information such as decimal numbers how much tokens will be published on the network, and most importantly how much Tokens can 1 Ethereum buy. Now since we have the code needed to create an ICO, the next step is to be headed to Remix IDE to compile our code into ByteCode in order to publish our ServiceCoin (SRV).

22.5.4 REMIX IDE

Remix IDE as an open-source program has to power to convert Solidity code into bytecode that the ethereum network can understand inside the browser. It is written in JavaScript. In Remix IDE, we can test, debug, and deploy smart contracts and more.[24] Now we go head to Remix Ide, which we can access, from the Ethereum website and copy our code in the editor like in Figure 22.20:

```
pragma solidity ^0.4.4;

contract Token {

    /// @return total amount of tokens
    function totalSupply() constant returns (uint256 supply) {}

    /// @param _owner The address from which the balance will be retrieved
    /// @return The balance
    function balanceOf(address _owner) constant returns (uint256 balance) {}

    /// @notice send `_value` token to `_to` from `msg.sender`
    /// @param _to The address of the recipient
    /// @param _value The amount of token to be transferred
    /// @return Whether the transfer was successful or not
    function transfer(address _to, uint256 _value) returns (bool success) {}

    /// @notice send `_value` token to `_to` from `_from` on the condition it is approved by `_from`
    /// @param _from The address of the sender
    /// @param _to The address of the recipient
    /// @param _value The amount of token to be transferred
    /// @return Whether the transfer was successful or not
    function transferFrom(address _from, address _to, uint256 _value) returns (bool success) {}

    /// @notice `msg.sender` approves `_addr` to spend `_value` tokens
    /// @param _spender The address of the account able to transfer the tokens
    /// @param _value The amount of wei to be approved for transfer
    /// @return Whether the approval was successful or not
    function approve(address _spender, uint256 _value) returns (bool success) {}
```

FIGURE 22.20 Copied code from the brackets editor for ServiceCoin (SRV).

By copying the code, we can see that there are warnings, since we did not get red warnings we are great to go. The yellow ones are concerns from the compiler about security issues with our token. Just to note that we are not building an actual Coin with actual ETH. Before we compile we need to make sure that we are using 0.4.24+commit.e67f0147 compiler since it will be the best for our token, the newer releases will not allow us to pass those security issues without reading and reediting the code. It is important that we have the 0.4.24+commit.e67f0147 compiler selected like in Figure 22.21:

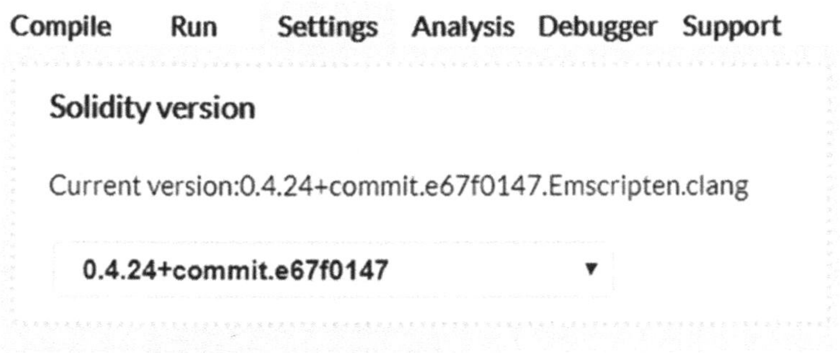

FIGURE 22.21 Compiler version for our code.

Now we are headed back to compile submenu that we can see in Figure 22.21. In addition, we select ServiceCoin and press on Details and we can see the info like in Figure 22.22.

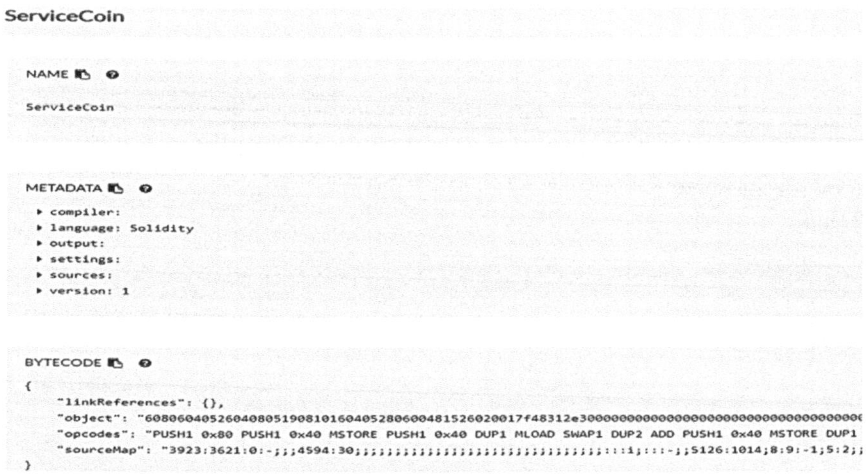

FIGURE 22.22 Pop Up after clicking on details.

The Pop Up appears with information generated from our code. We are interested in getting only the Bytecode. In this case we need to copy everything that is stated in "object" but without the. We copy everything and paste it into a text editor, and we add a 0x at the start. This is a unique sequence generated just for our code it will be important to note this in the next steps. Now comes the next step, taking this chunk of letters and numbers and putting them to use with MyEtherWallet contract publishing.

22.5.5 MYETHER WALLET CONTRACT PUBLISHING

MyEtherWallet is open-source and free-to-use interface that produces ethereum wallets and can publish contracts by converting the bytecode from Remix IDE. It is famous for its security and one of the first ethereum wallets available. The one contract we will talk about is sending our contract by utilizing byteCode with MyEtherWallet. To convey the byteCode, you have to open your record and send transaction. There are different potential techniques to open your record including utilizing Private key, KeyStore document, etc. We can pick any of the techniques. When your account is unlocked successfully, you see a Sign Transaction button.[27] But in our case we are using Metamask so the browser will auto-recognize the extension and will offer you to connect via MetaMask, you select that and you'll see two text boxes with Raw Transaction and Signed Transaction and a Deploy Contract button below that. Now we need to copy the byteCode that we got from Remix IDE like in Figure 22.23:

FIGURE 22.23 The view after copying the byteCode.

Now we can notice that the Gas Limit has automatically updated and when we press on Sign Transaction we can see Raw Transaction and Signed Transaction now that means we are good to go and press Deploy Contract.

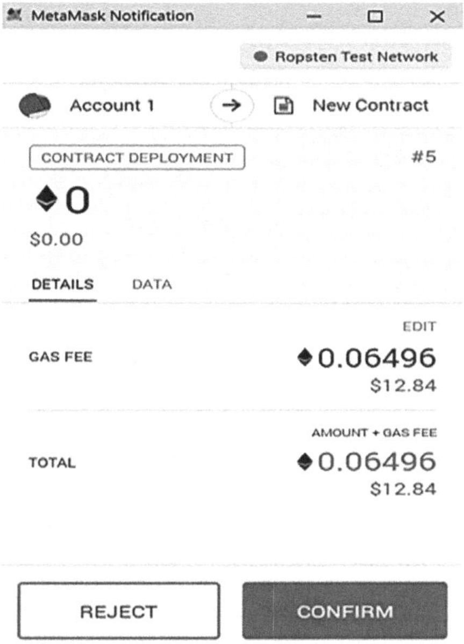

FIGURE 22.24 Notification from MetaMask extension.

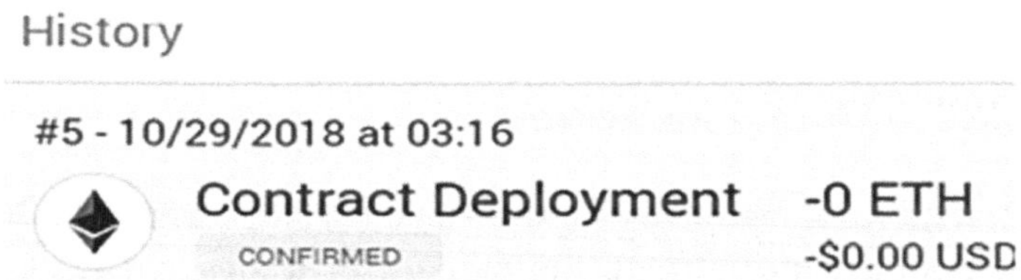

FIGURE 22.25 Contract Deployment transaction on MetaMask confirmed.

Now we can see that in Figure 22.24, mentioned above the MetaMask wallet extension is notifying us to confirm the transaction. Now when we press confirm the contract will be published and we can see in Figure 22.25. Now we can see on the MetaMask history that we successfully deployed a contract, now we need to check the contract on the Ropsten Etherscan Network. We are going to do with a click on the MetaMask arrow icon transaction; it will autolink us through the browser directly to the transaction as in Figure 22.26, shown to be able to see the transaction on the Ropsten Test Network.

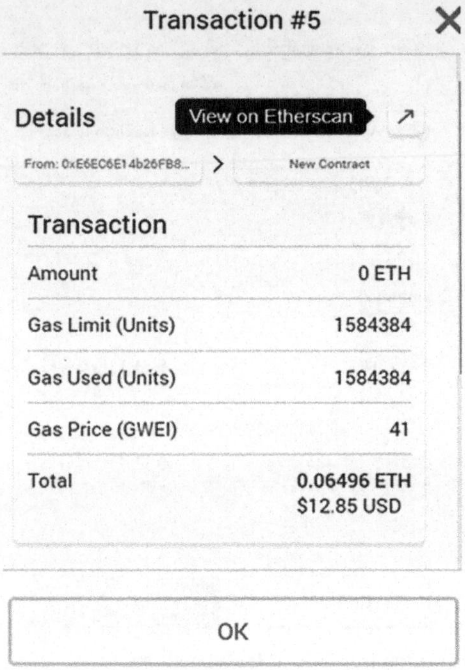

FIGURE 22.26 Transaction to Etherscan.

22.5.6 ROPSTEN TEST NETWORK

Ropsten test network represents a testing blockchain network; it is basically the same as Ethereum but with the purpose of testing contracts and for testing tokens. The usage of this network is that this network is free, you get your testing Ethereum from the given faucet of Ropsten, and you can have as many Ethereum as you want. Before publishing a contract or app, developers always test it on a test network. Because debugging is not possible for example, we have a token, which is unchangeable and after publishing, you cannot edit, and you spent some ETH for generating it. You can have trouble generating it due to compiler errors you get since you have so many similar variables, but that rarely happens. There were few more test networks but only 4–5 are left, the reason behind this is because there were some flaws in the build that they had, like spamming the network or not using the faucet or forcing the faucet so other users cannot claim the test Ethereum. Let us get back to our work, on the Etherscan.io after selecting the Ropsten Testnet you add the contract address and this window will pop up. Now your browser tab should look like Figure 22.27.

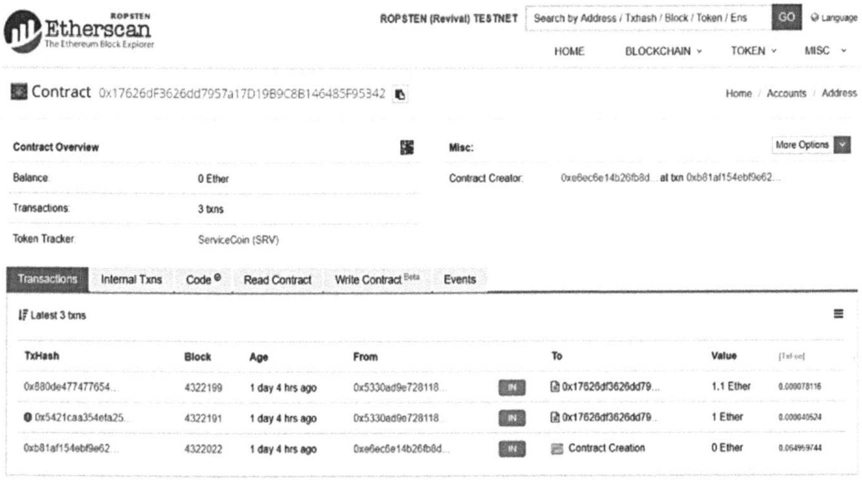

FIGURE 22.27 After clicking from Metamask to transfer you to the transaction.

We can see our transaction was confirmed, to Contract Creation row. And our contract address is 0x17626dF3626dd7957a17D19B9C8B146485F95 342. That means that we have successfully created our own token. Anyone who sends now ETH to this contract address will receive SRV. Now let us add our token to our wallet, meaning our wallet to recognize this token. We are going to do that in MetaMask by going in the menu and pressing add new Token and entering the contract address, in our case 0x17626dF3626d-d7957a17D19B9C8B146485F95342 like in Figure 22.28.

22.5.7 ACTUAL TRANSACTIONS TO OWN THE TOKEN

In this case I already added it and now it states that I own it. I will make a transaction from another account on the Ropsten Test Network, I will send 1.1 ETH and in return I will need to receive 110 SRV tokens. Sending will look like Figure 22.29.

FIGURE 22.28 Adding our Token to metamask.

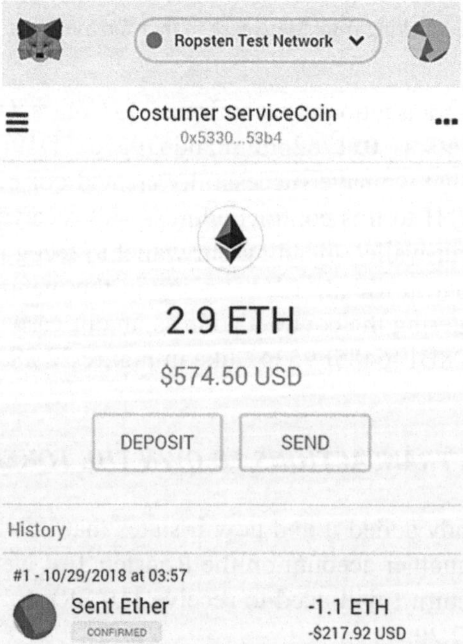

FIGURE 22.29 Sending ETH from another account to our contract.

Few minutes after I got a confirmation and by using the same steps I also added the token to MetaMask and on the next figure you can see that the "Costumer" got its SRV coin! In Figure 22.30, shown below you can see that the user got the SRV Tokens. By following the above-mentioned process and steps, a successful token is created privately which helps to start the marketing and start collecting funds for our project. It should be verified before deployment, so all of our future users can see the source code and to verify for themselves in the open environment. To do this we need to go again in Ropsten Network, but now we will open up the contract address. After opening up the address, we should enter our source code and to select our compiler and to press Verify. The logic behind this is when you put the source code and Etherscan is looking for the byte codes, if they receive the same bytecodes as I did when I entered the code into Remix IDE, then we have a successful verification. One of the most important factors of ICO's is their website. For our purpose I have built a page and the Contract address has been written and the price per 100 tokens. For making the simple view we are headed to Sandeep Panda ICO example on how to build a simple website for you token[22] at and we take from the source code and edit it for our purpose it could be checked in a live preview on the browser view.

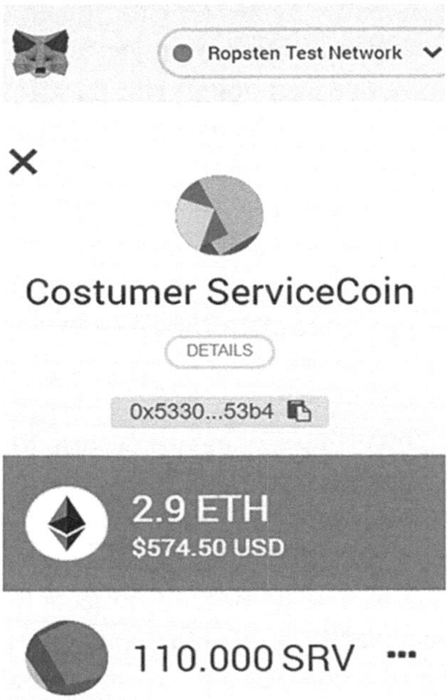

FIGURE 22.30 ServiceCoin (SRV) arrived at the costumer account.

22.5.8 MARKET CAPITALIZATION

Market Capitalization (Market Cap) is important in the cryptocurrency market, meaning that most of the attention and especially free marketing you get when you have a big amount of money invested into your currency. The laws of determining the market capitalization for a cryptocurrency are the same as any currency in the world backed by something real in value such as gold. Let us say that we have some coin named Tonium (TON), if TON has a supply of 10,000,000 coins in the network and $100,000,000 invested into the coin, 1 coin would cost $10, it is simple as that. If people are investing more into that coin (buying it from exchanges etc.), the price rises automatically by the same law we mentioned above. Now the status of the market can be checked at coimarketcap.com[7] here is a figure of the top cryptocurrencies right now.

Everything is automatically calculated as illustrated in the above process. The past months the prices got low but everyone hopes that the age of cryptocurrencies has just started. Ethereum is on the second place with 20 billion USD market capacity. If we calculate from Figure 22.31, the circulating supply and multiply with the present price, we are going to get our market capitalization of that coin, and vice versa by changing the parameters proportionally. Table 22.3 illustrates the basic parameters for concluding why we should build a project using the ICO—Ethereum network ERC20 token rather than going with the Crowdfunding on online platforms such as Kickstarter, GoFundMe, or Indiegogo. By looking at the data and considering the moral reasoning of average investors, the first parameter of Liquidity stands out as an important factor in today's world. In ICO, we get our tokens immediately, so we instantly "rewarded" for our investment and get that gratification feeling. On the opposite side, we have the Crowdfunding via the online platforms. As a business if we consider the costs for maintaining a marketing for gathering funds, a big commission can set us back; in this case we can conclude that the massive advantage is toward the ICO method, since investing and having 0 costs for having the offering. The amount raised parameter states that the online platforms are winning the race, but if we consider how long they are on the market compared to the ICO method of crowdfunding, by proportions the ICO method is really up in the trends right now. The number of projects being run currently on both methods cannot be exactly compared due to different types of offers to invest or to donate, but also considering the time frame of existence of both methods, the proportions will also suggest that ICO is going to take over the market in the following 3 years. At last the restrictions, they are really important due to the laws that every country has, for example third-world countries have some restrictions to invest in some online platforms, but that cannot happen in the ICO method, so again ICO is the better way to go if you are looking to build a worldwide project.

#	Name	Market Cap	Price	Volume (24h)	Circulating Supply	Change (24h)	Price Graph (7d)
1	Bitcoin	$110,137,540,720	$6,348.26	$4,235,475,997	17,349,262 BTC	-1.99%	
2	Ethereum	$20,337,086,622	$197.66	$1,500,377,317	102,886,874 ETH	-3.35%	
3	XRP	$17,872,752,742	$0.444535	$361,882,632	40,205,513,967 XRP *	-3.10%	
4	Bitcoin Cash	$7,281,877,060	$417.77	$299,336,534	17,430,150 BCH	-4.75%	
5	EOS	$4,661,698,429	$5.14	$621,462,600	906,245,118 EOS *	-4.62%	
6	Stellar	$4,270,193,783	$0.225775	$46,878,987	18,913,500,521 XLM *	-2.17%	
7	Litecoin	$2,909,604,997	$49.36	$361,488,475	58,944,027 LTC	-4.85%	
8	Tether	$1,817,360,671	$0.995039	$2,365,850,300	1,826,421,736 USDT *	0.06%	
9	Cardano	$1,809,530,671	$0.069793	$21,631,863	25,927,070,538 ADA *	-3.71%	
10	Monero	$1,687,827,139	$102.11	$13,329,527	16,529,343 XMR	-1.66%	

FIGURE 22.31 Current market capitalization.

TABLE 22.3 ICO Using Ethereum Blockchain Compared to Crowdfunding on Online Platforms Such as Kickstarter, Gofundme, Indiegogo.

	Liquidity	Commission per project	Amount raised	Number of projects	Restrictions
ICO on the Ethereum network	Immediate	0%	4.5 billion USD	4000+	No restrictions
Crowdfunding on online platforms	1–5 years	8%-20%	34 billion USD	12,000+	State law, investing law

Table 22.4 shows the analysis of data, gives out an example of parameters in different fundings of 100k, and approaches at the beginning and the end of the funding. Hugo Benedetti and Leonard Kostovetsky et al.[1] state that most of the successful ICO were creating big amounts of return on investments;

they also mentioned how few of them started and how they developed the business and what costs they had after having the funds. Gathering funds around $100,000 how much actually we are going to have at the end of the campaign and after few weeks/months, the results are shown in Table 15.4. The sum up of all the expenses to run an own project, even if you are successful and you want to go further, ICO with Ethereum ERC20 Token is 80% more efficient way. Since while having a $100,000 budget, you will end up spending more than $50,000 after the first month on team, marketing, and production while in the ICO you get to keep the funds and you have the time to gather the proper team and then to develop and apply for an appropriate project.

TABLE 22.4 Input of Factors to Consider When Looking to Fund $100,000 for a Project.

Block chain approach	ICO with Ethereum ERC20 Tokens	Crowdfunding with online platforms such as Kickstarter, Indiegogo, or GoFundme
Required team at the start with expenses and presentation	No team required, 0 costs	Yes, average starting team costs between $10,000 and $20,000
Development preparation costs	No, you can easily do it yourself, 0 costs	Yes, Video, Prototype, around $5000 costs
Commissions if there is successful funding	No, 0% commissions	Yes, from 5% to 20% of total funding
Restrictions to operate	No, it is global and free to use	Yes, for every country's law
Cost to maintain the idea	No, the blockchain keeps it free and published	Yes, after funding you spend on salaries, marketing production
Eco-friendly	Yes, it is not using any physical product	Maybe, could be eco-friendly, but if there is a physical product it is not
Return on investment	Yes, the price of the token fluctuates how the project develops	Most of the time no, you just get the product or service

22.6 CONCLUSION

At the end a fully operational cryptocurrency Ethereum ERC20 Token ICO was created. Now the idea is available for everyone to see at the Ropsten Test network. It does not mean that this process has unique identification of crowd financing. The technology is every time upgraded through myetherwallet.com, with the option to interact with the smart contract. They just need to send Ropsten Ethereum to the contract address and before that to add the token to their wallet using the same contract address and they will receive the ServiceCoin token. The Contract Address is 0x17626dF3626dd7957a17D19B9C8B146485F95342. That is not all right with potential investors since they cannot know what kind of changes will be made. In search engine started rising dramatically for the ICO with ethereum method at the start of 2017, while crowdfunding with online platforms such as Kickstarter, Indiegogo, and Gofundme is even going downward. That is why most of the great ICO's are building into the code a finite number of supply and restrictions about attacks. For the public, the cryptocurrency market is still not mainstream enough to get the attention of an everyday buyer, but updates and new valuable coins are developed daily. From the observation in minimum time, block chain technology allows us to start using cryptocurrency as a daily routine. By analyzing the data given on Datatoken.io the total amount raised so far for Ethereum ICO's is getting toward 4.5 billion USD; rapidly the amount was growing every year. A large number of devices exist to arm financial specialists with data. Digital money showcase capitalization lives in the weapons store of those apparatuses. It gives a long haul point of view as value alone changes for the time being.

22.7 FUTURE WORKS

To look forward to do something about the future of block chain approach for crowd financing, is going to spread around the world and the technology behind cryptocurrencies is being highlighted. Some refer it as the new golden age, some who are not informed enough they say it is a digital pyramid Ponzi fraud. This paper would be a milestone to educate fellow researchers and the industrialist to have an appropriate vision on the logic and concept behind the block chain technology, which would bring maximum benefits in all means.[2–4,10,14,16–19] In addition, an active crypto currency miner and trader is creating job vacancy for the relevant seekers. Concentrating on such opportunity would improve the crypto currencies' benefits and it is younger in the market. Moreover, it has a bright boom in the trade market with the crypto world. It is observed collectively pervasive users have started to give their comments after using such technology. The benefits and performance are more stunning with more efficient and safe with block chain technology. Also it takes the power of the rich and gives it to everyone.

KEYWORDS

- **cryptocurency**
- **Ethereum**
- **ERC20 Tokens**
- **Solidity**
- **Metamask cryptocurrency wallet**
- **transactions between wallets**
- **MyEtherWallet contracts**
- **Remix IDE**
- **initial coin offering**

REFERENCES

1. Benedetti, H.; Kostovetsky, L. Digital Tulips? Returns to Investors in Initial Coin Offerings. *J. Corp. Finan.* **2021,** *66*, 101786. https://doi.org/10.1016/j.jcorpfin.2020.101786.
2. Bose, J. S. C.; Mehrez, M.; Badawy, A. S.; Ghribi, W.; Bangali, H.; Basha, A. Development and Designing of Fire Fighter Robotics Using Cyber Security. In *2017 2nd International Conference on Anti-Cyber Crimes (ICACC)*, March 2017. https://doi.org/10.1109/anti-cybercrime.2017.7905275.
3. Bose, S. C.; Mubarakali, A.; Marina, N.; Reshmi. Design and Development of the Robot for Aquatic Debris Using a Smartphone. In *Innovative Data Communication Technologies and Application. Lecture Notes on Data Engineering and Communications Technologies*; Raj, J. S., Iliyasu, A. M., Bestak, R., Baig, Z. A., Eds.; Springer: Singapore; Vol 59, pp 41–55. https://doi.org/10.1007/978-981-15-9651-3_4.
4. Bose, S. C.; Veerasamy, M.; Mubarakali, A.; Marina, N.; Hadzieva, E. Analysis of Feature Extraction Algorithm Using Two Dimensional Discrete Wavelet Transforms in Mammograms to Detect Microcalcifications. *Comput. Vis. Bio-Inspired Comput.* **2020,** 26–39. https://doi.org/10.1007/978-3-030-37218-7_4.
5. brackets.io. A Modern, Open Source Text Editor That Understands Web Design. *Brackets*, 2016. http://brackets.io/.
6. Chrome Web Store. *MetaMask*, 2018. https://chrome.google.com/webstore/detail/metamask/nkbihfbeogaeaoehlefnkodbefgpgknn?hl=en.
7. CoinMarketCap. *Cryptocurrency Prices, Charts and Market Capitalizations*, 2018. https://coinmarketcap.com.
8. cointelegraph. *Latest Bitcoin, Blockchain and Ethereum News*; Maxwell William, 2018. https://cointelegraph.com/explained/erc-20-tokens-explained+may+12,+2018.
9. Dannen, C. *Introducing Ethereum and Solidity: Foundations of Cryptocurrency and Blockchain Programming for Beginners*, 1st ed. Apress: New York, NY, 2017.

10. Subash Chandra Bose. J.; Gopinath, G. A Survey Based on Image Encryption Then Compression Techniques for Efficient Image Transmission. *J. Ind. Eng. Res.* **2015,** *1* (1), 15–18. http://oaji.net/articles/2015/2164-1437300161.pdf.
11. Emsley, J. *10 Most Successful ICOs of All Time. Invest in Blockchain*; May 14, 2018. https://www.investinblockchain.com/10-most-successful-icos/.
12. ethdocs.org. *Ethereum Homestead Documentation—Ethereum Homestead 0.1 Documentation*, 2016. https://ethdocs.org/_/downloads/en/latest/pdf/. https://www.ethdocs.org/en/latest/.
13. Ethereum. *Home*; Ethereum.Org, 2021. https://ethereum.org/en/.
14. Subash Chandra Bose, J.; Shankar Kumar, K. R. Detection of Micro-Classification in Mammograms Using Soft Computing Techniques. *Eur. J. Sci. Res.* **2012,** *86* (1), 103–122.
15. https://www.academia.edu/8934493/Detection_of_Microcalcification_in_Mammograms_using_Soft_Computing_Techniques.
16. Subash Chandra Bose, J.; Gopinath, G. An ETC System Using Advanced Encryption Standard and Arithmetic Coding. *Middle-East J. Sci. Res.* **2015,** *23* (5), 932–935. https://www.idosi.org/mejsr/mejsr23(5)15/23.pdf.
17. Subash Chandra Bose, J.; Saranya, B.; Monisha, S. Optimization and Generalization of Lloyd's Algorithm for Medical Image Compression. *Middle-East J. Sci. Res.* **2015,** *23* (4), 647–651. https://www.idosi.org/mejsr/mejsr23(4)15/13.pdf.
18. Sebastian, L.; Subash Chandra Bose, J. Efficient Restoration of Corrupted Images and Data Hiding in Encrypted Images. *J. Ind. Eng. Res.* **2015,** *1* (2), 38–44. http://oaji.net/articles/2015/2164-1437300684.pdf.
19. Mubarakali, A.; Bose, S. C.; Srinivasan, K.; Elsir, A.; Elsier, O. Design a Secure and Efficient Health Record Transaction Utilizing Block Chain (SEHRTB) Algorithm for Health Record Transaction in Block Chain. *J. Amb. Intell. Hum. Comput.* **2019**. https://doi.org/10.1007/s12652-019-01420-0.
20. Neto, M. *Get Ropsten Ethereum—The Easy Way*; Medium, June 19, 2018. https://medium.com/bitfwd/get-ropsten-ethereum-the-easy-way-f2d6ece21763.
21. Neto, M. *How to Issue Your Own Token on Ethereum in Less than 20 Minutes*; Medium, February 16, 2020. https://medium.com/bitfwd/how-to-issue-your-own-token-on-ethereum-in-less-than-20-minutes-ac1f8f022793.
22. Panda, S. *Hashnode Test Coin ICO*, 2018. http://icoin20mins.s3-website-us-west-2.amazonaws.com/.
23. Panda, S. *How to Build Your Own Ethereum Based ERC20 Token and Launch an ICO in Next 20 Minutes*; Hashnode, 2018, March 14. https://hashnode.com/post/how-to-build-your-own-ethereum-based-erc20-token-and-launch-an-ico-in-next-20-minutes-cjbcpwzec01c93awtbij90uzn.
24. Remix IDE. *Maze Found | Read the Docs*, 2018. https://remix.readthedocs.io/en/latest/+2018+Revision+6d296658.
25. Rosic, A. *What Is Ethereum? [The Most Updated Step-by-Step-Guide!]*; Blockgeeks, 2020, September 8. https://blockgeeks.com/guides/ethereum/.
26. Rosic, A. *What Is Cryptocurrency? [Everything You Need To Know!]*; Blockgeeks, 2020, November 25. https://blockgeeks.com/guides/what-is-cryptocurrency/.
27. Singh, P. P. *Deploying Contract Using ByteCode (MyEtherWallet and Remix)*; Medium, 2019, January 29. https://medium.com/sofocle-technologies/deploying-contract-using-bytecode-myetherwallet-and-remix-10f643a82d40.

28. Yu, S.; Johnson, S.; Lai, C.; Cricelli, A.; Fleming, L. Crowdfunding and Regional Entrepreneurial Investment: An Application of the CrowdBerkeley Database. *Res. Policy* **2017,** *46* (10), 1723–1737. https://doi.org/10.1016/j.respol.2017.07.008.
29. Zhang, Y. *An Empirical Study into the Field of Crowdfunding*; Lund University, August 2012. https://lup.lub.lu.se/luur/download?func=downloadFile&recordOId=3049774&fileOId=30497.

Index

F

G

H

N

O

P

R